EDEXCEL ADVANCED MATHEMATICS

The Edexcel course is based on units in the four strands of Pure Mathematics, Mechanics, Statistics and Decision Mathematics. The first unit in each of these strands is designated AS and so is Pure Mathematics: Core 2; all others are A2.

The units may be aggregated as follows:

3 units   AS Mathematics
6 units   A Level Mathematics
9 units   A Level Mathematics + AS Further Mathematics
12 units   A Level Mathematics + A Level Further Mathematics

Core 1 and 2 are compulsory for AS Mathematics, and Core 3 and 4 must also be included in a full A Level award.

Examinations are offered by Edexcel twice a year, in January (most units) and in June (all units). All units are assessed by examination only; there is no longer any coursework in the scheme.

Candidates are not permitted to use electronic calculators in the Core 1 examination. In all other examinations candidates may use any legal calculator of their choice, including graphical calculators.

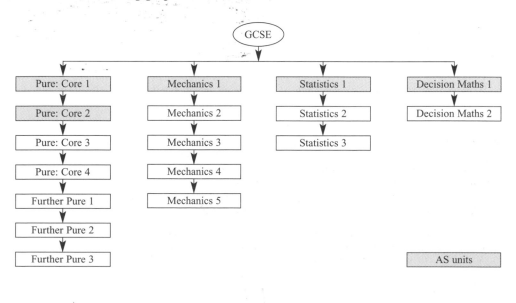

WITHDRAWN

INTRODUCTION

This is the first book written to support the Pure Mathematics units in the Edexcel Advanced Level Mathematics scheme. Much of the material is new and there has been a substantial re-write of previous material to provide complete coverage of the Edexcel Pure Mathematics Core 1 and Core 2 units.

Each book follows the order and structure of the Edexcel examination specifications. Throughout the series the emphasis is on understanding and applying a wide variety of mathematical skills. The style of writing provides clear explanations of the methods and techniques. There is a large number of examples illustrating the problem-solving processes. There are many questions from Edexcel question papers.

The Core 1 section of this book has five chapters relating to the five sections of the syllabus. It has been written to reflect the rule that no calculator is allowed in the Core 1 examination. The first chapter develops skills in the use of algebra, building upon GCSE topics. Chapter 2 illustrates methods used in coordinate geometry. In Chapter 3 arithmetic series and recurrence relations are covered. Finally, the essentials of basic calculus are introduced, with chapters 4 on differentiation and 5 on integration.

In the Core 2 section there are seven chapters numbered sequentially from the Core 1 section. Chapter 6 develops algebraic skills with the factor and remainder theorems. Chapter 7 covers the coordinate geometry of circles with Chapter 8 explaining the geometric series and the binomial expansion. Chapter 9 introduces the reader to radians and trigonometric functions and identities. Chapter 10 covers exponentials and logarithms, then the last two chapters build upon the calculus of Core 1.

I should like to thank the many people in the preparation and checking of material. Special thanks go to Val Hanrahan and Peter Secker, who wrote the original MEI edition. Further thanks go to my colleagues at Sedbergh School.

John Sykes

CONTENTS

Core 1

C1

Chapter one

ALGEBRA AND FUNCTIONS

A good notation has a subtlety and suggestiveness which at times makes it seem almost like a live teacher.

Bertrand Russell

· · · · · · · · · · · · · · · ·

The graph in Figure 1.1 refers to the moons of the planet Jupiter. For six of its moons, the time, T, that each one takes to orbit Jupiter is plotted against the average radius, r, of its orbit. (The remaining ten moons of Jupiter would be either far off the scale or bunched together near the origin.)

The curves $T = kr^1$ and $T = kr^2$ (using suitable units and constant k) are also drawn on the graph. You will see that the curve defined by Jupiter's moons lies somewhere between the two.

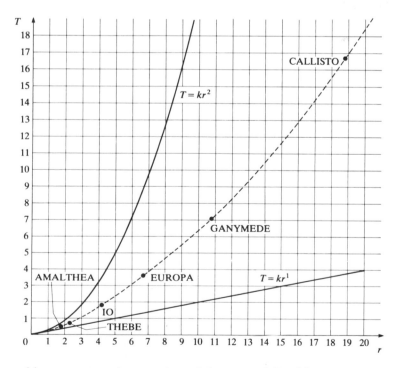

FIGURE 1.1

How could you express the equation of the curve defined by Jupiter's moons?

INDICES

You may have suggested that the equation of the curve defined by Jupiter's moons could be written as $T = kr^{1\frac{1}{2}}$. If so, well done! This is correct, but what would a power, or *index*, of $1\frac{1}{2}$ mean?

Before answering this question it is helpful to review the language and laws relating to positive whole-number indices.

Terminology

In the expression a^m the number represented by a is called the *base* of the expression; m is the *index*, or the *power* to which the base is raised. (The plural of index is *indices*.)

Multiplication

Multiplying 3^6 by 3^4 gives:

$$3^6 \times 3^4 = (3 \times 3 \times 3 \times 3 \times 3 \times 3) \times (3 \times 3 \times 3 \times 3)$$
$$= 3^{10}$$

Clearly it is not necessary to write down all the 3s like that. All you need to do is to add the powers.

$$6 + 4 = 10$$

and so $\qquad 3^6 \times 3^4 = 3^{6+4} = 3^{10}$

This can be written in general form as:

$$a^m \times a^n = a^{m+n}$$

Another important multiplication rule arises when a base is successively raised to one power and then another, as for example in $(3^4)^2$.

$$(3^4)^2 = (3^4) \times (3^4)$$
$$= (3 \times 3 \times 3 \times 3) \times (3 \times 3 \times 3 \times 3)$$
$$= 3^8$$

In this case the powers to which 3 is raised are multiplied.

$$4 \times 2 = 8$$

Written in general form this becomes:

$$(a^m)^n = a^{mn}$$

Division

In the same way, dividing 3^6 by 3^4 gives:

$$3^6 \div 3^4 = \frac{3 \times 3 \times 3 \times 3 \times 3 \times 3}{3 \times 3 \times 3 \times 3}$$

$$= 3^2$$

In this case you subtract the powers.

$$6 - 4 = 2$$

and so $\qquad\qquad\qquad 3^6 \div 3^4 = 3^{6-4} = 3^2$

This can be written in general form as:

$$a^m \div a^n = a^{m-n}$$

EXTENDING THE USE OF INDICES

Using these rules you can now go on to give meanings to indices which are not positive whole numbers.

INDEX 0

If you divide 3^4 by 3^4 the answer is 1, since this is a number divided by itself. However you can also carry out the division using the rules of indices to get:

$$3^4 \div 3^4 = 3^{4-4}$$
$$= 3^0$$

and so it follows that $\quad 3^0 = 1$

The same argument applies to 5^0, 2.9^0 or any other (non-zero) number raised to the power 0; they are all equal to 1.

In general:

$$a^0 = 1 \qquad a \neq 0$$

NEGATIVE INDICES

Dividing 3^4 by 3^6 gives:

$$3^4 \div 3^6 = \frac{3 \times 3 \times 3 \times 3}{3 \times 3 \times 3 \times 3 \times 3 \times 3}$$

$$3^{4-6} = \frac{1}{3 \times 3}$$

and so $\qquad 3^{-2} = \frac{1}{3^2}$

This can be generalised to: $\quad a^{-m} = \dfrac{1}{a^m}$

FRACTIONAL INDICES

What number multiplied by itself gives the answer 3? The answer, as you will know, is the square root of 3, usually written $\sqrt{3}$. Suppose instead that the square root of 3 is written 3^p; what then is the value of p?

Since $\qquad 3^p \times 3^p = 3^1$

it follows that $\qquad p + p = 1$

and so $\qquad p = \frac{1}{2}$

In other words, the square root of a number can be written as that number raised to the power $\frac{1}{2}$.

$$\sqrt{a} = a^{\frac{1}{2}}$$

The same argument may be extended to other roots, so that the cube root of a number may be written as that number raised to the power $\frac{1}{3}$, the fourth root corresponds to power $\frac{1}{4}$ and so on.

This can be generalised to:

$$\sqrt{a} = a^{\frac{1}{n}}$$

Using the previous result that $(a^m)^n = a^{mn}$, with the fact that $a^{\frac{1}{n}} = \sqrt[n]{a}$, gives the general rule that:

$$a^{\frac{m}{n}} = \sqrt[n]{a^m}$$

and this can be worked out as:

either $\quad a^{\frac{m}{n}} = \left(\sqrt[n]{a}\right)^m \quad$ or $\quad a^{\frac{m}{n}} = \sqrt[n]{(a^m)}$

In the example of Jupiter's moons, or indeed the moons or planets of any system, the relationship between T and r is of the form:

$$T = kr^{\frac{1}{12}}$$

Squaring both sides gives:

$$T \times T = kr^{\frac{1}{12}} \times kr^{\frac{1}{12}}$$

which may be written as $T^2 = cr^3$ (where the constant $c = k^2$).

This is one of Kepler's laws of planetary motion, first stated in 1619.

The use of indices not only allows certain expressions to be written more simply, but also, and this is more important, it makes it possible to carry out arithmetical and algebraic operations (such as multiplication and division) on them. These processes are shown in the following examples.

EXAMPLE 1.1

Write the following numbers as the base 5 raised to a power.

(a) 625   **(b)** 1   **(c)** $\dfrac{1}{125}$ **(d)** $5\sqrt{5}$

Solution   Notice that $5^1 = 5$, $5^2 = 25$, $5^3 = 125$, $5^4 = 625$, ...

(a) $625 = 5^4$ **(b)** $1 = 5^0$

(c) $\dfrac{1}{125} = \dfrac{1}{5^3}$ **(d)** $5\sqrt{5} = 5^1 \times 5^{\frac{1}{2}}$

$\qquad\qquad = 5^{-3}$ $= 5^{1\frac{1}{2}}$

EXAMPLE 1.2

Simplify these numbers.   **(a)** $(2^3)^4$ **(b)** $27^{\frac{1}{3}}$ **(c)** $4^{-2\frac{1}{2}}$

Solution   **(a)** $(2^3)^4 = 2^{12}$ **(b)** $27^{\frac{1}{3}} = \sqrt[3]{27}$ **(c)** $4^{-2\frac{1}{2}} = (2^2)^{-\frac{5}{2}}$

$\qquad\qquad\quad = 4096$ $= 3$ $= 2^{2 \times (-\frac{5}{2})}$

$\qquad\qquad\qquad\qquad\qquad\qquad\qquad\qquad\qquad\qquad\quad = 2^{-5}$

$\qquad\qquad\qquad\qquad\qquad\qquad\qquad\qquad\qquad\qquad\quad = \dfrac{1}{32}$

EXAMPLE 1.3

Simplify $8^{\frac{2}{3}}$.

Solution   There are two ways to approach this.

(a) $8^{\frac{2}{3}} = \left(\sqrt[3]{8}\right)^2$ **(b)** $8^{\frac{2}{3}} = \sqrt[3]{8^2}$

$\qquad\quad = 2^2$ $= \sqrt[3]{64}$

$\qquad\quad = 4$ $= 4$

Both give the same answer, and both are correct.

EXAMPLE 1.4

Simplify $\dfrac{32^3 \times 8^2}{4^4}$.

Solution   32, 8 and 4 are all powers of 2, hence:

$$\frac{32^3 \times 8^2}{4^4} = \frac{(2^5)^3 \times (2^3)^2}{(2^2)^4}$$

$$= \frac{2^{15} \times 2^6}{2^8}$$

$$= 2^{13}$$

EXAMPLE 1.5

Simplify $\left(4\sqrt{2} \times \frac{1}{16} \times \sqrt[5]{32}\right)^2$.

Solution $\left(4\sqrt{2} \times \frac{1}{16} \times \sqrt[5]{32}\right)^2 = (2^2 \times 2^{\frac{1}{2}} \times 2^{-4} \times (2^5)^{\frac{1}{5}})^2$

$$= (2^{2\frac{1}{2}} \times 2^{-4} \times 2^1)^2$$

$$= (2^{2\frac{1}{2} - 4 + 1})^2$$

$$= (2^{-\frac{1}{2}})^2$$

$$= 2^{-1}$$

$$= \tfrac{1}{2}$$

Note

In Examples 1.4 and 1.5 all the terms are expressed as a power of the base 2. When using the laws of indices you should always make sure that all the terms you are multiplying or dividing are written to the same base.

SIMPLIFYING SUMS AND DIFFERENCES OF FRACTIONAL POWERS

The next two examples illustrate a type of simplication which you will often find you need to do.

EXAMPLE 1.6

Simplify $5^{\frac{1}{2}} - 5^{\frac{3}{2}} + 5^{\frac{5}{2}}$.

Solution   This can be written as:

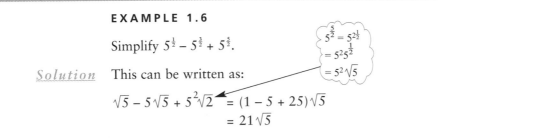

$$\sqrt{5} - 5\sqrt{5} + 5^2\sqrt{2} = (1 - 5 + 25)\sqrt{5}$$

$$= 21\sqrt{5}$$

$$5^{\frac{5}{2}} = 5^{2\frac{1}{2}}$$
$$= 5^2 5^{\frac{1}{2}}$$
$$= 5^2\sqrt{5}$$

EXAMPLE 1.7

Simplify $3(x + 7)^{\frac{1}{2}} - \frac{2}{3}(x + 7)^{\frac{3}{2}}$.

Solution   The expression is $3(x + 7)^{\frac{1}{2}} - \frac{2}{3}(x + 7)(x + 7)^{\frac{1}{2}}$.

$$\{3 - \tfrac{2}{3}(x - 7)\}(x + 7)^{\frac{1}{2}} = \frac{(9 - 2x - 14)}{3}(x + 7)^{\frac{1}{2}}$$

$$= -\frac{(5 + 2x)}{3}(x + 7)^{\frac{1}{2}}$$

$$\dots \text{or} -\frac{(5 + 2x)}{3}\sqrt{x + 7}$$

MIXED BASES

If a problem involves a combination of different bases, you may well need to split them, using the rule:

$$(a \times b)^n = a^n \times b^n$$

before going on to work with each base in turn.

For example: $6^3 = (2 \times 3)^3$
$$= 2^3 \times 3^3$$

which is another way of saying $216 = 8 \times 27$.

EXERCISE 1A   Answer these questions without using your calculator, but do use it to check your answers. Remember, you will not be allowed a calculator in your C1 examination.

1   Write the following numbers as powers of 3.

   (a) 81   (b) 1   (c) 27   (d) $\dfrac{1}{27}$ (e) $\sqrt{3}$ (f) 3

2   Write the following numbers as powers of 4.

   (a) 16   (b) 2   (c) $\dfrac{1}{4}$ (d) $\dfrac{1}{2}$ (e) 8   (f) $\dfrac{1}{8}$

3   Write the following numbers as powers of 10.

   (a) 1000   (b) 0.0001 (c) $\dfrac{1}{1000}$ (d) $\sqrt{1000}$ (e) $\dfrac{1}{\sqrt{10}}$

   (f) one millionth

4   Write the following as whole numbers or fractions.

   (a) 2^{-5} (b) $27^{\frac{1}{3}}$ (c) $27^{-\frac{1}{3}}$ (d) $625^{\frac{1}{4}}$ (e) $625^{-\frac{1}{4}}$ (f) 19^0

5   Simplify the following.

   (a) $3\sqrt{3} \times 3^{\frac{1}{2}}$ (b) 1000×10^{-2} (c) $\sqrt{25} \times 5^{-2}$
   (d) $7^{\frac{1}{2}} \times 7^{-\frac{1}{2}}$ (e) $\sqrt{5^4 \times 5^6}$ (f) $\sqrt[3]{243 \times 81}$

6   Simplify the following.

   (a) $\dfrac{7^{-2}}{7^{-4}}$ (b) $(3^{\frac{1}{2}})^2$ (c) $(3^2)^{\frac{1}{2}}$ (d) $(\sqrt{27})^4$ (e) $\sqrt[4]{16^3}$ (f) $(4^{-\frac{1}{2}})^{-2}$

7   Simplify the following.

   (a) $64^{\frac{1}{2}} - 64^{\frac{1}{3}}$ (b) $(11^2)^3 - (11^3)^2$

   (c) $25^{\frac{1}{2}} - \dfrac{5^{-4}}{5^{-5}}$ (d) $2^0 + 2^{-1} + 2^{-2} + 2^{-3} + 2^{-4}$

   (e) $(3^{\frac{1}{2}})^2 - (3^2)^{\frac{1}{2}}$ (f) $3^1 - 3^{-1}$

8 Simplify the following.

(a) $2^{\frac{1}{2}} + 2^{\frac{3}{2}}$

(b) $4^{\frac{1}{3}} - 4^{\frac{4}{3}}$

(c) $7^{\frac{1}{2}} - 7^{\frac{3}{2}} + 7^{\frac{5}{2}}$

(d) $(1 + x)^{\frac{1}{2}} + (1 + x)^{\frac{3}{2}}$

(e) $4(2 + 3x)^{\frac{1}{2}} + 3(2 + 3x)^{\frac{3}{2}}$

(f) $6(x + y)^{\frac{1}{2}} - 5(x + y)^{\frac{3}{2}}$

(g) $x^{\frac{1}{2}} + x^{-\frac{1}{2}}$

(h) $(x^2 - 2x + 3)^{\frac{1}{2}} - (x^2 - 2x + 3)^{\frac{3}{2}}$

9 Simplify the following.

(a) $\dfrac{4^2 \times 16^3}{8^4}$

(b) $\dfrac{\sqrt{27} \times \sqrt[4]{81}}{\sqrt{3}}$

(c) $\dfrac{625 \times 5^3}{25^2}$

(d) $\dfrac{2^{\frac{1}{2}} \times 4^{\frac{1}{3}}}{8^{\frac{1}{4}}}$

10 Simplify the following.

(a) $(2^3 \times 3^6)^{\frac{1}{3}}$

(b) $(\sqrt{2}ab^2)^4$

(c) $\left(\dfrac{x^{\frac{1}{3}} \times y^{\frac{1}{2}}}{z}\right)^6$

(d) $(\sqrt{a} + \sqrt{b})^2$

SURDS

You often meet square roots while doing mathematics, and there are times when it is helpful to be able to manipulate them, rather than just find their values from your calculator. This ensures that you are working with the exact value of the square root, rather than a rounded-off version. It may also keep your work tidier, without long numbers appearing at every line. Sometimes you find that the square roots cancel out, and this may indicate a relationship that you would otherwise have missed.

A number which is partly rational and partly square root, such as $(\frac{1}{2} + \sqrt{5})$, is an example of a *surd*. Surds may also involve cube roots, etc.

The basic fact about the square root of a number is that when multiplied by itself, it gives the original number.

$$\sqrt{5} \times \sqrt{5} = 5$$

This can be rearranged to give:

$$\sqrt{5} = \frac{5}{\sqrt{5}} \qquad \text{and} \qquad \frac{1}{\sqrt{5}} = \frac{\sqrt{5}}{5}$$

and both of these are useful results.

Surds may be added or subtracted just like algebraic expressions, keeping the rational numbers and the square roots separate.

Note

The symbol $\sqrt{}$ means 'the positive square root of'. Thus $\sqrt{8} = +2.828\ldots$ and the solution of $x^2 = 8$ is $x = +\sqrt{8}$ or $x = -\sqrt{8}$, often written as $x = \pm\sqrt{8}$.

EXAMPLE 1.8 Simplify $\sqrt{50}$.

Solution   Look for the largest square number that is a factor of 50, i.e. 25.

$$\sqrt{50} = \sqrt{25 \times 2} = \sqrt{25} \times \sqrt{2} = 5\sqrt{2}$$

Note

In general, the rules are:

$$\sqrt{ab} = \sqrt{a} \times \sqrt{b}$$

$$\sqrt{\frac{a}{b}} = \frac{\sqrt{a}}{\sqrt{b}}$$

However, you cannot combine $\sqrt{a} + \sqrt{b}$ into a single term.

EXAMPLE 1.9

Add $(2 + 3\sqrt{5})$ to $(3 - \sqrt{5})$.

Solution $2 + 3\sqrt{5} + 3 - \sqrt{5} = 2 + 3 + 3\sqrt{5} - \sqrt{5}$

$$= 5 + 2\sqrt{5}$$

EXAMPLE 1.10

Simplify $4(\sqrt{7} - 2\sqrt{2}) + 3(2\sqrt{7} + \sqrt{2})$.

Solution $4(\sqrt{7} - 2\sqrt{2}) + 3(2\sqrt{7} + \sqrt{2}) = 4\sqrt{7} - 8\sqrt{2} + 6\sqrt{7} + 3\sqrt{2}$

$$= 10\sqrt{7} - 5\sqrt{2}$$

When multiplying two surds, do so term by term, as in the following examples.

EXAMPLE 1.11

Simplify $(2 + \sqrt{3})^2$.

Solution $(2 + \sqrt{3})^2 = (2 + \sqrt{3}) \times (2 + \sqrt{3})$

$$= 2 \times 2 + 2 \times \sqrt{3} + \sqrt{3} \times 2 + \sqrt{3} \times \sqrt{3}$$

$$= 4 + 2\sqrt{3} + 2\sqrt{3} + 3$$

$$= 7 + 4\sqrt{3}$$

EXAMPLE 1.12

Simplify $(\sqrt{2} + \sqrt{3})(\sqrt{2} + 2\sqrt{3})$.

Solution
$$\begin{aligned}(\sqrt{2} + \sqrt{3})(\sqrt{2} + 2\sqrt{3}) &= \sqrt{2} \times \sqrt{2} + \sqrt{2} \times 2\sqrt{3} + \sqrt{3} \times \sqrt{2} + \sqrt{3} \times 2\sqrt{3} \\ &= 2 + 2\sqrt{6} + \sqrt{6} + 2 \times 3 \\ &= 8 + 3\sqrt{6}\end{aligned}$$

Notice that in the last example we used the fact that $\sqrt{2} \times \sqrt{3} = \sqrt{2 \times 3} = \sqrt{6}$.

EXAMPLE 1.13

Simplify $(2 + \sqrt{3}) \times (2 - \sqrt{3})$.

Solution
$$\begin{aligned}(2 + \sqrt{3}) \times (2 - \sqrt{3}) &= 2 \times 2 - 2 \times \sqrt{3} + \sqrt{3} \times 2 - \sqrt{3} \times \sqrt{3} \\ &= 4 - 2\sqrt{3} + 2\sqrt{3} - 3 \\ &= 1\end{aligned}$$

You will notice that in the last example the terms involving square roots disappear, leaving an answer that is a rational number. This is because the two expressions to be multiplied together are the factors of the *difference of two squares*.

$$(a + b)(a - b) = a^2 - b^2$$

In this case $a = 2$ and $b = \sqrt{3}$, so that:

$$\begin{aligned}(2 + \sqrt{3})(2 - \sqrt{3}) &= 2^2 - (\sqrt{3})^2 \\ &= 4 - 3 \\ &= 1\end{aligned}$$

This is the basis of a useful technique for simplifying a fraction of which the bottom line is a surd. The technique, called *rationalising the denominator*, is illustrated in the next examples.

EXAMPLE 1.14

Rationalise $\dfrac{3}{\sqrt{5}}$.

Solution   Multiply the top and bottom line by $\sqrt{5}$.

$$\frac{3}{\sqrt{5}} = \frac{3 \times \sqrt{5}}{\sqrt{5} \times \sqrt{5}} = \frac{3\sqrt{5}}{5} = \tfrac{3}{5}\sqrt{5}$$

EXAMPLE 1.15

Simplify the fraction $\dfrac{(4 + \sqrt{5})}{(3 + \sqrt{5})}$ by rationalising the denominator.

Solution   Multiply both the top and the bottom by $(3 - \sqrt{5})$. You can see that this will make the bottom of the fraction into an expression of the form $(a + b)(a - b)$, which is $(a^2 - b^2)$, and this is a rational number.

$$\frac{(4 + \sqrt{5})}{(3 + \sqrt{5})} = \frac{(4 + \sqrt{5}) \times (3 - \sqrt{5})}{(3 + \sqrt{5}) \times (3 - \sqrt{5})}$$

$$= \frac{12 - 4\sqrt{5} + 3\sqrt{5} - 5}{9 - 3\sqrt{5} + 3\sqrt{5} - 5}$$

$$= \frac{7 - \sqrt{5}}{4}$$

$$= 1\tfrac{3}{4} - \tfrac{1}{4}\sqrt{5}$$

Note

This technique involves multiplying the top line and the bottom line of a fraction by the same factor. This does not change the value of the fraction: it is equivalent to multiplying the fraction by 1.

EXERCISE 1B   Leave all answers in this exercise in surd form, where appropriate.

1 Write the following in the form $a\sqrt{b}$ where a and b are integers.

(a) $\sqrt{8}$ (b) $\sqrt{27}$ (c) $\sqrt{200}$

(d) $\sqrt{288}$ (e) $\sqrt{63}$ (f) $\sqrt{98}$

(g) $\sqrt{320}$ (h) $\sqrt{\tfrac{9}{49}}$ (i) $\sqrt{\tfrac{16}{36}}$

(j) $\sqrt{\tfrac{64}{25}}$

2 Simplify the following.

(a) $\dfrac{7}{\sqrt{7}}$ (b) $\dfrac{9}{3\sqrt{3}}$ (c) $\dfrac{(2 + \sqrt{2})}{\sqrt{2}}$ (d) $\dfrac{5\sqrt{5}}{\sqrt{125}}$ (e) $\dfrac{6}{\sqrt{3}} - \dfrac{8}{2\sqrt{2}}$

3 Simplify the following.

(a) $(3 + \sqrt{2}) + (5 + 2\sqrt{2})$ (b) $(5 + 3\sqrt{2}) - (5 - 2\sqrt{2})$

(c) $4(1 + \sqrt{5}) + 3(5 + 4\sqrt{5})$ (d) $3(\sqrt{2} - \sqrt{5}) + 5(\sqrt{2} + \sqrt{5})$

(e) $3(\sqrt{2} - 1) + 3(\sqrt{2} + 1)$

4 Simplify the following.

(a) $(3 + \sqrt{2})(4 - \sqrt{2})$

(b) $(2 - \sqrt{3})^2$

(c) $(3 - \sqrt{5})(\sqrt{5} + 5)$

(d) $(1 + \sqrt{5})(1 - \sqrt{5})$

(e) $(\frac{1}{2} + \frac{1}{2}\sqrt{3})^2$

(f) $(2 - \sqrt{7})(\sqrt{7} - 1)$

(g) $(4 + \sqrt{2})(4 - \sqrt{2})$

(h) $(5 + \sqrt{30})(6 - \sqrt{30})$

(i) $(\sqrt{7} + \sqrt{5})(\sqrt{7} - \sqrt{5})$

5 Simplify each fraction by rationalising the denominator.

(a) $\dfrac{(3 + \sqrt{2})}{(4 - \sqrt{2})}$

(b) $\dfrac{(2 - \sqrt{3})}{(1 + \sqrt{3})}$

(c) $\dfrac{(3 - \sqrt{5})}{(\sqrt{5} + 5)}$

(d) $\dfrac{1}{(1 - \sqrt{5})}$

(e) $\dfrac{\sqrt{3}}{(\frac{1}{2} - \frac{1}{2}\sqrt{3})}$

(f) $\dfrac{2}{(\sqrt{7} - 1)}$

(g) $\dfrac{(4 + \sqrt{2})}{(4 - \sqrt{2})}$

(h) $\dfrac{(6 + \sqrt{30})}{(6 - \sqrt{30})}$

Historical note

The word algebra is thought to have derived from the Arabic word *al-jabr*. During the Renaissance Arabic texts were translated into Latin. As there was no direct translation the word *algebra* was invented. The term *al-jabr* is thought to mean 'restoring'. It refers to the process of changing a quantity from being subtracted on one side of an equation to being added on the other side.

One of the earliest Arabic texts was written by Muhammad ibn Musa al-Khwarizmi, *Al-kitab al-muhtasar fi hisab al-jabr wa'l-muqabala* (*The Condensed Book on the Calculation of Algebra*).

Algebra was first used by the ancient Greeks and developed most notably by Diophantus in the mid-third century.

QUADRATIC EQUATIONS

An equation such as $2x^2 - 5x + 6 = 0$ in which the highest power of x is x^2 is called a *quadratic equation*. It is usual to write the equation with the right-hand side being 0.

EXAMPLE 1.16

The length of a field is 40 m greater than its width, and its area is 6000 m². Form a quadratic equation involving the length, x m, of the field.

Solution   Since the length of the field is 40 m greater than the width, the width, in m, must be $x - 40$ and the area, in m², is $x(x - 40)$.

So the required quadratic equation is:

$$x(x - 40) = 6000$$
$$\Rightarrow x^2 - 40x - 6000 = 0$$

QUADRATIC FACTORISATION

The process of factorisation requires the quadratic to be written using brackets.

EXAMPLE 1.17

Factorise $x^2 - 5x + 6$.

Solution The first term is x^2 so each bracket will begin with an x.

So $x^2 - 5x + 6 = (x \quad)(x \quad)$.

The two numbers in the brackets must multiply together to make $+6$ and add together to make -5.

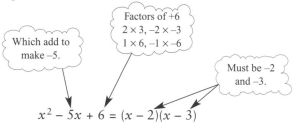

Which add to make -5.

Factors of $+6$
$2 \times 3, -2 \times -3$
$1 \times 6, -1 \times -6$

Must be -2 and -3.

$$x^2 - 5x + 6 = (x - 2)(x - 3)$$

Note Always check your answer by multiplying it out to get back to the original expression.

EXAMPLE 1.18

Factorise $x^2 - 20x + 100$.

Solution $\begin{aligned} x^2 - 20x + 100 &= (x - 10)(x - 10) \\ &= (x - 10)^2 \end{aligned}$

Factors of 100 which add to make -20 are -10 and -10.

Note Example 1.18 was a *perfect square*. It is helpful to recognise the form of such expressions:

$$(x + a)^2 = x^2 + 2ax + a^2$$

and

$$(x - a)^2 = x^2 - 2ax + a^2$$

EXAMPLE 1.19

Factorise $x^2 - 49$.

Solution $x^2 - 49 = (x - 7)(x + 7)$

> Factors of –49
> which add to make
> 0 are 7 and –7.

Note

The example above was the *difference of two squares*. It is helpful to recognise this type of expression in its general form:

$x^2 - a^2 = (x + a)(x - a)$

The quadratic expression may not begin with just x^2 as example 1.20 illustrates.

EXAMPLE 1.20

Factorise $6x^2 + x - 12$.

Solution

> 6 has factors
> $1 \times 6, 2 \times 3.$

> –12 has factors $1 \times -12,$
> $-12 \times 1, 2 \times -6, -2 \times 6,$
> $3 \times -4, -3 \times 4.$

$6x^2 + x - 12 = (2x + 3)(3x - 4)$

> $-8x + 9x = x$

This becomes easier with practice!

Note

The method used in the earlier examples is really the same as this. It is just that in those cases the coefficient of x^2 was 1 and so multiplying the constant term by it had no effect.

 Before starting the procedure for factorising a quadratic, you should always check that the terms do not have a common factor as for example in:

$2x^2 - 8x + 6$

which can be written as $2(x^2 - 4x + 3)$ and factorised to give $2(x - 3)(x - 1)$.

EXAMPLE 1.21

Factorise $(x - 3)^2 - 1$.

Solution Expand then factorise the quadratic.

$$(x - 3)^2 - 1 = x^2 - 6x + 9 - 1$$
$$= x^2 - 6x + 8$$
$$= (x - 2)(x - 4)$$

EXAMPLE 1.22

Factorise $3x^2 - 6x$.

Solution $3x$ is a common factor.

$$\therefore \ 3x^2 - 6x = 3x(x - 2)$$

EXERCISE 1C   Factorise the following quadratic expressions.

1 $x^2 + 6x + 8$ 2 $x^2 - 8x + 15$
3 $x^2 - 10x + 16$ 4 $x^2 - x - 12$
5 $x^2 + 11x + 24$ 6 $x^2 - 10x - 24$
7 $x^2 - 5x - 36$ 8 $x^2 + 12x + 32$
9 $x^2 - 81$ 10 $x^2 - 10x + 25$
11 $x^2 + 12x + 36$ 12 $x^2 - 169$
13 $2x^2 - 5x - 3$ 14 $2x^2 - 7x + 6$
15 $2x^2 + 13x + 20$ 16 $2x^2 - 5x - 12$
17 $3x^2 - 14x + 8$ 18 $3x^2 + 20x + 25$
19 $6x^2 + 5x - 6$ 20 $6x^2 + 19x - 20$
21 $(x - 3)^2 - 4$ 22 $(x + 5)^2 - 1$
23 $3(x - 2)^2 + x - 6$ 24 $4(x - 3)^2 + 5x - 41$
25 $(2x + 3)^2 + x - 6$

SOLVING QUADRATIC EQUATIONS

FACTORISATION

It is a simple matter to solve a quadratic equation once the quadratic expression has been factorised. Since the product of the two factors is 0, it follows that one or other of them must equal 0, and this gives the solution.

EXAMPLE 1.23

Solve $x^2 - 40x - 6000 = 0$.

Solution   First factorise the left-hand side.

$$x^2 - 40x - 6000 = 0$$
$$\therefore (x + 60)(x - 100) = 0$$

$60 \times -100 = -6000$
$60 - 100 = -40$

$\Rightarrow$ either the first bracket is 0 $\therefore x + 60 = 0$
$$x = -60$$

or the second bracket is 0 $\therefore x - 100 = 0$
$$x = 100$$

The solution is $x = -60$ or $x = 100$.

Comparing this with example 1.16, where x is the length of a field, the only valid answer would have been 100.

Note

The *solution* of the equation in the example is $x = -60$ or $x = 100$.

The *roots* of the equation are the values of x which satisfy the equation, in this case one root is $x = -60$ and the other root is $x = 100$.

EXAMPLE 1.24

Solve $2x^2 - 11x + 5 = 0$.

Solution
$$2x^2 - 11x + 5 = 0$$
$$\Rightarrow (2x - 1)(x - 5) = 0$$
$$\Rightarrow \quad 2x - 1 = 0 \quad \text{or} \quad x - 5 = 0$$
$$\Rightarrow \quad x = \tfrac{1}{2} \quad \text{or} \quad x = 5$$

EXERCISE 1D   Solve the following equations by first factorising the expressions.

1 $x^2 - 3x + 2 = 0$ 2 $x^2 + 3x - 10 = 0$ 3 $x^2 - 10x + 24 = 0$
4 $x^2 + 8x + 15 = 0$ 5 $x^2 - 5x - 14 = 0$ 6 $x^2 + 2x - 48 = 0$
7 $x^2 + 16x + 48 = 0$ 8 $x^2 - 14x + 49 = 0$ 9 $x^2 - 49 = 0$
10 $x^2 - 4x - 32 = 0$ 11 $2x^2 - 9x + 4 = 0$ 12 $2x^2 + x - 15 = 0$
13 $3x^2 + 11x + 6 = 0$ 14 $6x^2 - 17x + 10 = 0$ 15 $4x^2 - 16x + 15 = 0$
16 $4x^2 + 5x - 6 = 0$ 17 $10x^2 + 9x + 2 = 0$ 18 $10x^2 - 26x + 12 = 0$
19 $6x^2 - 13x + 6 = 0$ 20 $x^2 + 15 = 8x$ 21 $x^2 + x = 20$
22 $(x - 6)^2 = x$ 23 $6(x - 1)^2 = 5x - 6$ 24 $x + 4 = \dfrac{36}{x - 1}$
25 $x - 8 = \dfrac{26}{x + 3}$

EQUATIONS THAT CANNOT BE FACTORISED

The method of quadratic factorisation is fine so long as the quadratic expression can be factorised, but not all of them can. In the case of $x^2 - 6x + 2$, for example, it is not possible to find two whole numbers which add to give −6 and multiply to give +2.

There are other techniques available for such situations, as you will see in the next few pages.

Graphical solution

If an equation has a solution, you can always find an approximate value for it by working out coordinates and drawing a graph. In the case of:

$$f(x) = x^2 - 6x + 2$$

draw the graph of $y = f(x)$ and solve $f(x) = 0$ by first finding where it cuts the x-axis.

Note	f(x) means a function of x.

x	0	1	2	3	4	5	6
x^2	0	1	4	9	16	25	36
$-6x$	0	−6	−12	−18	−24	−30	−36
$+2$	+2	+2	+2	+2	+2	+2	+2
y	+2	−3	−6	−7	−6	−3	+2

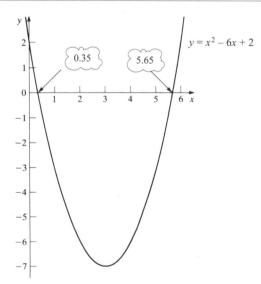

FIGURE 1.2

From figure 1.2, when $f(x) = 0$ x is approximately 0.35 or 5.65.

Clearly the accuracy of the answer is dependent on the scale of the graph, but however large a scale you use, your answer will never be completely accurate.

EXERCISE 1E   Make an accurate drawing of $y = f(x)$ and solve $f(x) = 0$ for each of the following.

1 $f(x) = x^2 - 5x + 2$ $0 \leqslant x \leqslant 5$ **2** $f(x) = x^2 + 4x - 4$ $-5 \leqslant x \leqslant 1$

3 $f(x) = x^2 - 6x + 3$ $0 \leqslant x \leqslant 6$ **4** $f(x) = x^2 - 8x + 13$ $-1 \leqslant x \leqslant 7$

5 $f(x) = 2 + 5x - x^2$ $-1 \leqslant x \leqslant 6$ **6** $f(x) = 3 + 4x - x^2$ $-1 \leqslant x \leqslant 5$

7 $f(x) = 2x^2 - 6x + 3$ $0 \leqslant x \leqslant 3$ **8** $f(x) = 3x^2 + x - 5$ $-2 \leqslant x \leqslant 2$

9 $f(x) = 4 + 3x - 2x^2$ $-1 \leqslant x \leqslant 3$ **10** $f(x) = 1 + 5x - 3x^2$ $-1 \leqslant x \leqslant 2$

COMPLETING THE SQUARE

It is possible to write quadratic equations in the form $a(x + b)^2 + c$.

The method used to obtain an expression of this form is called *completing the square*. To do this, you need to recall these results obtained earlier in the chapter:

$$(x + a)^2 = x^2 + 2ax + a^2 \quad \text{and} \quad (x - a)^2 = x^2 - 2ax + a^2$$

EXAMPLE 1.25

Express $x^2 + 8x + 5$ in the form $(x + a)^2 + b$, where a and b are integers.

Solution   Using $x^2 + 2ax + a^2 = (x + a)^2$ a is half the coefficient of the term in x.

$$\therefore x^2 + 8x + 5 = (x + 4)^2 - 16 + 5$$

$x^2 + 8x + 16$ Need -16 to maintain the original expression.

$$\therefore x^2 + 8x + 5 = (x + 4)^2 - 11$$

EXAMPLE 1.26

Use the method of completing the square to write $2x^2 - 5x + 11$ in the form $a(x + b)^2 + c$, where a, b and c are rational numbers.

Solution   First take out 2 as a factor.

$$2x^2 - 5x + 11 = 2\left[x^2 - \frac{5}{2}x + \frac{11}{2}\right]$$

Now complete the square on the contents of the square brackets.

$$2x^2 - 5x + 11 = 2\left[(x - \frac{5}{4})^2 - \frac{25}{16} + \frac{11}{2}\right]$$

$$= 2\left[(x - \frac{5}{4})^2 + \frac{63}{16}\right]$$

$$= 2(x - \frac{5}{4})^2 + \frac{63}{8}$$

EXERCISE 1F

1 Express the following in the form $(x + a)^2 + b$, where a and b are rational numbers.

  (a) $x^2 + 6x + 7$ (b) $x^2 + 10x + 30$ (c) $x^2 - 12x + 24$ (d) $x^2 - 8x - 3$

  (e) $x^2 + 16x + 70$ (f) $x^2 + 5x + 6$ (g) $x^2 - 9x + 21$ (h) $x^2 + x + 1$

  (i) $x^2 + \frac{1}{2}x - \frac{1}{4}$ (j) $x^2 + px + q$

2 Express the following in the form $a(x + b)^2 + c$, where a, b and c are rational numbers.

  (a) $2x^2 + 8x + 7$ (b) $2x^2 + 10x + 19$ (c) $3x^2 - 9x + 10$ (d) $5x^2 + 20x - 1$

  (e) $4x^2 + 10x + 5$ (f) $5 + 2x - x^2$ (g) $6 - 3x - x^2$ (h) $-3 + 5x - 2x^2$

  (i) $16 + 4x - 3x^2$ (j) $5x^2 - 40x + 68$

3 Show that $f(x) = 3x^2 - 9x + 1$ can be written as $f(x) = 3(x - \frac{3}{2})^2 - \frac{23}{4}$. What is the minimum value of $f(x)$ and what is the value of x that makes $f(x)$ a minimum?

4 By completing the square, show that $f(x) = 5 + 4x - x^2$ has a maximum value of 9. For what value of x is $f(x)$ a maximum?

5 Express $ax^2 + bx + c$ in the form $p(x + q)^2 + r$.

SOLVING QUADRATIC EQUATIONS BY COMPLETING THE SQUARE

The following examples illustrate how you can use the method of completing the square to solve quadratic equations.

EXAMPLE 1.27

Solve the equation $x^2 - 6x + 2 = 0$ by completing the square.

Solution $x^2 - 6x + 2 = 0$

First complete the square.

$$(x - 3)^2 - 9 + 2 = 0$$
$$\therefore \quad (x - 3)^2 - 7 = 0$$
$$\text{Solving gives: } (x - 3)^2 = 7$$
$$x - 3 = \pm \sqrt{7}$$
$$x = 3 \pm \sqrt{7}$$

EXAMPLE 1.28

By completing the square, solve $6 - 5x - 6x^2 = 0$.

Solution $6 - 5x - 6x^2 = 0$

The first step is to ensure that the expression always begins with x^2.

So $6x^2 + 5x - 6 = 0$

Then divide by 6.

$$x^2 + \frac{5}{6}x - 1 = 0$$

By inspection of $(x + a)^2 = x^2 + 2ax + a^2$, a has to be $\frac{5}{12}$ and a^2 is therefore $\frac{25}{144}$, so completing the square gives:

$$\left(x + \frac{5}{12}\right)^2 - \frac{25}{144} - 1 = 0 \qquad \text{and} -\frac{25}{144} - 1 = -\frac{169}{144}$$

so $$\left(x + \frac{5}{12}\right)^2 - \frac{169}{144} = 0$$

$$\Rightarrow \left(x + \frac{5}{12}\right)^2 = \frac{169}{144}$$

$$\Rightarrow x + \frac{5}{12} = \pm\frac{13}{12}$$

$$\Rightarrow x = -\frac{5}{12} \pm \frac{13}{12} = -\frac{18}{12} \text{ or } \frac{8}{12}$$

So the solution is $x = -1.5$ or $\frac{2}{3}$

EXERCISE 1G   Solve the following quadratic equations by completing the square.

1 $x^2 - 4x + 2 = 0$ 2 $x^2 + 4x - 4 = 0$

3 $x^2 - 6x + 3 = 0$ 4 $x^2 - 8x + 13 = 0$

5 $2 + 5x - x^2 = 0$ 6 $3 + 4x - x^2 = 0$

7 $2x^2 - 8x + 3 = 0$ 8 $2x^2 - 6x + 3 = 0$

9 $5 + 12x - 2x^2 = 0$ 10 $4 + 3x - 2x^2 = 0$

THE QUADRATIC FORMULA

To solve a general quadratic equation $ax^2 + bx + c = 0$ the *quadratic formula* may be used which gives the solutions as:

$$x = \frac{-b \pm \sqrt{b^2 - 4ac}}{2a}$$

The proof uses the method of completing the square:

$$ax^2 + bx + c = 0$$

divide throughout by a:

$$x^2 + \frac{b}{a}x + \frac{c}{a} = 0$$

Using the general result $(x + a)^2 = x^2 + 2ax + a^2$ gives:

$$\left(x + \frac{b}{2a}\right)^2 - \frac{b^2}{4a^2} + \frac{c}{a} = 0$$

$$\Rightarrow \qquad \left(x + \frac{b}{2a}\right)^2 = \frac{b^2}{4a^2} - \frac{c}{a} = \frac{b^2 - 4ac}{4a^2}$$

$$\Rightarrow \qquad x + \frac{b}{2a} = \pm \sqrt{\frac{b^2 - 4ac}{4a^2}} = \frac{\pm\sqrt{b^2 - 4ac}}{2a}$$

$$\Rightarrow \qquad x = \frac{-b \pm \sqrt{b^2 - 4ac}}{2a}$$

This equation should be learnt.

EXAMPLE 1.29

Solve $2x^2 - 6x + 3 = 0$.

Solution   Comparing $2x^2 - 6x + 3 = 0$ with $ax^2 + bx + c = 0$ gives $a = 2$, $b = -6$ and $c = 3$.
Substituting into the quadratic formula gives:

$$x = \frac{-(-6) \pm \sqrt{(-6)^2 - 4 \times 2 \times 3}}{2 \times 2}$$

$$\Rightarrow \quad x = \frac{6 \pm \sqrt{12}}{4}$$

$$\Rightarrow \quad x = \frac{6 \pm 2\sqrt{3}}{4}$$

$$\Rightarrow \quad x = \frac{3 \pm \sqrt{3}}{2}$$

These are the *exact* values of the roots and, given that you are not allowed to use a calculator in Core 1, this is the form in which your answer should be left.
If you use a calculator, you should obtain the answers $x = 2.366$ or $x = 0.634$ (to 3 decimal places).

Note   If $b^2 - 4ac$ is negative its square root cannot be found. In such cases the quadratic has *no real solutions*. You will find more about this on pages 25–27.

EXERCISE 1H   In questions 1 to 20 use the quadratic formula to find the exact values of the roots.

1 $x^2 + 20x + 96 = 0$ 2 $x^2 - 16x + 63 = 0$
3 $x^2 - 2x - 80 = 0$ 4 $x^2 - 24x + 144 = 0$
5 $x^2 - 2x + 3 = 0$ 6 $x^2 + 10x + 25 = 0$
7 $x^2 - 4x + 2 = 0$ 8 $x^2 + 4x - 4 = 0$
9 $x^2 - 6x + 3 = 0$ 10 $x^2 - 8x + 13 = 0$
11 $2 + 5x - x^2 = 0$ 12 $3 + 4x - x^2 = 0$
13 $2x^2 - 8x + 3 = 0$ 14 $2x^2 - 6x + 3 = 0$
15 $5 + 12x - 2x^2 = 0$ 16 $4 + 3x - 2x^2 = 0$
17 $3x^2 + 4x + 5 = 0$ 18 $3x^2 + 4x - 5 = 0$
19 $(2x - 3)^2 = 8$ 20 $(2x - 1)(3x - 1) = x - 5$

....

Note   Many calculators have built-in programs to solve quadratic equations. It is good practice to use the formula to solve the quadratic equation and then use the resident program to check your answer. However, remember that you will not be able to use a calculator in the Core 1 examination.

....

EXERCISE 1I   Use any appropriate method in the following questions.

1 Solve the following equations.

    (a) $x^2 - 11x + 24 = 0$ **(b)** $x^2 + 11x + 24 = 0$

    (c) $x^2 - 11x + 18 = 0$ **(d)** $x^2 - 6x + 9 = 0$

    (e) $x^2 - 64 = 0$

2 Solve the following equations.

    (a) $3x^2 - 5x + 2 = 0$ **(b)** $3x^2 + 5x + 2 = 0$

    (c) $3x^2 - 5x - 2 = 0$ **(d)** $25x^2 - 16 = 0$

    (e) $9x^2 - 12x + 4 = 0$

3 Solve the following equations, where possible.

    (a) $x^2 + 8x + 5 = 0$ **(b)** $x^2 + 2x + 4 = 0$

    (c) $x^2 - 5x - 19 = 0$ **(d)** $(x - 2)^2 = 5$

    (e) $(2x + 1)^2 = \frac{3}{4}$

4 The length of a rectangular field is 30 m greater than its width, w metres.

    (a) Write down an expression for the area, $A\,\mathrm{m}^2$, of the field, in terms of w.

    (b) The area of the field is $8800\,\mathrm{m}^2$. Find its width and perimeter.

5 The height h metres of a ball at time t seconds after it is thrown up in the air is given by the expression:

$$h = 1 + 15t - 5t^2$$

    (a) Find the times at which the height is $11\,\mathrm{m}$.

    (b) At what time does the ball hit the ground?

    (c) What is the greatest height the ball reaches?

6 A cylindrical tin, of height h cm and radius r cm, has surface area, including its top and bottom, $A\,\mathrm{cm}^2$.

    (a) Write down an expression for A in terms of r, h and π.

    (b) A tin of height $6\,\mathrm{cm}$ has surface area $54\pi\,\mathrm{cm}^2$. What is the radius of the tin?

    (c) Another tin has the same diameter as height. Its surface area is $150\pi\,\mathrm{cm}^2$. What is its radius?

7 When the first n positive integers are added together, their sum is given by $\frac{1}{2}n(n + 1)$.

   (a) Demonstrate that this result holds for the case $n = 5$.
   (b) Find the value of n for which the sum is 105.
   (c) What is the smallest value of n for which the sum exceeds 1000?

8 The shortest side AB of a right-angled triangle is x cm long. The side BC is 1 cm longer than AB and the hypotenuse, AC, is 29 cm long.

   Form an equation for x and solve it to find the lengths of the three sides of the triangle.

9 The cost of an anorak rose by £6. As a result a shop could buy five fewer anoraks for £600. If the cost of the anorak was £x before the rise, find expressions, in terms of x, for the number of anoraks which could be bought before and after the rise. Hence form an equation in x and show that it reduces to $x^2 + 6x - 720 = 0$. Solve this equation and state the original cost of the anorak.

[MEI]

10 A grower planned to plant 160 fruit trees in a new orchard. He planted n equal rows. Write down an expression for the number of trees in each row.

   He noticed that if he had had 10 more trees he could have planted two more rows, but with three fewer trees in each row. Write down an equation involving n.

   Show that the equation can be rewritten as:

   $$3n^2 + 16n - 320 = 0$$

   Solve this equation, and hence find the number of rows of trees that he actually planted.

[MEI]

DISCRIMINANT

The part of the quadratic formula that determines whether or not the roots are real and how many of them there are is the part which appears under the square root. This is called the *discriminant*.

In the case of $\quad x = \dfrac{-b \pm \sqrt{b^2 - 4ac}}{2a}$

$b^2 - 4ac$ is the discriminant.

If $b^2 - 4ac > 0$ then the equation has *two real distinct roots*. Graphically, this means the graph of the quadratic function will cut the x-axis in two places, as shown in figure 1.3.

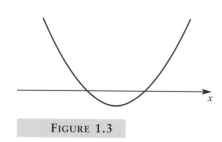

FIGURE 1.3

If $b^2 - 4ac = 0$ then the equation has *two equal real roots*. Graphically, this means the graph of the quadratic function will touch the x-axis at one place, as shown in figure 1.4.

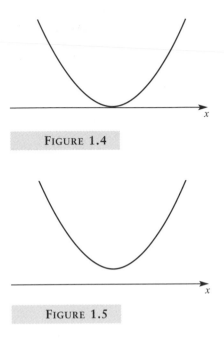

FIGURE 1.4

If $b^2 - 4ac < 0$ then the equation has *no real roots*. Graphically, this means the graph of the quadratic function will not cut the x-axis, as shown in figure 1.5.

FIGURE 1.5

Note

All of the above diagrams assume $a > 0$. If $a < 0$ then they would look like the curves shown in figure 1.6.

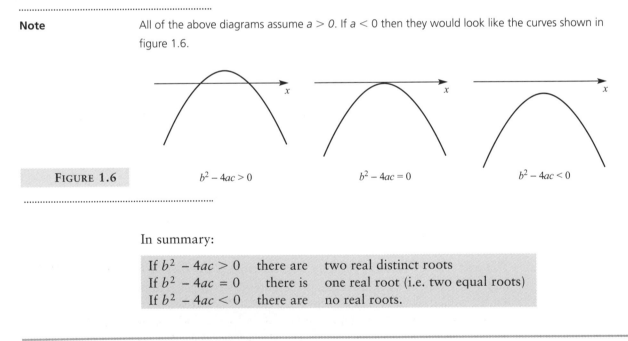

FIGURE 1.6 $b^2 - 4ac > 0$ $b^2 - 4ac = 0$ $b^2 - 4ac < 0$

In summary:

If $b^2 - 4ac > 0$ there are   two real distinct roots
If $b^2 - 4ac = 0$ there is   one real root (i.e. two equal roots)
If $b^2 - 4ac < 0$ there are   no real roots.

EXAMPLE 1.30

Determine the nature of the roots of $5x^2 - 6x - 7 = 0$.

Solution $b^2 - 4ac = 36 + 140 > 0$

There are two real distinct roots.

EXAMPLE 1.31

$4x^2 + kx + 81 = 0$ has equal roots. Find the values of k.

Solution   For equal roots $b^2 - 4ac = 0$ so:

$$k^2 - 4 \times 4 \times 81 = 0$$
$$k^2 = 1296$$
$$k = 36 \quad \text{or} \quad k = -36$$

Note   If $b^2 - 4ac$ is a perfect square then the quadratic can be factorised.

EXERCISE 1J

1 For each of the following state whether there are two real distinct roots, one real root or no real roots.

(a) $x^2 + 2x - 3 = 0$ (b) $x^2 - 2x + 3 = 0$
(c) $x^2 - 4x + 4 = 0$ (d) $2x^2 + 4x - 3 = 0$
(e) $3x^2 + 2x - 4 = 0$ (f) $5x^2 + 8x + 3 = 0$
(g) $4x^2 + 2x - 3 = 0$ (h) $4x^2 - 10x + 9 = 0$
(i) $4x^2 - 12x + 9 = 0$ (j) $6x^2 - 4x - 5 = 0$

2 If $3x^2 - kx + 48 = 0$ has two real distinct roots find the range of values that k can take.

3 $px^2 + 6x + 1 = 0$ has equal roots. What is the value of p?

4 Find the set of values that b can take if $5x^2 + bx + 45 = 0$ has no real roots.

5 Find the range of values that k can take if $kx^2 - 4x + k - 3 = 0$ has real roots.

THE GRAPHS OF QUADRATIC FUNCTIONS

Look at the curve in figure 1.7. It is the graph of $y = x^2 - 4x + 5$ and it has the characteristic shape of a quadratic; it is a parabola.

Notice that:

- it has a minimum turning point (or *vertex*) at $(2, 1)$;
- it has a line of symmetry, $x = 2$.

Is it possible to find the vertex and the line of symmetry without plotting the points? The answer is yes, by the technique of completing the square. The steps are as follows.

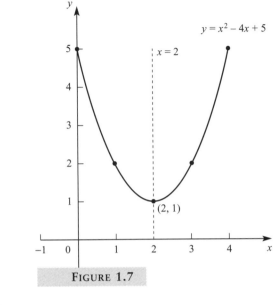

FIGURE 1.7

$$x^2 - 4x + 5 = (x - 2)^2 - 4 + 5$$
$$= (x - 2)^2 + 1$$

The minimum value is 1, so the vertex is $(2, 1)$.

The line of symmetry is $x - 2 = 0$ or $x = 2$.

EXAMPLE 1.32

Write $x^2 + 5x + 4$ in completed square form.

Hence state the equation of the line of symmetry and the coordinates of the vertex of the curve $y = x^2 + 5x + 4$, and sketch the curve.

Solution
$$x^2 + 5x + 4 = (x + \tfrac{5}{2})^2 - \tfrac{25}{4} + 4$$
$$= (x + \tfrac{5}{2})^2 - \tfrac{9}{4}$$
$$= (x + 2.5)^2 - 2.25$$

The line of symmetry is $x + 2.5 = 0$, or $x = -2.5$.

The vertex is at $(-2.5, -2.25)$.

The graph is shown in figure 1.8.

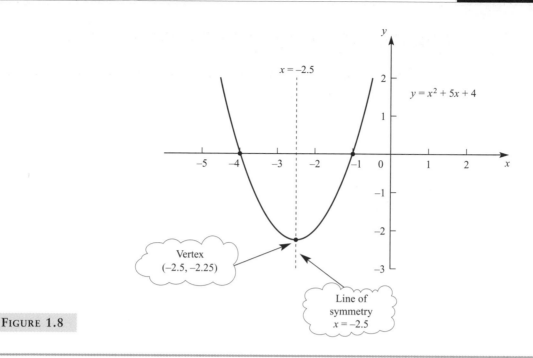

$x = -2.5$

$y = x^2 + 5x + 4$

Vertex
$(-2.5, -2.25)$

Line of
symmetry
$x = -2.5$

FIGURE 1.8

EXAMPLE 1.33

Write $f(x) = 2x^2 - 6x + 5$ in completed square form.

Sketch $y = f(x)$, showing its turning point.

Solution

$2x^2 - 6x + 5$

Take out 2 as a factor

$= 2\{x^2 - 3x + 2.5\}$

$= 2\{(x - 1.5)^2 - 2.25 + 2.5\}$

Complete the
square inside the
brackets

$= 2\{(x - 1.5)^2 + 0.25\}$

$= 2(x - 1.5)^2 + 0.5$

Removing the outside
brackets gives the
completed square form

The quadratic has a minimum
turning point at $(1.5, 0.5)$ and its
graph is shown in figure 1.9.

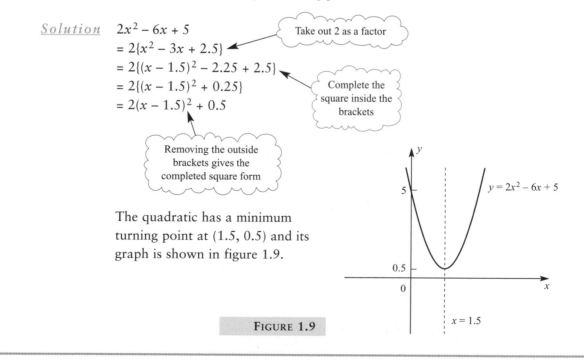

$y = 2x^2 - 6x + 5$

$x = 1.5$

FIGURE 1.9

EXERCISE 1K

1 For each of the following:
   (a) write down the equation of the line of symmetry and the coordinates of the vertex
   (b) write the expression as a polynomial in descending powers of x
   (c) sketch the curve.

   (i) $y = (x + 2)^2 - 3$ (ii) $y = (x + 4)^2 - 4$ (iii) $y = (x - 1)^2 + 2$
   (iv) $y = (x - 10)^2 + 12$ (v) $y = (x - \frac{1}{2})^2 + \frac{3}{4}$ (vi) $y = (x + 0.1)^2 + 0.99$

2 For each of the following equations:
   (a) write it in completed square form
   (b) hence write down the equation of the line of symmetry and the coordinates
       of the vertex
   (c) sketch the curve.

   (i) $y = x^2 + 4x + 9$ (ii) $y = x^2 - 4x + 9$
   (iii) $y = x^2 + 4x + 3$ (iv) $y = x^2 - 4x + 3$
   (v) $y = x^2 + x + 2$ (vi) $y = x^2 - 3x - 7$
   (vii) $y = x^2 - \frac{1}{2}x + 1$ (viii) $y = x^2 + 0.1x + 0.03$

3 Write the following in completed square form.
   (a) $2x^2 + 4x + 6$ (b) $3x^2 - 18x - 27$
   (c) $-x^2 - 2x + 5$ (d) $-2x^2 - 2x - 2$

4 The curves below all have equations of the form $y = x^2 + bx + c$. In each case
   find the values of b and c.

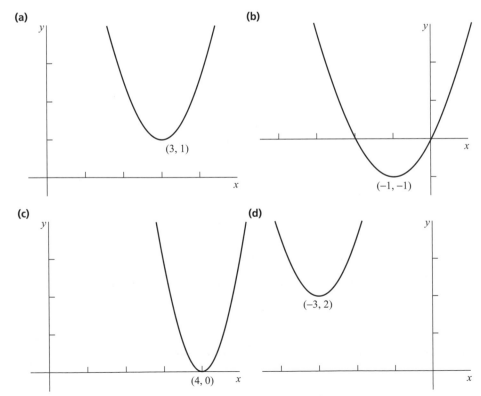

RELATED QUADRATICS

In some equations where the variable appears as a power and as that power squared then a suitable substitution will reduce the equation to a quadratic which can be solved.

EXAMPLE 1.34

Solve the equation $y^4 - 6y^2 - 27 = 0$.

Solution   This can be made into a quadratic equation by putting $x = y^2$ and so $x^2 = y^4$ to give:

$$x^2 - 6x - 27 = 0$$
$$(x - 9)(x + 3) = 0$$

so $x = 9$ or $x = -3$

then $y^2 = 9$ or $y^2 = -3$ (which has no real roots)

giving $y = \pm 3$

Both these values satisfy the original equation.

EXAMPLE 1.35

By using the substitution $x = \sqrt{t}$ solve $t - 3 = 2\sqrt{t}$.

Solution   If $x = \sqrt{t}$ then $x^2 = t$ giving:

$$x^2 - 3 = 2x$$
$$x^2 - 2x - 3 = 0$$

Factorising: $(x - 3)(x + 1) = 0$

Solving: $x = 3$ or $x = -1$

so $\sqrt{t} = 3$ or $\sqrt{t} = -1$

giving $t = 9$ or $t = 1$

Check your solutions in the original equation.

$t = 9$ satisfies the equation but $t = 1$ does not.

The final answer is $t = 9$.

EXERCISE 1L   By making a suitable substitution solve the following equations.

1 $y - 2y^{\frac{1}{2}} - 3 = 0$ **2** $t^4 - 13t^2 + 36 = 0$

3 $p - \sqrt{p} - 6 = 0$ **4** $z^6 - 35z^3 + 216 = 0$

5 $y - 7y^{\frac{1}{2}} - 18 = 0$ **6** $q^4 - 17q^2 + 16 = 0$

7 $\sqrt{t} - 4\sqrt[4]{t} + 3 = 0$ **8** $x^4 - 16x^2 - 225 = 0$

9 $2y^{\frac{2}{3}} + y^{\frac{1}{3}} - 1 = 0$ **10** $4z - 17\sqrt{z} + 4 = 0$

SIMULTANEOUS EQUATIONS

There are many situations which can only be described mathematically in terms of more than one variable. When you need to find the values of the variables in such situations, you need to solve two or more equations simultaneously (i.e. at the same time). Such equations are called *simultaneous equations*. If you need to find values of two variables, you will need to solve two simultaneous equations; if three variables, then three equations, and so on. The work here is confined to solving two equations to find the values of two variables, but most of the methods can be extended to more variables if required.

EXAMPLE 1.36

At a poultry farm, six hens and one duck cost £40, while four hens and three ducks cost £36. What is the cost of each type of bird?

Solution   Let the cost of 1 hen be £h and the cost of 1 duck be £d.
Then the information given can be written as:

$$6h + d = 40 \qquad ①$$
$$4h + 3d = 36 \qquad ②$$

There are several methods of solving this pair of equations.

Method 1: Elimination

Multiplying equation ① by 3	$\Rightarrow$	$18h + 3d = 120$
Leaving equation ②	$\Rightarrow$	$4h + 3d = 36$
Subtracting	$\Rightarrow$	$14h = 84$
Dividing both sides by 14	$\Rightarrow$	$h = 6$

Substituting $h = 6$ in equation ① gives $36 + d = 40$
$$\Rightarrow \qquad d = 4$$

Therefore a hen costs £6 and a duck £4.

Note

1   The first step was to multiply equation ① by 3 so that there would be a term $3d$ in both equations. This meant that when equation ② was subtracted, the variable d was eliminated and so it was possible to find the value of h.

2   The value $h = 6$ was substituted in equation ① but it could equally well have been substituted in the other equation. Check for yourself that this too gives the answer $d = 4$.

Before you look at other methods for solving this pair of equations, here is another example.

EXAMPLE 1.37

Solve

$$3x + 5y = 12 \quad ①$$
$$2x - 6y = -20 \quad ②$$

Solution

$$① \times 6 \implies 18x + 30y = 72$$
$$② \times 5 \implies 10x - 30y = -100$$
$$\text{Adding} \implies 28x = -28$$

$$\text{Giving} \quad x = -1$$

Substituting $x = -1$ in equation ① $\implies -3 + 5y = 12$

Adding 3 to each side $\implies 5y = 15$

$$\implies y = 3$$

Therefore $x = -1$, $y = 3$.

Note

In this example, both equations were multiplied, the first by 6 to give $+30y$ and the second by 5 to give $-30y$. Because one of these terms was positive and the other negative, it was necessary to add rather than subtract in order to eliminate y.

Returning now to the pair of equations giving the prices of hens and ducks:

$$6h + d = 40 \quad ①$$
$$4h + 3d = 36 \quad ②$$

here are two alternative methods of solving them.

Method 2: Substitution

Rearrange the equation $6h + d = 40$ to make d its subject:

$$d = 40 - 6h$$

Now substitute this expression for d in the other equation, $4h + 3d = 36$, giving:

$$4h + 3(40 - 6h) = 36$$
$$\Rightarrow \quad 4h + 120 - 18h = 36$$
$$\Rightarrow \qquad\qquad -14h = -84$$
$$\Rightarrow \qquad\qquad\quad h = 6$$

Substituting for h in $d = 40 - 6h$ gives $d = 40 - 36 = 4$

Therefore a hen costs £6 and a duck £4 (the same answer as before, of course).

Method 3: Intersection of the graphs of the equations

Figure 1.10 shows the graphs of the two equations, $6h + d = 40$ and $4h + 3d = 36$. As you can see, they intersect at the solution, $h = 6$ and $d = 4$.

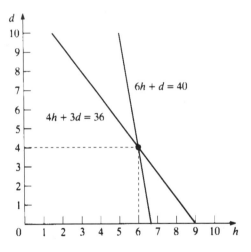

FIGURE 1.10

NON-LINEAR SIMULTANEOUS EQUATIONS

The simultaneous equations in the examples so far have all been *linear*, that is their graphs have been straight lines. A linear equation in, say, x and y contains only terms in x and y and a constant term. So $7x + 2y = 11$ is linear but $7x^2 + 2y = 11$ is not linear, since it contains a term in x^2.

You can solve a pair of simultaneous equations, one of which is linear and the other not, using the substitution method. This is shown in the next example.

EXAMPLE 1.38

Solve $\qquad x + 2y = 7 \qquad$ ①
$\qquad\qquad x^2 + y^2 = 10 \qquad$ ②

Solution   Rearranging equation ① gives $x = 7 - 2y$.

Substituting for x in equation ②:

$$(7 - 2y)^2 + y^2 = 10$$

Multiplying out the $(7 - 2y) \times (7 - 2y)$

gives $49 - 14y - 14y + 4y^2 = 49 - 28y + 4y^2$

so the equation is:

$$49 - 28y + 4y^2 + y^2 = 10$$

This is rearranged to give:

$$5y^2 - 28y + 39 = 0$$
$$\Rightarrow \qquad (5y - 13)(y - 3) = 0$$

> A quadratic in y which you can now solve using factorisation or the formula.

Either $5y - 13 = 0 \implies y = 2.6$
Or $\qquad y - 3 = 0 \implies y = 3$

Substituting in equation ①, $x + 2y = 7$:

$$y = 2.6 \implies x = 1.8$$
$$y = 3 \quad \implies x = 1$$

The solution is either $x = 1.8$, $y = 2.6$ or $x = 1$, $y = 3$.

Note

Always substitute into the linear equation. Substituting in the quadratic will give you extra answers which are not correct.

Historical note

Simultaneous equations have been around for a long time! Diophantus set the problem:
Find two numbers such that their sum is 20 and the sum of their squares is 208.

You could put $x + y = 20$ and $x^2 + y^2 = 208$ and solve the equations.
However, Diophantus set one number as $10 + a$ and the other as $10 - a$ so that:

$$(10 + a)^2 + (10 - a)^2 = 200 + 2a^2 = 208$$
$$a^2 = 4$$

giving the numbers as 8 and 12.

EXERCISE 1M   For questions 1–10 and 15–24, solve the pairs of simultaneous equations, and use a calculator or a computer algebra system to check your answers.

1 $2x + 3y = 8$
$3x + 2y = 7$

2 $x + 4y = 16$
$3x + 5y = 20$

3 $7x + y = 15$
$4x + 3y = 11$

4 $5x - 2y = 3$
$x + 4y = 5$

5 $8x - 3y = 21$
$5x + y = 16$

6 $8x + y = 32$
$7x - 9y = 28$

7 $4x + 3y = 5$
$2x - 6y = -5$

8 $3u - 2v = 17$
$5u - 3v = 28$

9 $4l - 3m = 2$
$5l - 7m = 9$

10 $3t_1 + 2t_2 = 8$
$5t_1 - 7t_2 = -28$

11 A student wishes to spend exactly £10 at a second-hand bookshop. All the paperbacks are one price, all the hardbacks another. She can buy five paperbacks and eight hardbacks. Alternatively she can buy ten paperbacks and six hardbacks.
(a) Write this information as a pair of simultaneous equations.
(b) Solve your equations to find the cost of each type of book.

12 In an examination there are short questions worth 2 marks each and long ones worth 5 marks each. A correct answer is given full marks, a wrong answer is given 0. Rajinder gets 70 marks from 17 correct answers.

(a) Write this information as a pair of simultaneous equations.
(b) Solve your equations to find how many of each type of question Rajinder gets right.

13 The cost of a pear is 5p greater than that of an apple. Eight apples and nine pears cost £1.64.
(a) Write this information as a pair of simultaneous equations.
(b) Solve your equations to find the cost of each type of fruit.

14 A car journey of 380 km lasts 4 hours. Part of this is on a motorway at an average speed of $110 \, \text{km h}^{-1}$, the rest on country roads at an average speed of $70 \, \text{km h}^{-1}$.
(a) Write this information as a pair of simultaneous equations.
(b) Solve your equations to find the distances travelled on each type of road.

15 $x^2 + y^2 = 5$
$2x - y = 4$

16 $x^2 + 4y^2 = 37$
$x - 2y = 7$

17 $x^2 + y^2 = 10$
$x + y = 4$

18 $x^2 - 2y^2 = 8$
$x + 2y = 8$

19 $2x^2 + 3y = 12$
$x - y = -1$

20 $k^2 + km = 8$
$m = k - 6$

21 $t_1{}^2 - t_2{}^2 = 75$
$t_1 = 2t_2$

22 $p + q + 5 = 0$
$p^2 = q^2 + 5$

23 $k(k - m) = 12$
$k(k + m) = 60$

24 $p_1{}^2 - p_2{}^2 = 0$
$p_1 + p_2 = 2$

INEQUALITIES

The symbols that are used for inequalities are $<, \leqslant, >$ and $\geqslant$ and these are defined as follows.

$a < b$ a is less than b	$a > b$ a is greater than b
$a \leqslant b$ a is less than or equal to b	$a \geqslant b$ a is greater than or equal to b.

LINEAR INEQUALITIES

EXAMPLE 1.39

Solve and illustrate on the number line the inequality $x + 7 > 3x - 5$.

Solution   Collecting terms together:

$$12 > 2x$$

or

$$x < 6$$

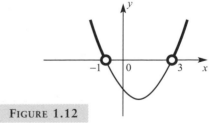

FIGURE 1.11

Note   If the end point is included in the solution it is shown as a solid circle ● but if it is not included then it is shown as a hollow circle ○.

QUADRATIC INEQUALITIES

These are illustrated by example 1.40.

EXAMPLE 1.40

Solve and illustrate $x^2 - 2x - 3 > 0$.

Solution   First solve the quadratic equation $x^2 - 2x - 3 = 0$.

Factorising gives:

$$(x + 1)(x - 3) = 0$$

so $x = -1$ or $x = 3$

Sketching gives the curve shown in figure 1.12.

FIGURE 1.12

From the sketch, for $x^2 - 2x - 3 > 0$ the solution is:

$$x < -1 \quad \text{or} \quad x > 3$$

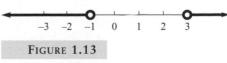

FIGURE 1.13

EXERCISE 1N　　　In each of the following solve the inequality and illustrate your answer with a diagram.

1 $x + 5 > 2$　　　　　　　　　　2 $2x - 7 \leqslant 10$

3 $2x - 4 < x + 1$　　　　　　　4 $5x + 2 - 3x \geqslant 7$

5 $2(x + 1) \geqslant 5(x - 2)$　　　6 $15 > 2(x + 3)$

7 $5(x - 5) - 2(x - 8) < 0$　　8 $4 + 3x \leqslant 2 - x$

9 $4(x + 3) \geqslant 6(x - 2)$　　10 $3(x - 5) - 5(x - 3) < 0$

11 $x^2 - 4x + 3 < 0$　　　　　12 $x^2 - x - 2 \geqslant 0$

13 $x^2 + 7x + 12 > 0$　　　　14 $4 - x^2 \geqslant 0$

15 $3 + 2x - x^2 < 0$　　　　　16 $2x^2 - x - 3 > 0$

17 $3x^2 - 11x - 4 \leqslant 0$　　18 $6x^2 + 5x - 4 > 0$

19 $6 - 5x - 4x^2 \geqslant 0$　　　20 $16 - 9x^2 < 0$

POLYNOMIALS

Polynomials are expressions involving powers of a variable (usually x). For example, $x^2 - 5x + 6$ is a polynomial of order 2. Polynomials of this type, that is $ax^2 + bx + c$ ($a \neq 0$), are known as *quadratics*.

An expression of the form $ax^3 + bx^2 + cx + d$, which includes a term in x^3, is called a *cubic* in x.

Examples of cubic expressions are:

$$2x^3 + 3x^2 - 2x + 11$$

$$3y^3 - 1$$

$$4z^3 - 2z$$

Similarly:

- a *quartic* expression in x, such as $x^4 - 4x^3 + 6x^2 - 4x + 1$, contains a term in x^4

- a *quintic* expression contains a term in x^5

and so on.

All these expressions are called *polynomials*.

The *order* of a polynomial is the highest power of the variable it contains. So a quadratic is a polynomial of order 2, a cubic is a polynomial of order 3 and $3x^8 + 5x^4 + 6x$ is a polynomial of order 8 (an *octic*).

Notice that a polynomial does not contain terms involving $\sqrt{x}$, $\frac{1}{x}$, etc. Apart from the constant term, all the others are multiples of x raised to a positive integer power.

ADDITION OF POLYNOMIALS

Polynomials are added by adding like terms, for example you add the *coefficients* of x^3 together (i.e. the numbers multiplying x^3), the coefficients of x^2 together, the coefficients of x together and the numbers together. You may find it easiest to set this out in columns.

EXAMPLE 1.41

Add $(5x^4 - 3x^3 - 2x)$ to $(7x^4 + 5x^3 + 3x^2 - 2)$.

Solution

$$\begin{array}{rrrrr} 5x^4 & -3x^3 & & -2x & \\ +\ (7x^4 & +5x^3 & +3x^2 & & -2) \\ \hline 12x^4 & +2x^3 & +3x^2 & -2x & -2 \end{array}$$

Note

This may alternatively be set out as follows.

$$\begin{aligned} (5x^4 - 3x^3 - 2x) + (7x^4 + 5x^3 + 3x^2 - 2) &= (5+7)x^4 + (-3+5)x^3 + 3x^2 - 2x - 2 \\ &= 12x^4 + 2x^3 + 3x^2 - 2x - 2 \end{aligned}$$

SUBTRACTION OF POLYNOMIALS

Similarly polynomials are subtracted by subtracting like terms.

EXAMPLE 1.42

Simplify $(5x^4 - 3x^3 - 2x) - (7x^4 + 5x^3 + 3x^2 - 2)$.

Solution

$$\begin{array}{rrrrr} 5x^4 & -3x^3 & & -2x & \\ -\ (7x^4 & +5x^3 & +3x^2 & & -2) \\ \hline -2x^4 & -8x^3 & -3x^2 & -2x & +2 \end{array}$$

Be careful with the signs when subtracting. You may find it easier to change the signs on the bottom line and then go on as if you were adding.

Note

This, too, may be set out as follows.

$$\begin{aligned} (5x^4 - 3x^3 - 2x) - (7x^4 + 5x^3 + 3x^2 - 2) &= (5-7)x^4 + (-3-5)x^3 - 3x^2 - 2x + 2 \\ &= -2x^4 - 8x^3 - 3x^2 - 2x + 2 \end{aligned}$$

MULTIPLICATION OF POLYNOMIALS

When you multiply two polynomials, you multiply each term of one by each term of the other, and all the resulting terms are added. Remember that when you multiply powers of x, you add the indices: $x^5 \times x^7 = x^{12}$.

EXAMPLE 1.43

Multiply $2x(3x + 4y)$.

Solution $2x(3x + 4y) = 2x \times 3x + 2x \times 4y$
$= 6x^2 + 8xy$

EXAMPLE 1.44

Multiply $(2x - 3y)(3x + 4y)$.

Solution $(2x - 3y)(3x + 4y) = 6x^2 + 8xy - 9xy - 12y^2$

Cover up $-3y$ and multiply by $2x$.

Cover up $2x$ and multiply by $-3y$.

$\therefore (2x - 3y)(3x + 4y) = 6x^2 - xy - 12y^2$

EXAMPLE 1.45

Expand $(2x + 5)^3$.

Solution $(2x + 5)^3 = (2x + 5)(2x + 5)(2x + 5)$
$= (2x + 5)(4x^2 + 20x + 25)$
$= 8x^3 + 40x^2 + 50x$
$\qquad\qquad + 20x^2 + 100x + 125$
$= 8x^3 + 60x^2 + 150x + 125$

EXAMPLE 1.46

Multiply $x^2 + 3x - 2$ by $x^2 - 2x - 4$.

Solution $(x^2 + 3x - 2)(x^2 - 2x - 4) = x^4 - 2x^3 - 4x^2$
$\qquad\qquad\qquad\qquad\qquad + 3x^3 - 6x^2 - 12x$
$\qquad\qquad\qquad\qquad\qquad\qquad - 2x^2 + 4x + 8$
$\qquad\qquad\qquad\qquad = x^4 + x^3 - 12x^2 - 8x + 8$

Multiply by x^2 first then by $+3x$ then by -2.

FACTORISING CUBIC EQUATIONS

You will learn more about factorising cubic equations in Chapter 6 of *Pure Mathematics: Core 2*.

Below are some examples of factorising cubics of the form $ax^3 + bx^2 + cx$, or when one factor is known.

EXAMPLE 1.47

Factorise $x^3 + 4x^2 + 3x$.

Solution   By inspection, x is a factor.

$$\therefore x^3 + 4x^2 + 3x = x(x^2 + 4x + 3)$$
$$= x(x + 1)(x + 3)$$

EXAMPLE 1.48

Find the values of a, b and c, given that $f(x) = 2x^3 + 9x^2 + 10x + 3$ can be factorised as $(x + 1)(ax^2 + bx + c)$ where a, b and c are integers.

Hence solve $f(x) = 0$.

Solution $2x^3 + 9x^2 + 10x + 3 = (x + 1)(ax^2 + bx + c)$
$$= ax^3 + (a + b)x^2 + (b + c)x + c$$

By inspection, $a = 2$, $c = 3$ and $a + b = 9$
$$\therefore b = 7$$

Hence $f(x) = (x + 1)(2x^2 + 7x + 3)$.

Factorising completely:

$$f(x) = (x + 1)(2x + 1)(x + 3)$$

If $f(x) = 0$ then $x = -1$, $x = -\frac{1}{2}$ or $x = -3$.

EXERCISE 1O

1 State the orders of each of the following polynomials.
  (a) $x^3 + 3x^2 - 4x$ (b) x^{12}
  (c) $5 - 3x - x^2$ (d) $2 + 6x^2 + 3x^7 - 8x^5$

2 Add $(x^3 + x^2 + 3x - 2)$ to $(x^3 - x^2 - 3x - 2)$.

3 Add $(x^3 - x)$, $(3x^2 + 2x + 1)$ and $(x^4 + 3x^3 + 3x^2 + 3x)$.

4 Subtract $(3x^2 + 2x + 1)$ from $(x^3 + 5x^2 + 7x + 8)$.

5 Subtract $(x^3 - 4x^2 - 8x - 9)$ from $(x^3 - 5x^2 + 7x + 9)$.

6 Subtract $(x^5 - x^4 - 2x^3 - 2x^2 + 4x - 4)$ from $(x^5 + x^4 - 2x^3 - 2x^2 + 4x + 4)$.

7 Multiply $(x^3 + 3x^2 + 3x + 1)$ by $(x + 1)$.

8 Multiply $(x^3 + 2x^2 - x - 2)$ by $(x - 2)$.

9 Multiply $(x^2 + 2x - 3)$ by $(x^2 - 2x - 3)$.

10 Multiply $(x^{10} + x^9 + x^8 + x^7 + x^6 + x^5 + x^4 + x^3 + x^2 + x^1 + 1)$ by $(x - 1)$.

11 Simplify $(x^2 + 1)(x - 1) - (x^2 - 1)(x - 1)$.

12 Simplify $(x^2 + 1)(x^2 + 4) - (x^2 - 1)(x^2 - 4)$.

13 Simplify $(x + 1)^2 + (x + 3)^2 - 2(x + 1)(x + 3)$.

14 Simplify $(x^2 + 1)(x + 3) - (x^2 + 3)(x + 1)$.

15 Simplify $(x^2 - 2x + 1)^2 - (x + 1)^4$.

16 Factorise $15x^2 + 12x$.

17 Factorise $8x^3 + 12x^2y + 4xz$.

18 Simplify and factorise $(2x + 1)^3 + (2x - 1)^3$.

19 Remove the brackets and simplify $(3x + 2)^3$.

20 Remove the brackets and simplify $(3 - 2x)^4$.

21 Factorise $x^3 + 8x^2 + 15x$.

22 Factorise $x^3 - 5x^2 - 14x$.

23 Factorise $6x^3 - 5x^2 - 4x$.

24 Express $f(x) = 2x^3 + 3x^2 - 5x - 6$ as the product $(x + 1)(ax^2 + bx + c)$.
  Factorise $f(x)$ completely and hence solve $f(x) = 0$.

25 Given that $f(x) = x^3 - 3x^2 - 2x + 4$ and $(x - 1)$ is a factor of $f(x)$, express $f(x)$ as a product of a linear and a quadratic function. Hence find the exact roots of $f(x) = 0$.

GRAPHS OF FUNCTIONS

You have already learned how to sketch the graphs of quadratic functions. You will now extend your curve-sketching skills to the graphs of cubic functions and reciprocal functions.

The key features are knowledge of the general shape of each function, and where they cross the x- and y-axes.

GRAPHS OF CUBIC FUNCTIONS

Consider the graph of $y = x^3 - 4x^2 - x + 4$.

You can evaluate values of y for various values of x and plot the points in a table, as follows.

x	−2	−1	0	1	2	3	4	5
y	−18	0	4	0	−6	−8	0	24

Plotting these points gives figure 1.14.

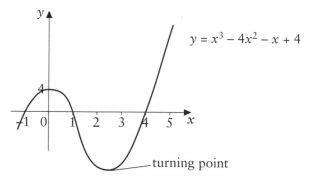

$y = x^3 - 4x^2 - x + 4$

turning point

FIGURE 1.14

This graph illustrates the general shape of a cubic function which has three factors. There are two *turning points* (or *stationary points*) on the curve.

So, if you can factorise the cubic equation you can then work out where the curve cuts the x-axis. By putting $x = 0$ you can also find where it cuts the y-axis. Given that you now know the general shape of the curve you have enough information to make a sketch of the curve.

EXAMPLE 1.49

Sketch the graph of $y = x^3 - 2x^2 - 3x$.

Solution   Factorising gives $y = x(x^2 - 2x - 3)$

$$\therefore y = x(x + 1)(x - 3)$$

Hence, when $y = 0$, $x = 0$, -1 or 3 $\Rightarrow$ where the graph cuts the x-axis.

When $x = 0$, $y = 0$ $\Rightarrow$ where the graph cuts the y-axis.

The sketch is shown in figure 1.15.

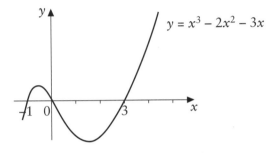

FIGURE 1.15

Note   You will learn how to factorise more complicated cubic functions in Chapter 6 of *Pure Mathematics: Core 2*.

You will learn how to find the coordinates of the turning points in Chapter 11 of *Pure Mathematics: Core 2*.

If the coefficient of the term in x^3 is positive then the graph usually looks like this.

(a)

FIGURE 1.16a

If the coefficient of the term in x^3 is negative then the graph will be reflected to look like this.

(b)

FIGURE 1.16b

EXAMPLE 1.50

Sketch the graph of $y = (2 - x)(3 - x)^2$.

Solution   Factorising gives $y = (2 - x)(3 - x)(3 - x)$.

Hence, when $y = 0$, $x = 2, 3$ or 3 and this is where the graph cuts the x-axis.

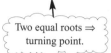

Two equal roots $\Rightarrow$ turning point.

When $x = 0$, $y = 2 \times 3 \times 3 = 18$ and this is where the graph cuts the y-axis.

The sketch looks like this.

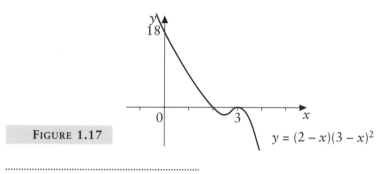

$y = (2 - x)(3 - x)^2$

FIGURE 1.17

Note

If you had multiplied out the brackets you would have seen that the coefficient of x^3 was negative and so you would have known that the graph would be this way round.

The graph in Example 1.50 illustrates the case where there are *two repeated roots*. In such cases the turning point occurs at the repeated root.

If there are *three repeated roots*, such as when $y = x^3$ or when $y = (x - 1)^3$ then the two turning points coincide at the repeated root. In such cases the graphs have the shape shown in figure 1.18.

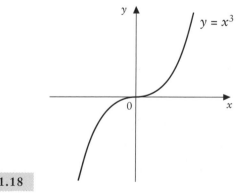

$y = x^3$

FIGURE 1.18

EXERCISE 1P

Sketch the graphs of the following functions, marking the points where the curve intersects the axes.

1 $y = x^3 + x^2 - 2x$

2 $y = 2x^3 - 9x^2 + 10x$

3 $y = 6x - x^2 - x^3$

4 $y = -x^3 + 6x^2 - 8x$

5 $y = 4x - x^3$

6 $y = (x - 1)(x - 2)(x - 3)$

7 $y = (4 + x)(2 + x)(2 - x)$

8 $y = (x + 1)(x - 1)^2$

9 $y = (x - 1)^3$

10 $y = x^3 - 1$

GRAPH OF THE RECIPROCAL FUNCTION $y = \dfrac{1}{x}$

It will be helpful to analyse the function $y = \frac{1}{x}$ before trying to sketch it.

Clearly, as x becomes large $\frac{1}{x}$ becomes small, i.e. as $x \to \infty$, $\frac{1}{x} \to 0$.

If $x = 0$ then $\frac{1}{x}$ is undefined (your calculator gives 'Error').

Considering what happens as x becomes closer to zero then:

if $x = \frac{1}{10}$ $\Rightarrow$ $\frac{1}{x} = 10$

if $x = \frac{1}{100}$ $\Rightarrow$ $\frac{1}{x} = 100$

if $x = \frac{1}{1000}$ $\Rightarrow$ $\frac{1}{x} = 1000$ i.e. as $x \to 0$ (taking positive values) then $\frac{1}{x} \to \infty$.

Using this information, plus any other points for which you may wish to calculate the coordinates, you can now sketch the graph. It should look like that shown in figure 1.19.

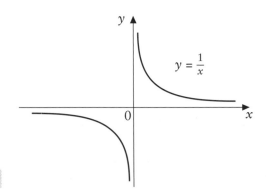

FIGURE 1.19

The curve has two *asymptotes*. These are the lines to which the curve becomes closer and closer but never quite reaches.

For $y = \frac{1}{x}$ the asymptotes are $y = 0$ (x-axis) and $x = 0$ (y-axis).

EXAMPLE 1.51

Sketch the graph of $y = 1 + \frac{1}{x}$, showing clearly any asymptotes.

Solution   The graph of $y = 1 + \frac{1}{x}$ is the same as that of $y = \frac{1}{x}$ with 1 added to all the y-values.

When $y = 0$, $x = -1$.
When $x = 0$, y is undefined (∞).

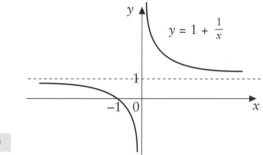

FIGURE 1.20

EXERCISE 1Q   In each of the following, plot points to draw accurately the graph of $y = \frac{1}{x}$ and the given graph.

In each case describe what has happened to the graph of $y = \frac{1}{x}$ to give the other graph.

1 $y = \frac{2}{x}$ **2** $y = \frac{3}{x}$

3 $y = \frac{1}{2x}$ **4** $y = -\frac{1}{x}$

5 $y = \frac{1}{(x-1)}$ **6** $y = \frac{1}{x} - 1$

GEOMETRIC SOLUTIONS OF EQUATIONS

You saw earlier in the chapter how to solve simultaneous equations. In example 1.36, method 3 referred to solving two linear equations by sketching the straight lines and finding the point of intersection.

This method can be applied to any type of curve, not just linear ones. The following examples illustrate how to solve equations by finding the points of intersection of the graphs.

EXAMPLE 1.52

By drawing the graphs of the lines solve the simultaneous equations:

$$x + 2y = 7$$
$$4x - y = 1$$

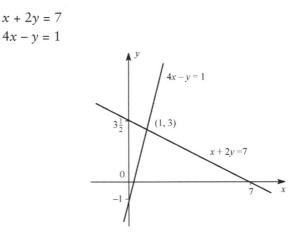

FIGURE 1.21

Solution $x + 2y = 7$ passes through $(0, 3\frac{1}{2})$ and $(7, 0)$.
$4x - y = 1$ passes through $(0, -1)$ and $(\frac{1}{4}, 0)$.

The lines cross at $(1, 3)$ giving the solution to the simultaneous equations as $x = 1$ and $y = 3$.

EXAMPLE 1.53

Draw the graph of $y = x^3 - x^2 - 2x$.

Use your graph to solve these equations.

(a) $x^3 - x^2 - 2x = 0$ (b) $x^3 - x^2 - 2x = 2x$

Solution   Plot points on $y = x^3 - x^2 - 2x$.

x	-2	-1	0	1	2	3
y	-8	0	0	-2	0	12

Now sketch the graph of $y = x^3 - x^2 - 2x$.

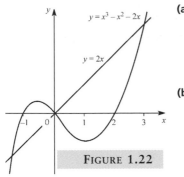

FIGURE 1.22

(a) To solve $x^3 - x^2 - 2x = 0$ you need to find the x-values of the points of intersection between $y = x^3 - x^2 - 2x$ and $y = 0$, i.e. where the curve cuts the x-axis (see figure 1.22). So the solutions are $x = -1$, $x = 0$ and $x = 2$

(b) To solve $x^3 - x^2 - 2x = 2x$ you need to find the x-values of the points of intersection between $y = x^3 - x^2 - 2x$ and $y = 2x$. To do this draw the line $y = 2x$ on your diagram (see figure 1.22). From the diagram the solutions are approximately $x = -1.6$, $x = 0$ and $x = 2.5$.

You should check these answers with a graphical calculator by drawing the graphs and using the trace function.

EXAMPLE 1.54

By drawing the graphs of $y = \frac{4}{x}$ and $y = x^2 - 9$, solve the equation $x^3 - 9x - 4 = 0$.

Solution $y = \frac{4}{x}$ has points:

x	−4	−2	−1	1	2	4
y	−1	−2	−4	4	2	1

$y = x^2 - 9$ has points:

x	−4	−2	−1	0	1	2	4
y	7	−5	−8	−9	−8	−5	7

Plotting the points and drawing the graphs gives the result sketched in figure 1.23.

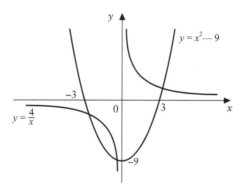

FIGURE 1.23

Where these graphs intersect gives $x^2 - 9 = \frac{4}{x}$

so $x^3 - 9x = 4$

or $x^3 - 9x - 4 = 0$

and the solution of this equation is found from the intersection of the graphs.

That is $x = -2.7$, $x = -0.5$ or $x = 3.2$.

EXERCISE 1R

1 Draw the graph of $y = x^2 - 1$ from $x = -3$ to $x = 3$.
   From your graph solve these equations.
   (a) $x^2 - 1 = 0$
   (b) $x^2 - 1 = 3$

2 Draw the graph of $y = x^2 - x - 1$.
   (a) Use your graph to solve $x^2 - x - 1 = 0$.
   (b) On the same diagram draw $y = x + 2$.
   (c) Use your graph to solve $x^2 - x - 1 = x + 2$.
   (d) For what values of x is $x^2 - x - 1 < x + 2$?

3 (a) On the same axes draw the graphs of $y = x(x + 3)$ and $y = x + 3$.
   (b) Show that $x(x + 3) = x + 3$ can be re-written as $x^2 + 2x - 3 = 0$.
   (c) Use your graph to find the solutions to $x^2 + 2x - 3 = 0$.

4 On the same diagram draw the graphs of $y = x^2 - 2x + 6$ and $y = 6x - x^2$.
   Use your diagram to solve $6x - x^2 > x^2 - 2x + 6$.

5 Draw the graph of $y = \dfrac{1}{x}$ for $-4 < x < 2$.
   On the same diagram draw the graph of $y = x + 3$.
   Show that $\dfrac{1}{x} = x + 3$ can be re-written as $x^2 + 3x - 1 = 0$.
   From your graph, state the approximate solutions to $x^2 + 3x - 1 = 0$.

6 Show that $x^3 - 9x + 4 = 0$ can be written as $\dfrac{4}{x} = 9 - x^2$.
   Draw the graphs of $y = \dfrac{4}{x}$ and $y = 9 - x^2$ and use them to find the solutions to
   $x^3 - 9x + 4 = 0$.

7 Find the x values of the points of intersection between the graphs of
   $y = 6 - x - x^2$ and $y = 2x + 1$.

8 (a) Express $f(x) = 2x^2 + 4x - 7$ in the form $a(x + b)^2 + c$ where a, b and c are
       integers.
   (b) Sketch the graph of $y = f(x)$ and state the coordinates of the minimum
       turning point.
   (c) Given that $2x^2 + 4x - 7 = k$, state the range of values of k for the equation
       to have real and distinct roots. Illustrate such a case on your diagram.

9 On the same graph draw $y = x^3 - 3x^2 - x + 3$ and $y = 2x^2 - 3x - 2$.
   Use your graphs to find the solutions of $x^3 - 5x^2 + 2x + 5 = 0$.

10 The cubic equation $x^3 + 2x^2 - 5x + 2 = 0$ has three solutions, α, β and γ, in the
   interval $-4 \leqslant x \leqslant 2$. Find these solutions by graphical means.
   (a) Draw the graph of $y = x^3 + 2x^2 - 5x + 2$ and hence find the values of α, β
       and γ.
   (b) Now draw the graph of $y = x^2 + 2x - 5$ and another graph. Determine the
       equation of the other graph and, after sketching the graphs of both
       equations, verify that your solutions for α, β and γ are correct.

TRANSFORMATIONS

You already know how to sketch the curves of many functions. Frequently, curve sketching can be made easier by relating the equation of the function to that of a standard function of the same form. This allows you to map the points on the standard curve to equivalent points on the curve you need to draw.

The mappings you will use for curve sketching are called *transformations*. There are several types of transformation, each with different effects, and you will find that by using them singly or in combination you can sketch a large variety of curves much more quickly than by plotting them.

TRANSLATIONS

Figure 1.24 shows the graphs of $y = x^2$ and $y = x^2 + 3$. You could draw this yourself, either on a graphics calculator or by hand. For any given value of x, the y-coordinate for the second curve is 3 units more than the y-coordinate for the first curve.

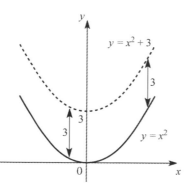

FIGURE 1.24

Although the curves appear to get closer together as you move away from the line of symmetry, their vertical separation is, in fact, constant. Each of the vertical arrows is 3 units long.

You can see that the graphs of $y = x^2 + 3$ and $y = x^2$ are exactly the same shape, but $y = x^2$ has been translated through 3 units in the positive y-direction to obtain $y = x^2 + 3$.

Similarly, $y = x^2 - 2$ could be obtained by translating $y = x^2$ through 2 units in the negative y-direction (i.e. −2 units in the positive y-direction).

In general, for any function $f(x)$, the curve $y = f(x) + b$ can be obtained from that of $y = f(x)$ by translating it through b units in the positive y-direction.

What about the relationship between the graphs of $y = x^2$ and $y = (x - 2)^2$? Figure 1.25 shows the graphs of these two functions. Again, these curves have exactly the same shape, but this time they are separated by a constant 2 units in the x-direction.

You may find it surprising that $y = x^2$ moves in the positive x-direction when 2 is subtracted from x. It happens because x must be correspondingly larger if $(x - 2)$ is to give the same output that x did in the first mapping.

Notice that the axis of symmetry of the curve $y = (x - 2)^2$ is the line $x = 2$.

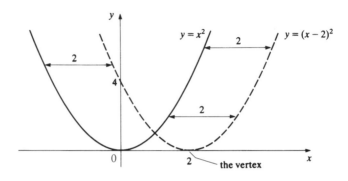

FIGURE 1.25

In general, the curve with equation $y = f(x - a)$ can be obtained from the curve with equation $y = f(x)$ by a translation of a units in the positive x-direction.

Combining these results, $y = f(x - a) + b$ is obtained from $y = f(x)$ by a translation of b units in the positive y-direction and a units in the positive x-direction. This translation is represented by the vector $\begin{pmatrix} a \\ b \end{pmatrix}$.

> $y = f(x)$ mapped on to $y = f(x - a) + b$
> $\Rightarrow$ translation $\begin{pmatrix} a \\ b \end{pmatrix}$

Note

You would usually write $y = f(x - a) + b$ with y as the subject, but this is equivalent to $y - b = f(x - a)$. This form emphasises that subtracting a number from x or y moves the graph in the positive x- or y-direction.

EXAMPLE 1.55

Sketch the curve $y = x^3$ and show how it can be used to obtain the graph of $y = (x - 2)^3 + 1$.

Solution   You can see that the graph of $y = (x - 2)^3 + 1$ is a translation of the graph of $y = x^3$ by the vector $\begin{pmatrix} 2 \\ 1 \end{pmatrix}$ (see figure 1.26).

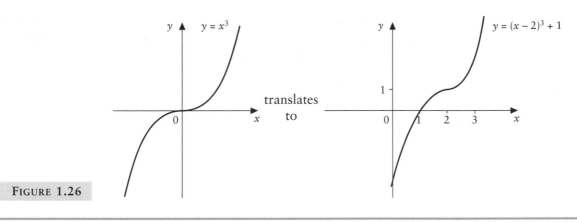

FIGURE 1.26

Earlier in this chapter you met the technique of completing the square of a quadratic function. Using this technique, any quadratic expression of the form $y = x^2 + bx + c$ can be written as $y = (x - p)^2 + q$, so its graph can be sketched by relating it to the graph of $y = x^2$.

EXAMPLE 1.56

(a) Find values of a and b such that $x^2 - 2x + 5 \equiv (x - a)^2 + b$.

(b) Sketch the graph of $y = x^2 - 2x + 5$ and state the position of its vertex and the equation of its axis of symmetry.

Solution

(a) Completing the square gives:
$$x^2 - 2x + 5 = (x - 1)^2 - 1 + 5$$
$$\therefore \quad x^2 - 2x + 5 = (x - 1)^2 + 4$$

(b) Re-writing the equation as $y = (x - 1)^2 + 4$ shows that the curve can be obtained from the graph of $y = x^2$ by a translation of 1 unit in the positive x-direction, and 4 units in the positive y-direction, i.e. a translation $\binom{1}{4}$.

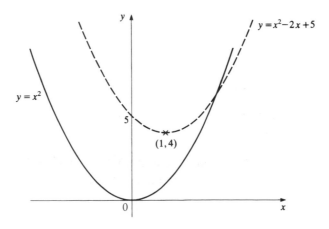

FIGURE 1.27

The vertex is at $(1, 4)$ and the axis of symmetry is the line $x = 1$ (see figure 1.27).

EXAMPLE 1.57

Figure 1.28 shows part of the scalloped bottom of a rollerblind.

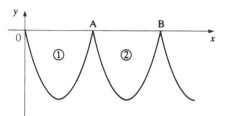

FIGURE 1.28

The equation of curve ① with respect to the x- and y-axes shown is $y = x^2 - 4x$ for

$0 \leqslant x \leqslant 4$.

Find the equation of curve ②.

Solution The equation of curve ① can be factorised to give $y = x(x - 4)$,
so when $y = 0$, $x = 0$ or 4.

Clearly, O is $(0, 0)$ and A is $(4, 0)$.

Curve ② is therefore obtained from curve ① by a translation of 4 units in the
positive x-direction.

To find the equation of curve ②, you replace x by $(x - 4)$ in the equation of curve ①.

Using the factorised form, the equation of curve ② is:

$$y = (x - 4)[(x - 4) - 4] \text{ for } 0 \leqslant x - 4 \leqslant 4$$
$$\Rightarrow \quad y = (x - 4)(x - 8) \qquad \text{for } 4 \leqslant x \leqslant 8$$

EXAMPLE 1.58

Sketch the graph of $y = \frac{1}{x}$.
On a separate diagram, marking any asymptotes, sketch the graph of
$y = \frac{1}{x + 1} + 2$.

Solution The graph of $y = \frac{1}{x}$ looks like figure 1.29.

Its asymptotes are the x-axis and the y-axis.

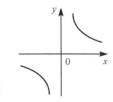

FIGURE 1.29

To obtain $y = \frac{1}{x+1} + 2$ the graph of $y = \frac{1}{x}$ is translated by $\begin{pmatrix} -1 \\ 2 \end{pmatrix}$.

This gives:

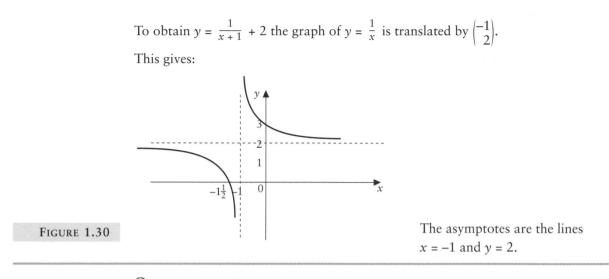

FIGURE 1.30

The asymptotes are the lines
$x = -1$ and $y = 2$.

ONE-WAY STRETCHES

Figure 1.31 shows the graphs of $y = x^2$ and $y = 3x^2$ on the same axes.

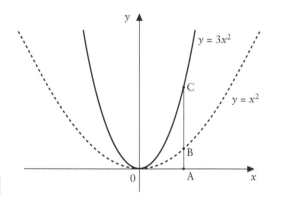

FIGURE 1.31

You should notice that for any three points A, B and C in a line parallel to the y-axis (as shown in the diagram), AC = $3 \times$ AB.

Thus, the transformation that maps $y = x^2$ onto $y = 3x^2$ is equivalent to the curve $y = x^2$ being *stretched* by a scale factor 3, parallel to the y-axis, with the x-axis *invariant* (i.e. not changing).

In general, for any curve $y = f(x)$ and any value of a ($a > 0$), then $y = af(x)$ is obtained from $y = f(x)$ by a stretch scale factor a, parallel to the y-axis, with the x-axis invariant.

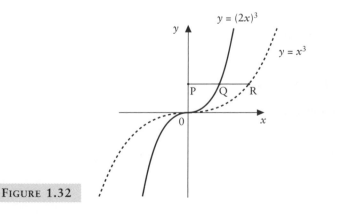

FIGURE 1.32

Now consider the graphs of $y = x^3$ and $y = (2x)^3$ on the same axes, as shown in figure 1.32.

This time, taking any three points P, Q and R in a line parallel to the x-axis (as shown) you should find that PQ = $\frac{1}{2}$ PR.

The graph of $y = x^3$ is compressed by a factor $\frac{1}{2}$ in a direction parallel to the x-axis, with the y-axis invariant.

In general, for any curve $y = f(x)$ and any value of a greater than 0, $y = f(ax)$ is obtained from $y = f(x)$ by a stretch, scale factor $\frac{1}{a}$, parallel to the x-axis, with the y-axis invariant.

Similarly $y = f(\frac{x}{a})$ corresponds to a stretch, scale factor a, parallel to the x-axis.

Note

This is as you would expect: dividing x by a gives a stretch of scale factor a in the x-direction, just as dividing y by a gives a stretch of scale factor a in the y-direction.

The rules are:

> $y = f(x)$ mapped onto $y = f(ax)$ $\Rightarrow$ stretch of scale factor $\frac{1}{a}$ parallel to the x-axis, y-axis invariant.
>
> $y = f(x)$ mapped onto $y = af(x)$ $\Rightarrow$ stretch of scale factor a parallel to the y-axis, x-axis invariant.

EXAMPLE 1.59

Start with the curve $y = x^2$ and sketch $y = \frac{1}{2}x^2$.

Solution   Noting that $y = af(x)$ is a stretch, scale factor a, parallel to the y-axis, then you can find $y = \frac{1}{2}x^2$ by stretching $y = x^2$ by scale factor $\frac{1}{2}$, parallel to the y-axis, as in figure 1.33.

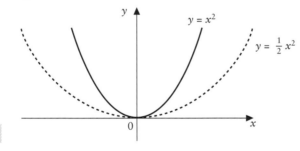

FIGURE 1.33

EXAMPLE 1.60

Figure 1.34 shows the graph of the function $y = f(x)$.

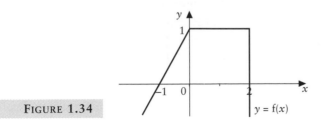

FIGURE 1.34

On separate diagrams, sketch: **(a)** $y = 2f(x)$ **(b)** $y = f(2x)$.

Solution **(a)** $y = 2f(x)$
Stretch, scale factor 2,
parallel to the y-axis.

(b) $y = f(2x)$
Stretch, scale factor $\frac{1}{2}$,
parallel to the x-axis.

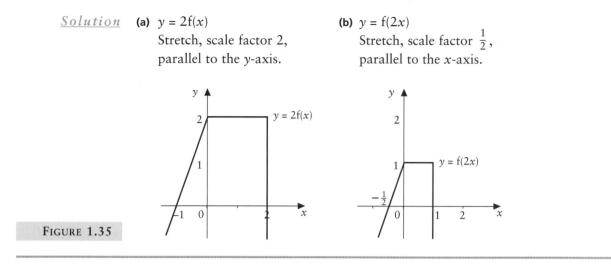

FIGURE 1.35

REFLECTIONS

You know that the graph of $y = (x - 1)^3$ looks like the curve shown in figure 1.36.

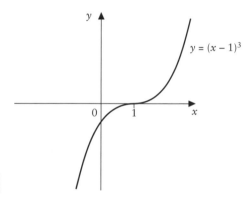

FIGURE 1.36

If you now draw the graph of $y = -(x - 1)^3$ it should look like the curve shown in figure 1.37.

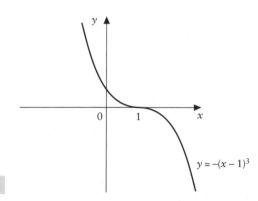

FIGURE 1.37

You can see that $y = -(x - 1)^3$ is a *reflection* of $y = (x - 1)^3$ in the x-axis.

If you now draw the graph of
$y = (-x - 1)^3$ it should look like
the curve shown in figure 1.38.

You can see that $y = (-x - 1)^3$ is a
reflection of $y = (x - 1)^3$ in the y-axis.

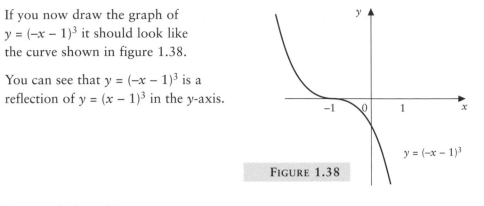

FIGURE 1.38

In general, the rules are:

> $y = f(x)$ mapped onto $y = f(-x) \Rightarrow$ reflection in the y-axis.
>
> $y = f(x)$ mapped onto $y = -f(x) \Rightarrow$ reflection in the x-axis

EXAMPLE 1.61

Figure 1.39 shows the graph of $y = f(x)$ where $f(x) = x^3 - 4x^2 + 3x$.

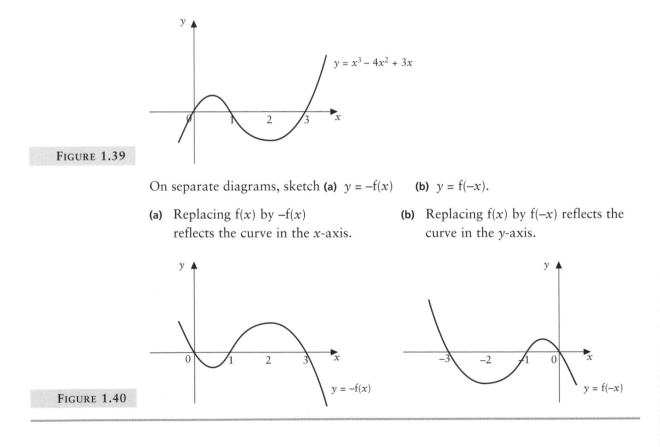

FIGURE 1.39

On separate diagrams, sketch **(a)** $y = -f(x)$ **(b)** $y = f(-x)$.

(a) Replacing $f(x)$ by $-f(x)$
reflects the curve in the x-axis.

(b) Replacing $f(x)$ by $f(-x)$ reflects the
curve in the y-axis.

FIGURE 1.40

EXERCISE 1S

1 Starting with the graph of $y = x^2$, state transformations which can be used to sketch the following curves. Specify the transformations in the order in which they are used, and where there is more than one stage in the sketching of the curve, state each stage. State the equation of the line of symmetry.

   (a) $y = x^2 - 2$
   (b) $y = (x + 4)^2$
   (c) $y = 4x^2$
   (d) $3y = x^2$
   (e) $y = (x - 3)^2 - 5$
   (f) $y = x^2 - 2x$
   (g) $y = x^2 - 4x + 3$
   (h) $y = -x^2$

2 (a) Show that $x^2 + 6x + 5$ can be written in the form $(x + 3)^2 + a$ where a is a constant to be determined.
   (b) Sketch the graph of $y = x^2 + 6x + 5$, giving the equation of the axis of symmetry and the coordinates of the vertex.

3 Given that $f(x) = x^2 - 6x + 11$, find values of p and q such that $f(x) \equiv (x - p)^2 + q$. On the same set of axes, sketch the curves
   (a) $y = f(x)$ and
   (b) $y = f(x + 4)$, labelling clearly which is which.

4 The diagram shows the graph of $y = f(x)$ which has a maximum point at $(-2, 2)$, a minimum point at $(2, -2)$, and passes through the origin.

   Sketch the following graphs, using a separate set of axes for each graph, and indicating the coordinates of the turning points.

   (a) $y = 2f(x)$
   (b) $y = f(x - 2)$
   (c) $y = f(2x)$
   (d) $y = 2 + f(x)$
   (e) $y = f(x + 2) - 2$
   (f) $y = 2f\left(\frac{x}{2}\right)$

5 The diagram shows the graph of $y = f(x)$. Sketch, on separate diagrams, the graphs of these functions.

   (a) $y = f(x + a)$
   (b) $y = f(2x)$
   (c) $y = f(-x)$

6 Sketch the graph of $y = \frac{1}{x}$.
   On separate diagrams, showing any asymptotes with dotted lines, sketch the graphs of these equations.

   (a) $y = \frac{1}{x - 1}$
   (b) $y = -\frac{1}{x}$
   (c) $y = \frac{1}{x - 2} + 3$

7 Starting with the graph of $y = x^2$, state transformations which can be used to sketch the following curves. Specify the transformations in the order in which they are used, and where there is more than one stage in the sketching of the curve, state each stage. State the equation of the line of symmetry.

   (a) $y = -2x^2$
   (b) $y = 4 - x^2$
   (c) $y = 2x - 1 - x^2$

8 (a) Write the expression $x^2 - 6x + 14$ in the form $(x - a)^2 + b$ where a and b are numbers which you are to find.

(b) Sketch the curves $y = x^2$ and $y = x^2 - 6x + 14$ and state the transformation which maps $y = x^2$ onto $y = x^2 - 6x + 14$.

(c) The curve $y = x^2 - 6x + 14$ is reflected in the x-axis. Write down the equation of the image.

9 (a) Sketch the curve with equation $y = x^2$.

(b) Given that $f(x) = (x - 2)^2 + 1$, sketch the curves with the following equations on separate diagrams. Label each curve and give the coordinates of its vertex and the equation of its axis of symmetry.

(i) $y = f(x)$ (ii) $y = -f(x)$ (iii) $y = f(x + 1) + 2$

[MEI]

10 Write the expression $2x^2 + 4x + 5$ in the form $a(x + b)^2 + c$ where a, b and c are numbers to be found.

Use your answer to *write down* the coordinates of the minimum point on the graph of $y = 2x^2 + 4x + 5$.

[OCR]

11 The diagram shows the graph of $y = f(x)$. The curve passes through the origin and has a maximum point at $(1, 1)$.

Sketch, on separate diagrams, the graphs of:

(a) $y = f(x) + 2$

(b) $y = f(x + 2)$

(c) $y = f(2x)$

giving the coordinates of the maximum point in each case.

[OCR]

12 In each of the diagrams below, the curve drawn with a dashed line is obtained as a mapping of the curve $y = f(x)$ using a single transformation. It could be a translation, a one-way stretch or a reflection. In each case, write down the equation of the image (dashed) in terms of $f(x)$.

(a)   (b)   (c)

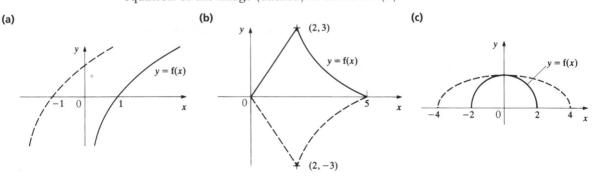

(d)　　　　　　　**(e)**　　　　　　　**(f)**

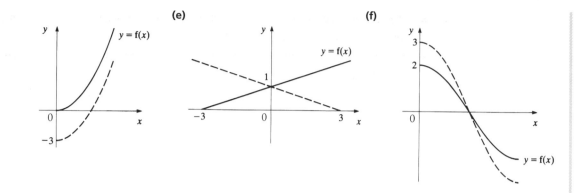

13 Sketch the graph of $y = f(x)$ where $f(x) = x^3 - 3x^2 - 4x$.

On separate diagrams sketch the graphs of:

(a)　$y = f(-x)$　　　(b)　$y = -f(x)$

describing, in each case, how $y = f(x)$ has been transformed.

14 (a)　Given that $f(x) = x^3 - 4x^2 + x + 6$ and that $f(x) = (x - 2)(ax^2 + bx + c)$, find a, b and c.

(b)　Factorise $f(x)$ completely.

(c)　Sketch, on separate diagrams, the following graphs.

(i)　　$y = f(x)$　　(ii)　　$y = -f(x)$　　(iii)　　$y = f(\frac{1}{2}x)$

(iv)　　$y = f(-x)$　　(v)　　$y = f(x + 3)$

15 The diagram shows the graph of $y = f(x)$ where $f(x) = \dfrac{1}{x - 1} + 2$.

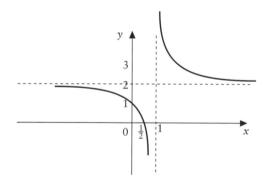

On separate diagrams sketch, showing any asymptomes, the graphs of these functions.

(a)　$y = f(x + 1) - 2$　　(b)　$y = f(\frac{1}{2}x)$　　(c)　$y = -f(x)$　　(d)　$y = 3f(x)$

EXERCISE 1T   **Examination-style questions**

1 Write $\left[\dfrac{125p^6}{27q^9}\right]^{-\frac{1}{3}}$ as an algebraic fraction in its simplest form.

2 If $p^m = 16$ and $p^n = 10$ find:

   (a) p^{m+n}

   (b) $8p^{-m}$

   (c) $p^{\frac{m}{2}-2n}$

3 Express $\dfrac{6+\sqrt{2}}{2+\sqrt{2}}$ in the form $a + b\sqrt{2}$.

4 Express $(3 - 2\sqrt{5})^2$ in the form $a + b\sqrt{5}$.

5 **(a)** Remove the brackets from $(x^2 + 2x + a)(x^2 + bx + 3)$.

   (b) Given that:

$$(x^2 + 2x + a)(x^2 + bx + 3) \equiv x^4 + cx^3 + 11x^2 + 14x + 12$$

   find a, b and c.

6 The cubic polynomial $2x^3 + x^2 + kx - 10$ has a factor $x - 2$. Find the value of k.

7 The cubic polynomial $x^3 + 5x^2 - 4x - 20$ is denoted by $f(x)$.

   (a) Show that $x - 2$ is a factor of $f(x)$.

   (b) Factorise $f(x)$ completely.

   (c) Sketch the graph of $y = f(x)$.

   (d) Solve $f(x) \geqslant 0$.

8 Given that $f(x) = 2x^3 - x^2 - 10x + 8$ find a linear factor and express $f(x)$ as the product of a linear factor and a quadratic factor.

Hence find the two positive roots of $f(x) = 0$, expressing one of your answers in the form $\dfrac{a+\sqrt{b}}{4}$ where a and b are integers.

9 **(a)** Factorise $(x - 2)^2 - 25$.

   (b) Solve $(x - 2)^2 - 25 = 0$.

10 **(a)** Factorise $25x^2 - 20x + 3$.

   (b) By making a suitable substitution solve:

$$25y - 20y^{\frac{1}{2}} + 3 = 0$$

   leaving your answers as fractions.

11 Solve these simultaneous equations.

$$x + 5y = 13$$
$$x^2 + 3y^2 = 21$$

12 Solve the inequality:

$$2x^2 + 5x - 3 \geqslant 0$$

13 By substituting $t = x^{\frac{1}{2}}$, or otherwise, find the values of x for which:

$$4x + 8 = 33x^{\frac{1}{2}}$$

[Edexcel]

14 (a) Use algebra to solve $(x - 1)(x + 2) = 18$.

   (b) Hence, or otherwise, find the set of values of x for which $(x - 1)(x + 2) > 18$.

[Edexcel]

15 (a) Show that $(x - 2)$ is a factor of $f(x) = x^3 + x^2 - 5x - 2$.

   (b) Hence, or otherwise, find the exact solutions of the equation $f(x) = 0$.

[Edexcel]

16 (a) Given that $8 = 2^k$, write down the value of k.

   (b) Given that $4^x = 8^{2 - x}$, write down the value of x.

[Edexcel]

17 The equation $x^2 + 5kx + 2k = 0$, where k is a constant, has real roots.

   (a) Prove that $k(25k - 8) \geqslant 0$.

   (b) Hence find the set of possible values of k.

   (c) Write down the values of k for which the equation $x^2 + 5kx + 2k = 0$ has equal roots.

[Edexcel]

18 Given that $2^x = \dfrac{1}{\sqrt{2}}$ and $2^y = 4\sqrt{2}$:

   (a) find the exact value of x and the exact value of y

   (b) calculate the exact value of $2^{y - x}$.

[Edexcel]

19 (a) By completing the square, find, in terms of k, the roots of the equation $x^2 + 2kx - 7 = 0$.

   (b) Prove that, for all real values of k, the roots of $x^2 + 2kx - 7 = 0$ are real and different.

   (c) Given that $k = \sqrt{2}$, find the exact roots of the equation.

[Edexcel]

20 Find the set of values of x for which:

   (a) $6x - 7 < 2x + 3$

   (b) $2x^2 - 11x + 5 < 0$

   (c) both $6x - 7 < 2x + 3$ and $2x^2 - 11x + 5 < 0$.

[Edexcel]

KEY POINTS

1 Rules of indices

$$a^m \times a^n = a^{m+n} \qquad a^m \div a^n = a^{m-n} \qquad a^0 = 1 \ (a \neq 0)$$
$$(a^m)^n = a^{mn} \qquad a^{\frac{m}{n}} = \sqrt[n]{a^m} = \left(\sqrt[n]{a}\right)^m \qquad a^{-m} = \frac{1}{a^m}$$

2 Rules of surds

$$\sqrt{ab} = \sqrt{a} \times \sqrt{b} \qquad \sqrt{\frac{a}{b}} = \frac{\sqrt{a}}{\sqrt{b}} \qquad \frac{1}{\sqrt{a}} = \frac{\sqrt{a}}{a}$$

3 Useful equations

$$(x + a)^2 = x^2 + 2ax + a^2 \qquad (x - a)^2 = x^2 - 2ax + a^2$$

4 Completing the square

$$x^2 + bx + c = (x + \frac{b}{2})^2 - \frac{b^2}{4} + c$$

5 Quadratic equation

$$x = \frac{-b \pm \sqrt{b^2 - 4ac}}{2a}$$

6 Discriminant

If $b^2 - 4ac > 0$ there are two real and distinct roots.
If $b^2 - 4ac = 0$ there is one real root (i.e. two equal roots).
If $b^2 - 4ac < 0$ there are no real roots.

7 Simultaneous equations

Solve algebraically by substitution (or by elimination).

8 Graphs of functions

General shapes

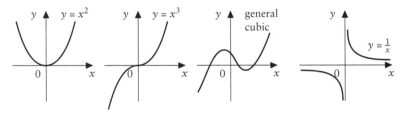

9 Transformations

$y = f(x)$ mapped onto $y = f(x - a) + b \quad \Rightarrow$ translation with vector $\binom{a}{b}$.

$y = f(x)$ mapped onto $y = f(ax) \quad \Rightarrow$ stretch, scale factor $\frac{1}{a}$, parallel to the x-axis, y-axis invariant.

$y = f(x)$ mapped onto $y = af(x) \quad \Rightarrow$ stretch, scale factor a parallel to the y-axis, x-axis invariant

$y = f(x)$ mapped onto $y = f(-x) \quad \Rightarrow$ reflection in the y-axis

$y = f(x)$ mapped onto $y = -f(x) \quad \Rightarrow$ reflection in the x-axis

Chapter two

COORDINATE GEOMETRY IN THE (x, y) PLANE

A place for everything, and everything in its place.

Samuel Smiles

COORDINATES

Coordinates are a means of describing a position relative to some fixed point, or *origin*. In two dimensions you need two pieces of information; in three dimensions, you need three pieces of information.

In the Cartesian system (named after René Descartes), position is given in perpendicular directions: x, y in two dimensions; x, y, z in three dimensions (see figure 2.1). This chapter concentrates exclusively on two dimensions.

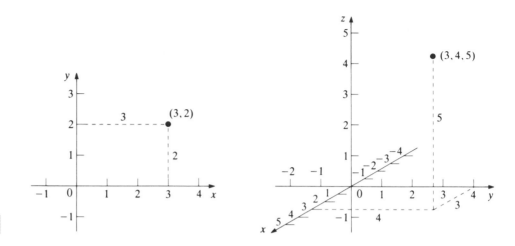

Figure 2.1

Plotting, sketching and drawing

In two dimensions, the coordinates of points are often marked on paper and joined up to form lines or curves. Several words are used to describe this process.

Plot (a line/curve) means mark the points and join them up as accurately as you can. You would expect to do this on graph paper and be prepared to read information from the graph.

Sketch means mark points in approximately the right positions and join them up in the right general shape. You would not expect to use graph paper for a sketch and would not read precise information from one. You would, however, mark on the coordinates of important points, such as intersections with the x- and y-axes and points at which the curve changes direction.

Draw means that you are to use a level of accuracy appropriate to the circumstances, and this could be anything between a rough sketch and a very accurately plotted graph.

The gradient of a line

In everyday English the word *line* is used to mean a straight line or a curve. In mathematics it is understood to mean a straight line. If you know the coordinates of any two points on a line, then you can draw the line.

The slope of a line is measured by its *gradient*. It is often denoted by the letter m.

In figure 2.2, A and B are two points on the line. The gradient of the line AB is given by the increase in value of the y-coordinate from A to B divided by the increase in value of the x-coordinate from A to B.

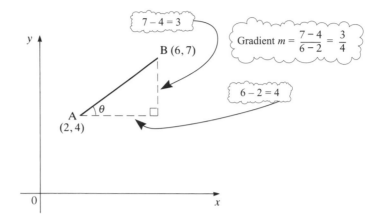

FIGURE 2.2

In general, when A is the point (x_1, y_1) and B is the point (x_2, y_2), the gradient is:

$$m = \frac{y_2 - y_1}{x_2 - x_1}$$

When the same scale is used on both axes, $m = \tan\theta$ (figure 2.2).

Figure 2.3 shows four lines. Looking at them in turn, from left to right: line A goes uphill and its gradient is positive; line B goes downhill and its gradient is negative. Line C is horizontal and its gradient is 0; the vertical line D has an infinite gradient.

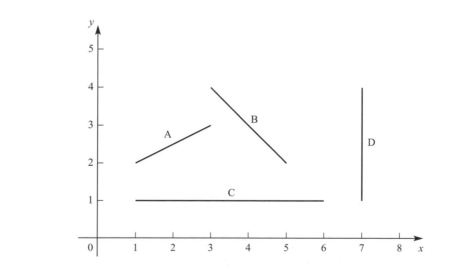

FIGURE 2.3

PARALLEL AND PERPENDICULAR LINES

If you know the gradients m_1 and m_2 of two lines then:

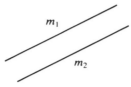 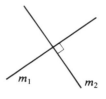

FIGURE 2.4

If the lines are parallel:

$$m_1 = m_2$$

If the lines are perpendicular:

$$m_1 m_2 = -1$$

This result can be shown in figure 2.5 which has perpendicular lines l_1 and l_2. The gradient m_1 of l_1 can be obtained from triangle PQR so that:

$$m_1 = \frac{b}{a}$$

Rotating triangle PQR 90° clockwise onto PQ'R' gives the gradient of l_2 to be:

$$m_2 = -\frac{a}{b}$$

from which:

$$m_1 m_2 = \frac{b}{a} \times -\frac{a}{b} = -1$$

FIGURE 2.5

Note You can only talk about lines being perpendicular if the same scale has been used for both axes.

DISTANCE BETWEEN TWO POINTS

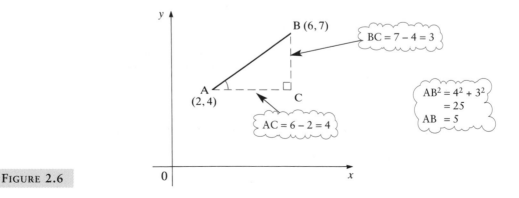

FIGURE 2.6

When you know the coordinates of two points, you can use Pythagoras' theorem to calculate the distance between them, as shown in figure 2.6.

This method can be generalised to find the distance between any two points, $A(x_1, y_1)$ and $B(x_2, y_2)$, as in figure 2.7.

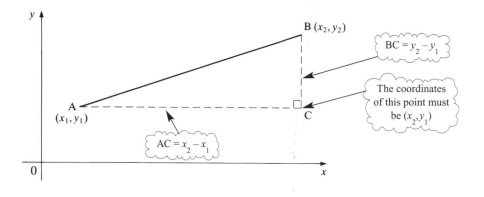

FIGURE 2.7

The length of the line AB is $\sqrt{(x_2 - x_1)^2 + (y_2 - y_1)^2}$

MID-POINT OF A LINE JOINING TWO POINTS

Look at the line joining the points A(2, 1) and B(8, 5) in figure 2.8.
The point M(5, 3) is the mid-point of AB.

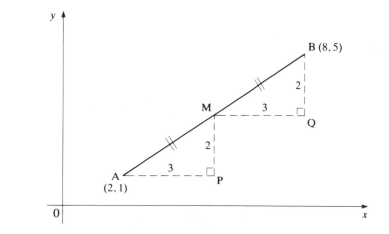

FIGURE 2.8

Notice that the coordinates of M are the *means* of the coordinates of A and B.

$$5 = \tfrac{1}{2}(2 + 8) \qquad 3 = \tfrac{1}{2}(1 + 5)$$

This result can be generalised as follows. For any two points $A(x_1, y_1)$ and $B(x_2, y_2)$, the coordinates of the mid-point of AB are the means of the coordinates of A and B so the mid-point is:

$$\left(\frac{x_1 + x_2}{2}, \frac{y_1 + y_2}{2} \right)$$

EXAMPLE 2.1

A and B are the points (2, 5) and (6, 3) respectively (see figure 2.9). Find:

(a) the gradient of AB

(b) the length of AB

(c) the mid-point of AB

(d) the gradient of the line perpendicular to AB.

Solution Taking A(2, 5) as the point (x_1, y_1), and B(6, 3) as the point (x_2, y_2) gives $x_1 = 2, y_1 = 5, x_2 = 6, y_2 = 3$.

(a) Gradient $= \dfrac{y_2 - y_1}{x_2 - x_1}$

$= \dfrac{3 - 5}{6 - 2}$

$= -\dfrac{1}{2}$

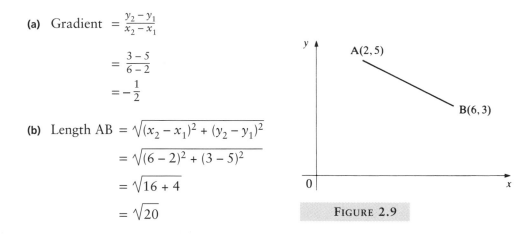

(b) Length AB $= \sqrt{(x_2 - x_1)^2 + (y_2 - y_1)^2}$

$= \sqrt{(6 - 2)^2 + (3 - 5)^2}$

$= \sqrt{16 + 4}$

$= \sqrt{20}$

FIGURE 2.9

(c) Mid-point $= \left(\dfrac{x_1 + x_2}{2}, \dfrac{y_1 + y_2}{2} \right)$

$= \left(\dfrac{2 + 6}{2}, \dfrac{5 + 3}{2} \right)$

$= (4, 4)$

(d) Gradient of AB $= m_1 = -\dfrac{1}{2}$.

If m_2 is the gradient of the line perpendicular to AB, then $m_1 m_2 = -1$

$\Rightarrow -\dfrac{1}{2} m_2 = -1$

$m_2 = 2$

EXAMPLE 2.2

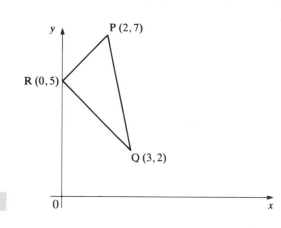

Using two different methods, show that the lines joining P(2, 7), Q(3, 2) and R(0, 5) form a right-angled triangle (see figure 2.10).

FIGURE 2.10

Solution   **Method 1**

Gradient of RP $= \dfrac{7-5}{2-0} = 1$

Gradient of RQ $= \dfrac{2-5}{3-0} = -1$

$\Rightarrow$ product of gradients $= 1 \times (-1) = -1$

$\Rightarrow$ sides RP and RQ are at right angles.

Method 2

Pythagoras' theorem states that for a right-angled triangle with hypotenuse of length c and in which the other sides have lengths a and b, $a^2 + b^2 = c^2$.

Conversely, if you can show that $a^2 + b^2 = c^2$ for a triangle with sides of lengths a, b and c, then the triangle has a right angle and the side of length a is the hypotenuse.

This is the basis for the alternative proof, in which you would use:

$$\text{length}^2 = (x_2 - x_1)^2 + (y_2 - y_1)^2$$
$$PQ^2 = (3-2)^2 + (2-7)^2 = 1 + 25 = 26$$
$$RP^2 = (2-0)^2 + (7-5)^2 = 4 + 4 = 8$$
$$RQ^2 = (3-0)^2 + (2-5)^2 = 9 + 9 = 18$$

Since $26 = 8 + 18,$ $PQ^2 = RP^2 + RQ^2$

$\Rightarrow$ sides RP and RQ are at right angles.

1 For the following pairs of points A and B, calculate:

    (a) the gradient of the line AB

    (b) the mid-point of the line joining A to B

    (c) the distance AB

    (d) the gradient of the line perpendicular to AB.

        (i) A(0, 1)   B(2, −3)   **(ii)** A(3, 2)   B(4, −1)

        (iii) A(−6, 3) B(6, 3)   **(iv)** A(5, 2)   B(2, −8)

        (v) A(4, 3)   B(2, 0)   **(vi)** A(1, 4)   B(1, −2)

2 The line joining the point P(3, −4) to Q(q, 0) has a gradient of 2. Find the value of q.

3 The three points X(2, −1), Y(8, y) and Z(11, 2) are collinear (i.e. they lie on the same straight line). Find the value of y.

4 The points A, B, C and D have coordinates (1, 2), (7, 5), (9, 8) and (3, 5).

    (a) Find the gradients of the lines AB, BC, CD and DA.

    (b) What do these gradients tell you about the quadrilateral ABCD?

    (c) Draw a diagram to check your answer to part (b).

5 The points A, B and C have coordinates (2, 1), (b, 3) and (5, 5), where $b > 3$ and $\angle ABC = 90°$. Find:

    (a) the value of b

    (b) the lengths of AB and BC

    (c) the area of triangle ABC.

6 The triangle PQR has vertices P(8, 6), Q(0, 2) and R(2, r). Find the values of r when the triangle:

    (a) has a right angle at P

    (b) has a right angle at Q

    (c) has a right angle at R

    (d) is isosceles with RQ = RP.

7 The points A, B, and C have coordinates (−4, 2), (7, 4) and (−3, −1).

    (a) Draw the triangle ABC.

    (b) Show by calculation that the triangle ABC is isosceles and name the two equal sides.

    (c) Find the mid-point of the third side.

    (d) By calculating appropriate lengths, calculate the area of the triangle ABC.

8 For the points P(x, y) and Q($3x$, $5y$), find in terms of x and y:

    (a) the gradient of the line PQ

    (b) the mid-point of the line PQ

    (c) the length of the line PQ.

9 A quadrilateral has vertices A(0, 0), B(0, 3), C(6, 6), and D(12, 6).

   (a) Draw the quadrilateral.
   (b) Show by calculation that it is a trapezium.
   (c) Find the coordinates of E when EBCD is a parallelogram.

10 Three points A, B and C have coordinates (1, 3), (3, 5), and (−1, y) respectively. Find the values of y when:

   (a) AB = AC
   (b) AC = BC
   (c) AB is perpendicular to BC
   (d) A, B and C are collinear.

11 The diagonals of a rhombus bisect each other at 90° and conversely, when two lines bisect each other at 90°, the quadrilateral formed by joining the end points of the lines is a rhombus.

   Use the converse result to show that the points with coordinates (1, 2), (8, −2), (7, 6) and (0, 10) are the vertices of a rhombus, and find its area.

THE EQUATION OF A STRAIGHT LINE

The word *straight* means going in a constant direction, that is with fixed gradient. This fact allows you to find the equation of a straight line from first principles.

EXAMPLE 2.3

Find the equation of the straight line with gradient 2 through the point (0, −5).

Solution

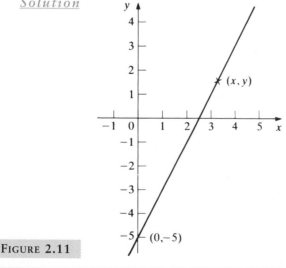

Take a general point (x, y) on the line, as shown in figure 2.11. The gradient of the line joining $(0, -5)$ to (x, y) is given by:

$$\text{gradient} = \frac{y - (-5)}{x - 0} = \frac{y + 5}{x}$$

Since you are told that the gradient of the line is 2, this gives:

$$2 = \frac{y + 5}{x}$$

$$\Rightarrow y = 2x - 5$$

Since (x, y) is a general point on the line, this holds for any point on the line and is therefore the equation of the line.

FIGURE 2.11

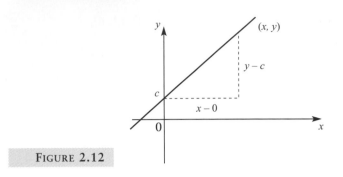

FIGURE 2.12

The result from example 2.3 can be generalised to give a well-known standard equation for a straight line with gradient m cutting the y-axis at the point $(0, c)$:

$$\text{gradient} = \frac{y - c}{x - 0} = m$$

which gives:

$$y = mx + c$$

EXAMPLE 2.4

What is the equation of the line passing through $(0, 1)$ and $(4, 3)$?

Solution   Gradient $m = \dfrac{3 - 1}{4 - 0} = \dfrac{1}{2}$

and $c = 1$

so using $y = mx + c$ gives:

$$y = \tfrac{1}{2}x + 1$$

or $2y = x + 2$

Note   It is good practice to avoid the use of fractions in your final equations.

THE EQUATION OF A STRAIGHT LINE GIVEN THE GRADIENT AND A POINT

If the gradient is m and the given point is (x_1, y_1), and (x, y) is taken to be a general point on the line then:

$$\text{gradient} = m = \frac{y - y_1}{x - x_1}$$

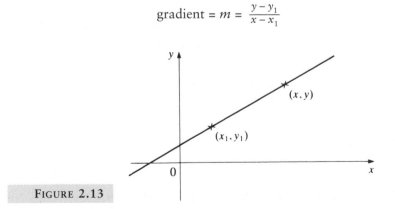

FIGURE 2.13

which gives the useful equation:

$$y - y_1 = m\,(x - x_1)$$

EXAMPLE 2.5

Find the equation of the line with gradient $\frac{1}{4}$ passing through $(1, -2)$.

Solution $m = \frac{1}{4}$, $x_1 = 1$ and $y_1 = -2$

So using $y - y_1 = m(x - x_1)$ gives:

$$y + 2 = \tfrac{1}{4}(x - 1)$$
$$4y + 8 = x - 1$$
$$4y = x - 9$$

THE EQUATION OF A STRAIGHT LINE GIVEN TWO POINTS

If the points are (x_1, y_1) and (x_2, y_2) then use $m = \frac{y_2 - y_1}{x_2 - x_1}$ to find the gradient and substitute into $y - y_1 = m\,(x - x_1)$ to find the equation of the line.

EXAMPLE 2.6

Find the equation of the line passing through $(1, 3)$ and $(3, 7)$.

Solution $m = \frac{y_2 - y_1}{x_2 - x_1} = \frac{7 - 3}{3 - 1} = 2$

and substituting into $y - y_1 = m(x - x_1)$ gives:

$$y - 3 = 2(x - 1)$$
$$\Rightarrow \qquad y = 2x + 1$$

Alternatively, given two points, you could use two expressions for the gradient, $m = \frac{y - y_1}{x - x_1}$ and $m = \frac{y_2 - y_1}{x_2 - x_1}$, to give a different equation for a straight line, that is

$$m = \frac{y - y_1}{x - x_1} = \frac{y_2 - y_1}{x_2 - x_1}$$

which may be rearranged into the form:

$$m = \frac{y - y_1}{y_2 - y_1} = \frac{x - x_1}{x_2 - x_1}$$

THE EQUATION OF A STRAIGHT LINE IN THE FORM $ax + by + c = 0$

It is sometimes more attractive, particularly to avoid using fractions, to rewrite an equation in the form $ax + by + c = 0$.

For example, the equation $y = -\frac{1}{3}x + 4$ may be rewritten as $x + 3y - 12 = 0$.

DRAWING A LINE, GIVEN ITS EQUATION

There are several standard forms for the equation of a straight line, as shown in figure 2.14.

When you need to draw the graph of a straight line, given its equation, the first thing to do is to look carefully at the form of the equation and see if you can recognise it.

(a), (b): Lines parallel to the axes

Lines parallel to the x-axis have the form $y = $ constant, those parallel to the y-axis have the form $x = $ constant. Such lines are easily recognised and drawn.

(a) Equations of the form $x = a$ **(b)** Equations of the form $y = b$

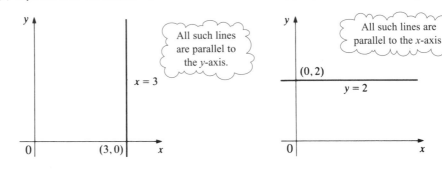

(c) Equations of the form $y = mx$ **(d)** Equations of the form $y = mx + c$

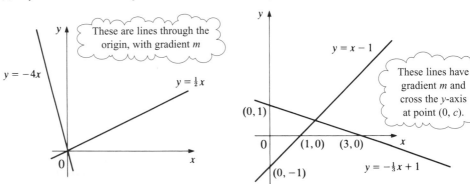

(e) Equations of the form $px + qy + r = 0$

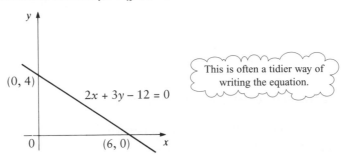

FIGURE 2.14

(c), (d): Equations of the form $y = mx + c$

The line $y = mx + c$ crosses the y-axis at the point $(0, c)$ and has gradient m. If $c = 0$, the line goes through the origin. In either case you know one point and can complete the line either by finding one more point, for example by substituting $x = 1$, or by following the gradient (e.g. 1 along and 2 up for gradient 2).

(e): Equations of the form $px + qy + r = 0$

In the case of a line given in this form, such as $2x + 3y - 6 = 0$, you can either rearrange it in the form $y = mx + c$ (in this example $y = -\frac{2}{3}x + 2$), or you can find the coordinates of two points that lie on it. Putting $x = 0$ gives the point where it crosses the y-axis, $(0, 2)$, and putting $y = 0$ gives its intersection with the x-axis, $(3, 0)$.

EXAMPLE 2.7

Sketch the lines $x = 5$, $y = 0$ and $y = x$ on the same axes. Describe the triangle formed by these lines.

Solution   The line $x = 5$ is parallel to the y-axis and passes through $(5, 0)$.
The line $y = 0$ is the x-axis.
The line $y = x$ has gradient 1 and goes through the origin.

The triangle obtained is an isosceles right-angled triangle, since OA = AB = 5 units, and $\angle$OAB = 90°.

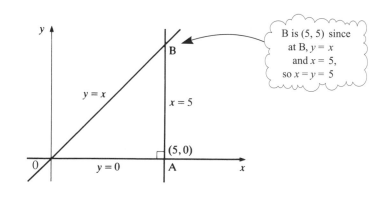

FIGURE 2.15

EXAMPLE 2.8

Sketch $y = x - 1$ and $3x + 4y = 24$ on the same axes.

Solution   The line $y = x - 1$ has gradient 1, and passes throught the point $(0, -1)$.
Substituting $y = 0$ gives $x = 1$, so the line also passes through $(1, 0)$.

Find two points on the line $3x + 4y = 24$.
Substituting $x = 0$ gives $4y = 24$ i.e. $y = 6$
substituting $y = 0$ gives $3x = 24$ i.e. $x = 8$

The line passes through $(0, 6)$ and $(8, 0)$.

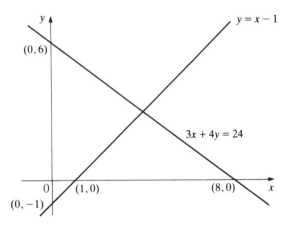

FIGURE 2.16

EXAMPLE 2.9

Two sides of a parallelogram are the lines $2y = x + 12$ and $y = 4x - 10$.
Sketch these lines on the same diagram.

The origin is a vertex of the parallelogram. Complete the sketch of the
parallelogram, and find the equations of the other two sides.

Solution   The line $2y = x + 12$ has gradient $\frac{1}{2}$ and passes through the point $(0, 6)$

(since dividing by 2 gives $y = \frac{1}{2}x + 6$).

The line $y = 4x - 10$ has gradient 4 and passes through the point $(0, -10)$.

The other two sides are lines with gradients $\frac{1}{2}$ and 4 which pass through $(0, 0)$, i.e.
$y = \frac{1}{2}x$ and $y = 4x$.

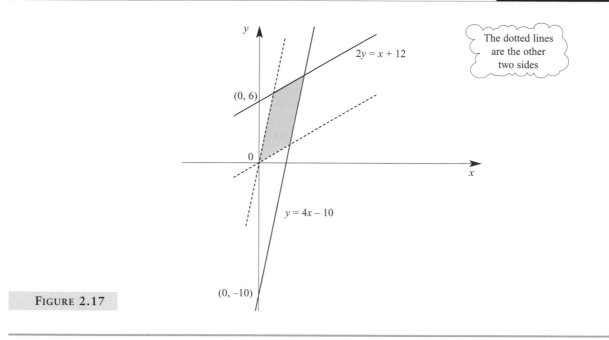

The dotted lines
are the other
two sides

FIGURE 2.17

EXAMPLE 2.10

Find the equation of the perpendicular bisector of the line joining P(-4, 5) to Q(2, 3).

Solution

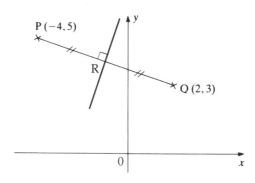

FIGURE 2.18

The gradient of the line PQ is:

$$\frac{3-5}{2-(-4)} = \frac{-2}{6} = -\frac{1}{3}$$

and so the gradient of the perpendicular bisector is $+3$.

The perpendicular bisector passes throught the mid-point, R, of the line PQ.

The coordinates of R are $\left(\frac{2+(-4)}{2}, \frac{3+5}{2}\right)$ i.e. (-1, 4).

Using $y - y_1 = m(x - x_1)$, the equation of the perpendicular bisector is:
$$y - 4 = 3(x - (-1))$$
$$y - 4 = 3x + 3$$
$$y = 3x + 7$$

EXAMPLE 2.11

The diameter of a snooker cue varies uniformly from 9 mm to 23 mm over its length of 140 cm. (*Varying uniformly* means that the graph of diameter against distance from the tip is a straight line.)

(a) Sketch the graph of diameter (y mm) against distance (x cm) from the tip.

(b) Find the equation of the line.

(c) Use the equation to find the distance from the tip at which the diameter is 15 mm.

Solution   (a) The graph passes through the points (0, 9) and (140, 23).

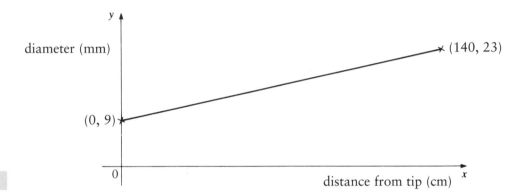

FIGURE 2.19

(b) Gradient $= \dfrac{y_2 - y_1}{x_2 - x_1}$

$\qquad = \dfrac{23 - 9}{140 - 0} = 0.1$

Using the form $y = mx + c$, the equation of the line is $y = 0.1x + 9$.

(c) Substituting $y = 15$ into the equation gives:

$$15 = 0.1x + 9$$
$$0.1x = 6$$
$$x = 60$$

$\Rightarrow$ the diameter is 15 mm at a point 60 cm from the tip.

EXERCISE 2B

In this exercise, you should sketch the lines by hand. If you have access to a graphics calculator or a computer graph-plotting package, you can use it to check your results.

1 Sketch the following lines.

(a) $y = -2$ (b) $x = 5$ (c) $y = 2x$

(d) $y = -3x$ (e) $y = 3x + 5$ (f) $y = x - 4$

(g) $y = x + 4$ (h) $y = \frac{1}{2}x + 2$ (i) $y = 2x + \frac{1}{2}$

(j) $y = -4x + 8$ (k) $y = 4x - 8$ (l) $y = -x + 1$

(m) $y = -\frac{1}{2}x - 2$ (n) $y = 1 - 2x$ (o) $3x - 2y = 6$

(p) $2x + 5y = 10$ (q) $2x + y - 3 = 0$ (r) $2y = 5x - 4$

(s) $x + 3y - 6 = 0$ (t) $y = 2 - x$

2 By calculating the gradients of the following pairs of lines, state whether they are parallel, perpendicular or neither.

(a) $y = -4$ $x = 2$

(b) $y = 3x$ $x = 3y$

(c) $2x + y = 1$ $x - 2y = 1$

(d) $y = 2x + 3$ $4x - y + 1 = 0$

(e) $3x - y + 2 = 0$ $3x + y = 0$

(f) $2x + 3y = 4$ $2y = 3x - 2$

(g) $x + 2y - 1 = 0$ $x + 2y + 1 = 0$

(h) $y = 2x - 1$ $2x - y + 3 = 0$

(i) $y = x - 2$ $x + y = 6$

(j) $y = 4 - 2x$ $x + 2y = 8$

(k) $x + 3y - 2 = 0$ $y = 3x + 2$

(l) $y = 2x$ $4x + 2y = 5$

3 Find the equations of the lines (a) to (e) in the diagram below.

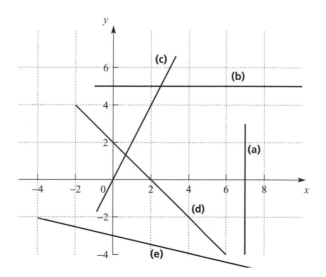

4 Find the equations of the lines (a) to (e) in the diagram below.

5 Find the equations of the following lines.

   (a) parallel to $y = 2x$ and passing through $(1, 5)$

   (b) parallel to $y = 3x - 1$ and passing through $(0, 0)$

   (c) parallel to $2x + y - 3 = 0$ and passing through $(-4, 5)$

   (d) parallel to $3x - y - 1 = 0$ and passing through $(4, -2)$

   (e) parallel to $2x + 3y = 4$ and passing through $(2, 2)$

   (f) parallel to $2x - y - 8 = 0$ and passing through $(-1, -5)$

   (g) parallel to $2y = 3x - 1$ and passing through $(4, 1)$

   (h) parallel to $5y = 2x + 3$ and passing through $(-5, 2)$

   (i) parallel to $4x + 2y = 3$ and passing through $(-1, -1)$

   (j) parallel to $2y - 4x = 5$ and passing through $(-3, 5)$

6 Find the equations of the following lines.

   (a) perpendicular to $y = 3x$ and passing through $(0, 0)$

   (b) perpendicular to $y = 2x + 3$ and passing through $(2, -1)$

   (c) perpendicular to $2x + y = 4$ and passing through $(3, 1)$

   (d) perpendicular to $2y = x + 5$ and passing through $(-1, 4)$

   (e) perpendicular to $2x + 3y = 4$ and passing through $(5, -1)$

   (f) perpendicular to $4x - y + 1 = 0$ and passing through $(0, 6)$

   (g) perpendicular to $3y = 2x + 4$ and passing through $(4, 3)$

   (h) perpendicular to $4y = 7 - x$ and passing through $(-2, 2)$

   (i) perpendicular to $3x + 2y = 5$ and passing through $(6, 0)$

   (j) perpendicular to $4x - y = 2$ and passing through $(2, 3)$

7 Find the equation of the line AB in each of the following cases.

    (a) A(0, 0)   B(4, 3)

    (b) A(2, −1)   B(3, 0)

    (c) A(2, 7)   B(2, −3)

    (d) A(3, 5)   B(5, −1)

    (e) A(−2, 4)   B(5, 3)

    (f) A(−4, −2)   B(3, −2)

    (g) A(1, 3)   B(3, −3)

    (h) A(5, −2)   B(8, −3)

    (i) $A(\frac{1}{2}, \frac{3}{2})$ B(3, 1)

    (j) $A(\frac{3}{12}, \frac{4}{12})$ $B(\frac{5}{12}, \frac{7}{12})$

8 Triangle ABC has an angle of 90° at B. Point A is on the y-axis, AB is part of the line $x - 2y + 8 = 0$, and C is the point (6, 2).

    (a) Sketch the triangle.

    (b) Find the equations AC and BC.

    (c) Find the lengths of AB and BC and hence find the area of the triangle.

    (d) Using your answer to (c), find the length of the perpendicular from B to AC.

9 A median of a triangle is a line joining one of the vertices to the mid-point of the opposite side.

In a triangle OAB, O is at the origin, A is the point (0, 6), and B is the point (6, 0).

    (a) Sketch the triangle.

    (b) Find the equations of the three medians of the triangle.

    (c) Show that the point (2, 2) lies on all three medians. (This shows that the medians of this triangle are concurrent.)

10 A quadrilateral ABCD has its vertices at the points (0, 0), (12, 5), (0, 10) and (−6, 8) respectively.

    (a) Sketch the quadrilateral.

    (b) Find the gradient of each side.

    (c) Find the length of each side.

    (d) Find the equation of each side.

    (e) Find the area of the quadrilateral.

INTERSECTION OF TWO LINES

In Chapter 1 you saw that the intersection of any two curves (or lines) can be found by solving their equations simultaneously. In the case of two distinct lines, there are two possibilities: (a) they are *parallel*, or (b) they *intersect at a single point*.

EXAMPLE 2.12

Sketch the lines $x + 2y = 1$ and $2x + 3y = 4$ on the same axes, and find the coordinates of the point where they intersect.

Solution The line $x + 2y = 1$ passes through $(0, \frac{1}{2})$ and $(1, 0)$.
The line $2x + 3y = 4$ passes through $(0, \frac{4}{3})$ and $(2, 0)$.

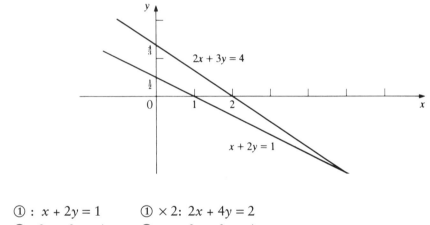

FIGURE 2.20

①$:\ x + 2y = 1$ ① × 2: $2x + 4y = 2$
②$:\ 2x + 3y = 4$ ②: $\underline{2x + 3y = 4}$
         Subtract: $y = -2$

Substituting $y = -2$ in ①: $x - 4 = 1$
               $\Rightarrow\ x = 5$

The coordinates of the point of intersection are $(5, -2)$.

EXAMPLE 2.13

Find the coordinates of the vertices of the triangle of which the sides have equations $x + y = 4$, $2x - y = 8$ and $x + 2y = -1$.

Solution A sketch will be helpful, so first find where each line crosses the axes.

① $x + y = 4$ crosses the axes at $(0, 4)$ and $(4, 0)$
② $2x - y = 8$ crosses the axes at $(0, -8)$ and $(4, 0)$
③ $x + 2y = -1$ crosses the axes at $(0, -\frac{1}{2})$ and $(-1, 0)$

Since two lines pass through the point $(4, 0)$ this is clearly one of the vertices. It has been labelled A on figure 2.21.

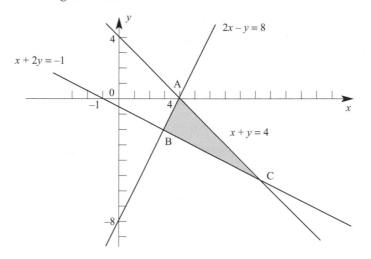

FIGURE 2.21

Point B is found by solving ② and ③ simultaneously.

$$
\begin{aligned}
\text{② × 2:} \qquad & 4x - 2y = 16 \\
\text{③:} \qquad & \underline{x + 2y = -1} \\
\text{Add:} \qquad & 5x = 15 \\
\Rightarrow \qquad & x = 3
\end{aligned}
$$

Back-substituting in ② gives $y = -2$, so B is the point $(3, -2)$.

Point C is found by solving ① and ③ simultaneously.

$$
\begin{aligned}
\text{①:} \qquad & x + y = 4 \\
\text{③:} \qquad & \underline{x + 2y = -1} \\
\text{Subtract:} \qquad & -y = 5 \qquad \text{so} \qquad y = -5
\end{aligned}
$$

Back-substitution gives $x = 9$, so C is the point $(9, -5)$.

Historical note

René Descartes was born near Tours in France in 1596. At the age of eight he was sent to a Jesuit boarding school where, because of his frail health, he was allowed to stay in bed until late in the morning. This habit stayed with him for the rest of his life and he claimed that he was at his most productive before getting up.

After leaving school he studied mathematics in Paris before becoming in turn a soldier, traveller and maker of optical instruments. Eventually he settled in Holland where he devoted his time to mathematics, science and philosophy, and wrote a number of books on these subjects. In an appendix, entitled *La Géométrie*, to one of his books, Descartes made the contribution to coordinate geometry for which he is particularly remembered. In 1649 he left Holland for Sweden at the invitation of Queen Christina but died there, of a lung infection, the following year.

EXERCISE 2C

1 (a) Find the vertices of the triangle ABC with sides given by the lines
AB: $x - 2y = -1$, BC: $7x + 6y = 53$ and AC: $9x + 2y = 11$.

(b) Show that the triangle is isosceles.

2 Two sides of a parallelogram are formed by parts of the lines
$2x - y = -9$ and $x - 2y = -9$.

(a) Show these two lines on a graph.

(b) Find the coordinates of the vertex where they intersect.

Another vertex of the parallelogram is the point $(2, 1)$.

(c) Find the equations of the other two sides of the parallelogram.

(d) Find the coordinates of the other two vertices.

3 A rhombus ABCD is such that the coordinates of A and C are $(0, 4)$ and $(8, 0)$ respectively.

(a) Show that the equation of the diagonal BD is $y = 2x - 6$.
(Hint: AC and BD bisect each other at an angle of 90°.)

The side AB has gradient -2.

(b) Find the coordinates of B and D.

(c) Show that the rhombus has an area of 30 square units.

[JMB]

4 The line with equation $5x + y = 20$ meets the x-axis at A and the line with equation $x + 2y = 22$ meets the y-axis at B. The two lines intersect at a point C.

(a) Sketch the two lines on the same diagram.

(b) Calculate the coordinates of A, B and C.

(c) Calculate the area of triangle OBC where O is the origin.

(d) Find the coordinates of the point E such that ABEC is a parallelogram.

5 Two rival taxi firms have the following fare structures.

Firm A: fixed charge of £1 plus 40p per kilometre
Firm B: 60p per kilometre, no fixed charge

(a) Sketch the graph of price (vertical axis) against distance travelled (horizontal axis) for each firm (on the same axes).

(b) Find the equation of each line.

(c) Find the distance for which both firms charge the same amount.

(d) Which firm would you use for a distance of 6 km?

6 The diagram (next page) shows the *supply* and *demand* of labour for a particular industry in relation to the wage paid per hour.

Supply is the number of people willing to work for a particular wage, and this increases as the wage paid increases. *Demand* is the number of workers that employers are prepared to employ at a particular wage: this is greatest for low wages.

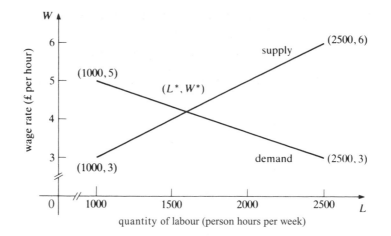

(a) Find the equation of each of the lines.

(b) Find the values of L^* and W^* at which the market 'clears', i.e. at which *supply* equals *demand*.

(c) Although economists draw the graph this way round, mathematicians would plot wage rate on the horizontal axis. Why?

7 A median of a triangle is a line joining a vertex to the mid-point of the opposite side. In any triangle, the three medians meet at a point. The centroid of a triangle is at the point of intersection of the medians.

Find the coordinates of the centroid for each triangle shown.

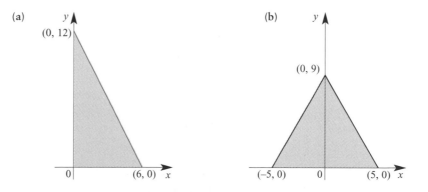

8 When the market price £p of an article sold in a free market varies, so does the number demanded, D, and the number supplied, S. In one case $D = 20 + 0.2p$ and $S = -12 + p$.

(a) Sketch both of these lines on the same graph.

The market reaches a state of equilibrium when the number demanded equals the number supplied.

(b) Find the equilibrium price and the number bought and sold in equilibrium.

9 You are given the coordinates of the four points

A(6, 2), B(2, 4), C(−6, −2) and D(−2, −4).

(a) Calculate the gradients of the lines AB, CB, DC and DA. Hence describe the shape of the figure ABCD.

(b) Show that the equation of the line DA is $4y - 3x = -10$, and find the length DA.

(c) Calculate the gradient of a line which is perpendicular to DA, and hence find the equation of the line l through B which is perpendicular to DA.

(d) Calculate the coordinates of the point P where l meets DA.

(e) Calculate the area of the figure ABCD.

[MEI]

10

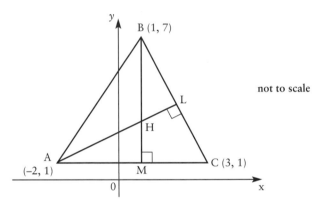

The diagram shows a triangle with vertices at A(−2, 1), B(1, 7) and C(3, 1). The point L is the foot of the perpendicular from A to BC, and M is the foot of the perpendicular from B to AC.

(a) Find the gradient of the line BC.

(b) Find the equation of the line AL.

(c) Write down the equation of the line BM.

The lines AL and BM meet at H.

(d) Find the coordinates of H.

(e) Show that CH is perpendicular to AB.

(f) Find the area of the triangle BLH.

[MEI]

EXERCISE 2D   **Examination-style questions**

1 A is the point $(-3, 8)$ and B is the point $(5, 2)$.

   (a) Find the length of AB.
   (b) Find the mid-point of AB.
   (c) Find the gradient of AB.
   (d) Find the equation of the perpendicular bisector of AB, expressing your
   answer in the form $ax + by + c = 0$, where a, b and c are integers.

2 **(a)** Find the equation of the line l which passes through the points A$(1, -3)$
   and B$(5, 5)$.
   (b) Find C, the point of intersection between l and the line m with equation
   $3x + 4y = 13$.
   (c) l cuts the y-axis at L and m cuts the y-axis at M. Show that the area of
   CLM is $\frac{99}{8}$.

3 Find the equations of the perpendicular bisectors of the lines between the
   following pairs of points.

   (a) $(3, 4)$ and $(7, 0)$
   (b) $(-3, -2)$ and $(1, 6)$

   Show that the point of intersection of these bisectors lies on the x-axis.

4 A is the point $(2, 3)$ and B is the point $(14, 8)$.

   (a) Find the distance between A and B.
   (b) Find the mid-point of the line between A and B.
   (c) Find the equation of the line passing through A and B.

   P and Q are points such that APBQ forms a square.

   (d) Find the coordinates of P and Q.
   (e) Show that the area of the square is 84.5 square units.

5 **(a)** Find the equation of the line l_1 which is parallel to $3x + 5y - 11 = 0$ and
   passes through the point $(-2, 4)$.
   (b) Find the equation of the line l_2 which is perpendicular to $2y = x - 3$ and
   passes through the point $(1, 5)$.
   (c) Show that the lines l_1 and l_2 intersect at the point $(3, 1)$.

6 The coordinates of the points P, Q and R are $(2, -1)$, $(1, 3)$ and $(4, 5)$ respectively.

   (a) Find the equation of the line parallel to PQ, passing through R.
   (b) Find the equation of the line parallel to PR, passing through Q.
   (c) These lines cross at S. Find the coordinates of S.
   (d) Show that the distance from the origin to S is $3\sqrt{10}$.

7 l_1 is a straight line with equation $x + 4y - 22 = 0$. A is the point with coordinates $(2, 5)$ and B is the point with coordinates $(4, 13)$. l_2 is a straight line parallel to l_1 passing through the point B.

(a) Show that A lies on l_1.

(b) Find the equation of l_2, giving your answer in the form $ax + by + c = 0$, where a, b and c are integers.

(c) Show that the line joining A and B is perpendicular to l_2.

(d) Find the shortest distance between the lines l_1 and l_2.

8 The straight line l passes through $A(1, 3\sqrt{3})$ and $B(2 + \sqrt{3}, 3 + 4\sqrt{3})$.

(a) Calculate the gradient of l, giving your answer as a surd in its simplest form.

(b) Give the equation of l in the form $y = mx + c$, where constants m and c are surds to be given in their simplest form.

(c) Show that l meets the x-axis at the point $C(-2, 0)$.

(d) Calculate the length AC.

(e) Find the size of the acute angle between the line AC and the x-axis, giving your answer in degrees.

[Edexcel]

9 (a) Find an equation of the straight line passing through the points with coordinates $(-1, 5)$ and $(4, -2)$, giving your answer in the form $ax + by + c = 0$, where a, b and c are integers.

The line crosses the x-axis at the point A and the y-axis at the point B, and O is the origin.

(b) Find the area of $\triangle OAB$.

[Edexcel]

10 The straight line l_1 passes through the points A and B with coordinates $(2, 2)$ and $(6, 0)$ respectively.

(a) Find an equation of l_1.

The straight line l_2 passes through the point C with coordinates $(-9, 0)$ and has gradient $\frac{1}{4}$.

(b) Find an equation of l_2.
The lines l_1 and l_2 intersect at the point D.

(c) Calculate, to 2 decimal places, the length of AD.

(d) Calculate the area of $\triangle DCB$.

[Edexcel]

11 The straight line l_1 has equation $4y + x = 0$.
The straight line l_2 has equation $y = 2x - 3$.

(a) On the same axes, sketch the graphs of l_1 and l_2. Show clearly the coordinates of all points at which the graphs meet the coordinate axes.

The lines l_1 and l_2 intersect at the point A.

(b) Calculate, as exact fractions, the coordinates of A.

(c) Find an equation of the line through A which is perpendicular to l_1. Give your answer in the form $ax + by + c = 0$, where a, b and c are integers.

[Edexcel]

12 The points A$(-1, -2)$, B$(7, 2)$ and C$(k, 4)$, where k is a constant, are the vertices of $\triangle$ABC. Angle ABC is a right angle.

(a) Find the gradient of AB.

(b) Calculate the value of k.

(c) Show that the length of AB may be written in the form $p\sqrt{5}$ where p is an integer to be found.

(d) Find the exact value of the area of $\triangle$ABC.

(e) Find an equation for the straight line l passing through B and C. Give your answer in the form $ax + by + c = 0$, where a, b and c are integers.

The line l crosses the x-axis at D and the y-axis at E.

(f) Calculate the coordinates of the mid-point of DE.

[Edexcel]

13 The points A$(-3, -2)$ and B$(8, 4)$ are the ends of a diameter of the circle shown in the diagram.

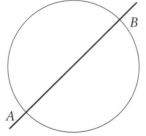

(a) Find the coordinates of the centre of the circle.

(b) Find an equation of the diameter AB, giving your answer in the form $ax + by + c = 0$, where a, b and c are integers.

(c) Find an equation of the tangent to the circle at B.

The line l passes through A and the origin.

(d) Find the coordinates of the point at which l intersects the tangent to the circle at B, giving your answers as exact fractions.

[Edexcel]

14 The points A and B have coordinates (4, 6) and (12, 2) respectively.
The straight line l_1 passes through A and B.

(a) Find an equation for l_1 in the form $ax + by + c = 0$, where a, b and c are integers.

The straight line l_2 passes through the origin and has gradient −4.

(b) Write down an equation for l_2.

The lines l_1 and l_2 intersect at the point C.

(c) Find the exact coordinates of the mid-point of AC.

[Edexcel]

KEY POINTS

1 The **gradient** of the straight line joining the points (x_1, y_1) and (x_2, y_2) is given by:

$$\text{gradient} = \frac{y_2 - y_1}{x_2 - x_1}$$

2 Two lines are **parallel** when their gradients are equal: $m_1 = m_2$.

3 Two lines are **perpendicular** when the product of their gradients is −1.
$m_1 m_2 = -1$

4 When the points A and B have coordinates (x_1, y_1) and (x_2, y_2) respectively, then:

$$\text{distance AB} = \sqrt{(x_2 - x_1)^2 + (y_2 - y_1)^2}$$

$$\text{mid-point of line AB is} \left(\frac{x_1 + x_2}{2}, \frac{y_1 + y_2}{2} \right)$$

5 The **equation of a straight line** may take any of the following forms.

- line parallel to the y-axis: $x = a$
- line parallel to the x-axis: $y = b$
- line through the origin with gradient m: $y = mx$
- line through $(0, c)$ with gradient m: $y = mx + c$
- line through (x_1, y_1) with gradient m: $y - y_1 = m(x - x_1)$
- line through (x_1, y_1) and (x_2, y_2)
$$\frac{y - y_1}{y_2 - y_1} = \frac{x - x_1}{x_2 - x_1} \text{ or } \frac{y - y_1}{x - x_1} = \frac{y_2 - y_1}{x_2 - x_1}$$

Chapter three

SEQUENCES AND SERIES

Again! Again! Again!

And the havoc did not slack.

Thomas Campbell

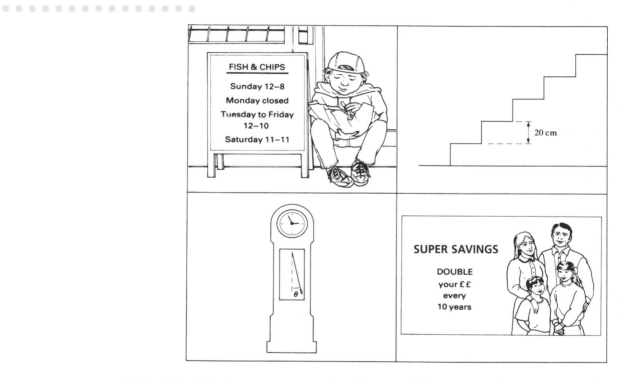

Each of the following sequences is related to one of the pictures above.

(a) 5000, 10 000, 20 000, 40 000, ...

(b) 8, 0, 10, 10, 10, 10, 12, 8, 0, ...

(c) 5, 3.5, 0, −3.5, −5, −3.5, 0, 3.5, 5, 3.5, ...

(d) 20, 40, 60, 80, 100, ...

(i) Identify which sequence goes with which picture.

(ii) Give the next few numbers in each sequence.

(iii) Describe the pattern of the numbers in each case.

(iv) Decide whether the sequence will go on for ever, or come to a stop.

DEFINITIONS AND NOTATION

A *sequence* is a set of numbers in a given order, such as:

$$\frac{1}{2}, \frac{1}{4}, \frac{1}{8}, \frac{1}{16}, \dots$$

Each of these numbers is called a *term* of the sequence.

When writing the terms of a sequence algebraically, it is usual to denote the position of any term in the sequence by a subscript, so that a general sequence might be written:

$$u_1, u_2, u_3, \dots, \text{with general term } u_n.$$

For the sequence above, the first term is $u_1 = \frac{1}{2}$, the second term is $u_2 = \frac{1}{4}$, and so on.

When the terms of a sequence are added together, like this:

$$\frac{1}{2} + \frac{1}{4} + \frac{1}{8} + \frac{1}{16} + \dots$$

the resulting sum is called a *series*.

The process of adding the terms together is called *summation* and indicated by the symbol $\sum$ (the Greek letter sigma), with the position of the first and last terms involved given as *limits*.

So $u_1 + u_2 + u_3 + u_4 + u_5$ is written $\sum_{r=1}^{5} u_r$.

In cases like this one, where there is no possibility of confusion, the sum may be written more simply as $\sum_{1}^{5} u_r$.

In general:

$$\sum_{r=1}^{n} f(r) = f(1) + f(2) + f(3) + \dots + f(n)$$

This is calculated by putting $r = 1$, evaluating the term, then putting $r = 2$, evaluating the term and continuing with r increasing in steps of 1 until $r = n$. All the terms are then added together.

A sequence may have an infinite number of terms, in which case it is called an *infinite sequence*.

The corresponding series is called an *infinite series*.

In mathematics, although the word *series* can describe the sum of the terms of any sequence, it is usually used only when summing the sequence provides some useful or interesting overall result. For example:

$$\pi = 2\sqrt{3}\left[1 + \left(\frac{-1}{3}\right) + 5\left(\frac{-1}{3}\right)^2 + 7\left(\frac{-1}{3}\right)^3 + ...\right]$$

$$(1 + x)^5 = 1 + 5x + 10x^2 + 10x^3 + 5x^4 + x^5$$

$$\sin x = x - \frac{x^3}{3!} + \frac{x^5}{5!} - ...\text{(where } x \text{ is measured in radians)}.$$

The phrase 'sum of a sequence' is often used to mean the sum of the terms of a sequence (i.e. the series).

PATTERNS IN SEQUENCES

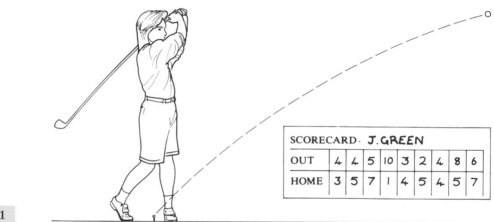

SCORECARD · J. GREEN								
OUT	4	4	5	10	3	2	4	8	6
HOME	3	5	7	1	4	5	4	5	7

FIGURE 3.1

Any ordered set of numbers, such as the scores of this golfer on an 18-hole round (figure 3.1) forms a *sequence*. In mathematics, we are particularly interested in those which have a well-defined pattern, often in the form of an algebraic formula linking the terms. The sequences you met at the start of this chapter show various types of pattern.

ARITHMETIC SEQUENCES

A sequence in which the terms increase by the addition of a fixed amount (or decrease by subtraction of a fixed amount), is described as *arithmetic*. The increase from one term to the next is called the *common difference*.

Thus the sequence 5   8   11   14 ... is arithmetic with

+3   +3   +3

common difference 3.

This sequence can be written algebraically in two quite different ways.

(a) $u_n = 2 + 3n$ for $n = 1, 2, 3, \ldots$

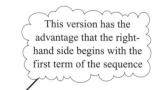

This version has the advantage that the right-hand side begins with the first term of the sequence

     When $n = 1$, $u_1 = 2 + 3 = 5$

             $n = 2$, $u_2 = 2 + 6 = 8$

             $n = 3$, $u_3 = 2 + 9 = 11$ and so on.

(An equivalent way of writing this is $u_n = 5 + 3(n - 1)$ for $n = 1, 2, 3, \ldots$.)

(b) $u_1 = 5$

$u_{n+1} = u_n + 3$ for $n = 1, 2, 3, \ldots$

Substituting $n = 1 \Rightarrow u_2 = u_1 + 3 = 5 + 3 = 8$

            $n = 2 \Rightarrow u_3 = u_2 + 3 = 8 + 3 = 11$

            $n = 3 \Rightarrow u_4 = u_3 + 3 = 11 + 3 = 14$

and so on.

The first form, which defines the value of u_n directly, is called a *deductive definition*. It is a *position-to-term* rule. The second, which defines each term by relating it to the previous one, is described as an *inductive* or a *term-to-term* rule. You should be prepared to use either.

EXAMPLE 3.1

A sequence is defined deductively by $u_n = 3n - 1$ for $n = 1, 2, 3, \ldots$.

(a) Write down the first six terms of the sequence, and say what kind of sequence it is.

(b) Find the value of the series $\sum_{r=1}^{6} u_r$

Solution   (a) Substituting $n = 1, 2, \ldots, 6$ in $u_n = 3n - 1$ gives:

$u_1 = 3 - 1 = 2$ $u_2 = 6 - 1 = 5$ $u_3 = 9 - 1 = 8$
$u_4 = 12 - 1 = 11$ $u_5 = 15 - 1 = 14$ $u_6 = 18 - 1 = 17$

so the sequence is   2, 5, 8, 11, 14, 17, ...

This is an arithmetic sequence with common difference 3.

(b) $\sum_{r=1}^{6} u_r = u_1 + u_2 + u_3 + u_4 + u_5 + u_6$

          $= 2 + 5 + 8 + 11 + 14 + 17$

          $= 57$

Arithmetic sequences are dealt with in more detail later in this chapter.

GEOMETRIC SEQUENCES

A sequence in which you find each term by multiplying the previous one by a fixed number is described as *geometric*; the fixed number is the *common ratio*.

Thus 10 20 40 80 ... is a geometric sequence with common ratio 2.

$\times 2$ $\times 2$ $\times 2$

It may be written algebraically as:

$u_n = 5 \times 2^n$ for $n = 1, 2, 3, ...$ (deductive definition)

or as: $u_1 = 10$

$u_{n+1} = 2u_n$ for $n = 1, 2, 3, ...$ (inductive definition).

You will study geometric sequences in more detail in Chapter 8 of *Pure Mathematics: Core 2*.

PERIODIC SEQUENCES

A sequence which repeats itself at regular intervals is called *periodic*. In the case of the fish bar at the start of this chapter, the number of hours it is open each day forms the sequence:

$u_1 = 8,$ $u_2 = 0,$ $u_3 = 10,$ $u_4 = 10,$ $u_5 = 10,$ $u_6 = 10,$ $u_7 = 12,$
(Sun)   (Mon)   (Tues)   (Wed)   (Thurs)   (Fri)   (Sat)

$u_8 = 8,$ $u_9 = 0$ ... .
(Sun)   (Mon)

There is no neat algebraic formula for the terms of this sequence but you can see that $u_8 = u_1$, $u_9 = u_2$, and so on.

In general:

$u_{n+7} = u_n$ for $n = 1, 2, 3, ...$.

This sequence is periodic, with period 7. Each term is repeated after 7 terms.

A sequence for which:

$u_{n+p} = u_n$ for $n = 1, 2, 3, ...$ (for a fixed integer, p)

is periodic. The period is the smallest positive value of p for which this is true.

OSCILLATING SEQUENCES

The terms of an oscillating sequence, such as:

   5, 6, 5, 4, 5, 6, 5, 4, ...

lie above and below a middle number, in this case 5 (see figure 3.2).

In this example, the sequence is also periodic, with period 4. However, some oscillating sequences, such as:

   $8, -4, 2, -1, \frac{1}{2}, \ldots$

are non-periodic. You will notice that the middle value, 0, is not itself a term in this sequence.

FIGURE 3.2

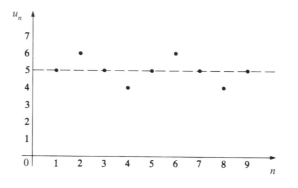

EXAMPLE 3.2

A sequence is defined by $u_n = (-1)^n$ for $n = 1, 2, 3, \ldots$.

(a) Write down the first six terms of this sequence and describe its pattern.

(b) Find the value of the series $\displaystyle\sum_{r=2}^{5} u_r$

(c) Describe the sequence defined by $t_n = 5 + (-1)^n \times 2$ for $n = 1, 2, 3, \ldots$.

Solution (a) $u_1 = (-1)^1 = -1$ $u_2 = (-1)^2 = +1$ $u_3 = (-1)^3 = -1$
           $u_4 = (-1)^4 = +1$ $u_5 = (-1)^5 = -1$ $u_6 = (-1)^6 = +1$

The sequence is $-1, +1, -1, +1, -1, +1, \ldots$.

It is oscillating, and periodic with period 2. It is also geometric, with common ratio -1.

(b) $\displaystyle\sum_{n=2}^{5} u_n = u_2 + u_3 + u_4 + u_5$

$= (+1) + (-1) + (+1) + (-1)$

$= 0$

(c) $t_1 = 5 + (-1) \times 2 = 3$ $\qquad$ $t_2 = 5 + (-1)^2 \times 2 = 7$

$t_3 = 5 + (-1)^3 \times 2 = 3$ $\qquad$ $t_4 = 5 + (-1)^4 \times 2 = 7$

and so on, giving the sequence 3, 7, 3, 7, ..., which is oscillating with period 2.

Note

The device $(-1)^n$ is very useful when writing the terms of oscillating sequences or series algebraically. It is used extensively in mathematics and so the ideas in Example 3.2 are important.

SEQUENCES WITH OTHER PATTERNS

There are many other possible patterns in sequences. Figure 3.3 shows a well-known children's toy in which squares are stacked to make a tower. The smallest square has sides 1 cm long, and the length of the sides increases in steps of 1 cm.

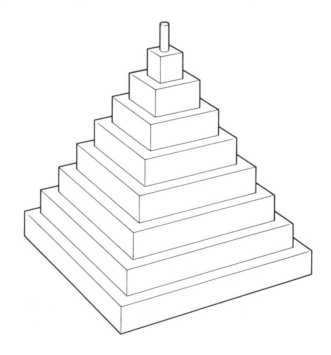

FIGURE 3.3

The areas of these squares, in cm², form the sequence:

$$1^2,\ 2^2,\ 3^2,\ 4^2,\ 5^2,\ ... \qquad \text{or} \qquad 1,\ 4,\ 9,\ 16,\ 25,\ ...\ .$$

This sequence does not fit any of the patterns described so far. If you subtract each term from the next, however, you will find that the differences form a pattern.

Sequence   1   4   9   16   25 ...

Difference   3   5   7   9 ...

These differences form an arithmetic sequence with common difference 2. The next difference in the sequence is $9 + 2 = 11$, and so the next term in the areas sequence is $25 + 11 = 36$, which is indeed 6^2.

Looking at the differences between the terms often helps you to spot the pattern within a sequence. Sometimes you may need to look at the differences of the differences, or go even further.

EXERCISE 3A

1 For each of the following sequences, write down the next four terms (assuming the same pattern continues) and describe its pattern as fully as you can.

(a) 7, 10, 13, 16, ...

(b) 8, 7, 6, 5, ...

(c) 10, 30, 90, 270, ...

(d) 64, 32, 16, 8, ...

(e) 2, 2, 2, 5, 2, 2, 2, 5, 2, 2, 2, 5, ...

(f) 4, 6, 7, 6, 4, 2, 1, 2, 4, 6, 7, 6, ...

(g) 1, –2, 4, –8, 16, ...

(h) 2, 4, 6, 8, 2, 4, 6, 8, 2, ...

(i) 4.1, 3.9, 3.7, 3.5, ...

(j) 1, 1.1, 1.21, 1.331, ...

2 Write down the first four terms of the sequences defined by the position-to-term rules. In each case, n takes the values 1, 2, 3, ...

(a) $u_n = 2n + 1$

(b) $u_n = 3 \times 2^n$

(c) $u_n = 2n + 2^n$

(d) $u_n = \dfrac{1}{n}$

(e) $t_n = \dfrac{1}{2}n(n + 1)$

(f) $t_n = n^n$

(g) $t_n = (n - 1)(n + 1)$

(h) $s_n = \dfrac{1}{n} - \dfrac{1}{n + 1}$

(i) $u_n = 5 + (-1)^n$

(j) $u_n = 1024(\dfrac{1}{2})^n$

3 Write down the first four terms of the sequences defined by the term-to-term rules. In each case, n takes the values 1, 2, 3, ...

(a) $u_{n+1} = u_n + 3,\ u_1 = 12$

(b) $u_{n+1} = -u_n,\ u_1 = -5$

(c) $u_{n+1} = \frac{1}{2}u_n,\ u_1 = 72$

(d) $u_{n+2} = u_n,\ u_1 = 4,\ u_2 = 6$

(e) $t_{n+1} = t_n^2 - 1,\ t_1 = 2$

(f) $t_{n+1} = t_n(t_n + 1),\ t_1 = 1$

(g) $t_{n+1} = 3t_n - 4,\ t_1 = 2$

(h) $t_{n+1} = 2t_n + 3,\ t_4 = 37$

(i) $t_{n+1} = \dfrac{1}{t_n + 1},\ t_1 = 3$

(j) $u_{n+1} = \sqrt{u_n},\ u_1 = 10^8$

4 The Fibonacci sequence is given by:

$$1, 1, 2, 3, 5, 8, \ldots$$

(a) Write down the sequence of the differences between the terms in this sequence, and comment on what you find.

(b) Write down the next three terms of the Fibonacci sequence.

5 The terms of a sequence are defined by:

$$u_n = 4 + (-1)^n \times 2$$

(a) Write down the first six terms of this sequence.

(b) Describe the sequence.

(c) Describe the effect of changing the sequence to:

(i) $u_n = 4 + (-2)^n$

(ii) $u_n = 4 + \left(-\dfrac{1}{2}\right)^n$

6 Calculate the first four terms of the sequence defined by:

$$u_n = \frac{1}{\sqrt{5}}\left\{\left(\frac{1+\sqrt{5}}{2}\right)^n - \left(\frac{1-\sqrt{5}}{2}\right)^n\right\}$$

The next two terms are 5 and 8. After whom is this series known?

7 A sequence of numbers $t_1, t_2, t_3, t_4, \ldots$ is formed by taking a starting value of t_1 and using the rule:

$$t_{n+1} = t_n^2 - 2, \qquad \text{for } n = 1, 2, 3, \ldots .$$

(a) If $t_1 = \sqrt{2}$, calculate t_2, t_3 and t_4. Show that $t_5 = 2$, and write down the value of t_{100}.

(b) If $t_1 = 2$, show that all terms of the sequence are the same.

Find the other value of t_1 for which all the terms of the sequence are the same.

(c) Determine whether the sequence converges, diverges or is periodic in the cases when:

(i) $t_1 = 3$ (ii) $t_1 = 1$ (iii) $t_1 = \dfrac{\sqrt{5}-1}{2}$

[MEI]

8 Work out each of the following by writing out all the terms and calculating their sum.

(a) $\displaystyle\sum_{r=1}^{5} r^2$ (b) $\displaystyle\sum_{r=1}^{6}(5r-3)$ (c) $\displaystyle\sum_{r=1}^{4}(2r+5)$ (d) $\displaystyle\sum_{r=1}^{6}(-1)^r r(r+1)$

9 Use the $\sum$ notation to describe each of these.

(a) $1 + 8 + 27 + \dots + 216$

(b) $1 \times 2 + 2 \times 3 + 3 \times 4 + \dots + 15 \times 16$

(c) $1 + \dfrac{1}{2} + \dfrac{1}{4} + \dots + \dfrac{1}{256}$

(d) $1 - 4 + 9 - 16 + \dots - 400$

10 You are given that $f(x) = \dfrac{5}{6-x}$.

(a) If $x_1 = 3$ and $x_{n+1} = f(x_n)$, find x_2, x_3 and x_4 and show that $x_5 = 1\dfrac{2}{313}$.

(b) What is the value of x_n as n becomes larger?

(c) When n is large, put $n = \alpha$ and $x_{n+1} = \alpha$. Hence prove that $\alpha = 1$ is an exact solution.

RECURRENCE RELATIONS $x_{n+1} = f(x_n)$

Recurrence relations are a form of term-to-term sequences. You start with a given term, usually the first term u_1, and use this to calculate the next term. The second term is used to find the third term, and so on. The process in which consecutive terms are calculated in this way is called *iteration*.

EXAMPLE 3.3

Find the first five terms of the recurrence relation:

$$x_{n+1} = \frac{2x_n + 10}{10}, \quad x_1 = 1$$

Solution $x_1 = 1$

$$x_2 = \frac{2 \times 1 + 10}{10} = 1.2$$

$$x_3 = \frac{2 \times 1.2 + 10}{10} = 1.24$$

$$x_4 = \frac{2 \times 1.24 + 10}{10} = 1.248$$

$$x_5 = \frac{2 \times 1.248 + 10}{10} = 1.2496$$

Note

In the Core 1 examination you are not allowed the use of a calculator. If you were (and it is useful to check your answers) then with this example you would start by keying in 1 = (or EXE) then (2×ANS + 10) ÷ 10 followed by repeated pressing of the = (or EXE key).

If you were to continue Example 3.3 on your calculator you would find it converged to a limit. What is that limit?

EXAMPLE 3.4

A recurrence relation is defined by $u_{n+1} = 2u_n - 1$.
- **(a)** What happens if $u_1 = 1$?
- **(b)** What happens if $u_1 = -1$?
- **(c)** A second recurrence relation is defined by $x_n = u_n + 1$. Show that $x_{n+1} = 2(x_n - 1)$.

Solution

- **(a)** $u_1 = 1$ gives the sequence 1, 1, 1, 1, 1, ...
  So 1 is a fixed point of the iteration.
- **(b)** $u_1 = -1$ gives the sequence $-1, -3, -7, -15, -31, ...$
  In this case the sequence diverges.
- **(c)** If $x_n = u_n + 1$ then:
$$x_{n+1} = u_{n+1} + 1$$
$$= 2u_n - 1 + 1$$
$$= 2u_n$$
So $x_{n+1} = 2(x_n - 1)$

EXERCISE 3B

1 A recurrence relation is given by $x_{n+1} = \frac{1}{4}x_n$ where $x_1 = 2$. Write the values of x_2, x_3, x_4 and x_5 leaving your answers as fractions. State the value of x_n as $n \to \infty$.

2 Given that $x_{n+1} = \sqrt{x_n - 2}$ and $x_1 = 123$ write down the values of x_2, x_3 and x_4 and explain why you cannot find x_5.

3 A recurrence relation has this equation.
$$x_{n+1} = \frac{x_n + 2}{2}$$

Given that $x_1 = 1$ find x_5 showing all intermediate iterations.

4 A sequence of terms $\{u_n\}$ is defined for $n \geqslant 1$, by the recurrence relation:
$$u_{n+1} = u_n^2$$

- **(a)** Determine what happens if $u_1 = 0$, $u_1 = 1$ and $u_1 = 2$.
- **(b)** Express u_2 and u_3 in terms of u_1.
- **(c)** Express u_n in terms of u_1.

5 A sequence of terms $u_1, u_2, ... u_n, ...$ is given by the formula:
$$u_{n+1} = (u_n - 1)^2$$

If $u_1 = 3$ find u_5.

6 **(a)** Write down the first five terms of $t_n = 2(n + 1)$.
- **(b)** Write down the first five terms of $u_{n+1} = u_n + 4$, $u_1 = 8$.
- **(c)** Using your answer to parts (a) and (b) express u_n in terms of n. Hence state the value of u_{100}.

7 A recurrence relation is $u_{n+1} = \frac{1}{2}u_n$.

    (a) Write down u_1, u_2, u_3, u_4 and u_5 given that $u_0 = 1$.

    (b) If an infinite number of terms of this sequence is taken what is their sum?

8 A sequence of terms x_1, x_2, ..., x_n, ... is given by the formula:

$$x_{n+1} = x_n^2 - 6$$

    (a) Find the iterations x_1 to x_4 in the case when $x_0 = 1$, $x_0 = 2$ and $x_0 = 3$.

    (b) When $x_{n+1} = x_n = x$ obtain a quadratic in x and solve it.

    (c) Comment on your answers to parts (a) and (b).

9 A recurrence relation has this equation.

$$u_{n+1} = \frac{5u_n + 2}{10}$$

    (a) With $u_1 = 0.2$ find u_5.

    (b) By putting $u_{n+1} = u_n = x$ obtain an equation in x.

    (c) Solve the equation in x. Hence state the limiting value of the iteration.

10 A sequence of terms $\{u_n\}$ is defined for $n \geqslant 1$, by the recurrence relation:

$$u_{n+1} = \frac{u_n}{3}$$

    (a) Find u_2, u_3, u_4 and u_5 in terms of u_1.

    (b) Express u_n in terms of u_1.

    (c) Given that $u_1 = 1$ find $\sum_{n=1}^{4} u_n$.

ARITHMETIC SEQUENCES AND SERIES

Earlier in this chapter, you learned that successive terms of an arithmetic sequence increase (or decrease) by a fixed amount called the common difference, d. So:

$$u_{n+1} = u_n + d.$$

When the terms of an arithmetic sequence are added together, the sum is called an *arithmetic series*. An alternative name is an *arithmetic progression*, often abbreviated to AP.

NOTATION

When describing arithmetic series and sequences in this book, the following conventions will be used:

- first term, $u_1 = a$
- number of terms $= n$
- last term $= l$
- common difference $= d$
- the general term, u_n, is that in position n (i.e. the nth term).

Thus in the arithmetic sequence 5, 7, 9, 11, 13, 15, 17:

$a = 5, l = 17, d = 2$ and $n = 7$

The terms are formed as follows:

$$u_1 = a \qquad\qquad\qquad = 5$$
$$u_2 = a + d \quad = 5 + 2 \quad = 7$$
$$u_3 = a + 2d = 5 + 2 \times 2 = 9$$
$$u_4 = a + 3d = 5 + 3 \times 2 = 11$$
$$u_5 = a + 4d = 5 + 4 \times 2 = 13$$
$$u_6 = a + 5d = 5 + 5 \times 2 = 15$$
$$u_7 = a + 6d = 5 + 6 \times 2 = 17$$

The 7th term is the 1st term (5) plus six times the common difference (2).

You can see that any term is given by the first term plus a number of differences. The number of differences is, in each case, one less than the number of the term. You can express this mathematically as:

$$u_n = a + (n - 1)d$$

If the nth term is the last term l, this becomes:

$$l = a + (n - 1)d$$

These are both general formulae which apply to any arithmetic sequence.

EXAMPLE 3.5

Find the 17th term in the arithmetic sequence 12, 9, 6, ...

Solution   In this case $a = 12$ and $d = -3$.

Using $u_n = a + (n - 1)d$ you obtain:
$$u_{17} = 12 + (17 - 1) \times (-3)$$
$$= 12 - 48$$
$$= -36$$

The 17th term is -36.

EXAMPLE 3.6

How many terms are there in this sequence?

11, 15, 19, ..., 643

Solution   This is an arithmetic sequence with:

first term, $a = 11$
last term, $l = 643$
common difference, $d = 4$.

Using the result $l = a + (n - 1)d$, you have:

$$643 = 11 + (n - 1) \times 4$$
$$\Rightarrow \quad 4n = 643 - 11 + 4$$
$$\Rightarrow \quad n = 159$$

There are 159 terms.

THE SUM OF THE TERMS OF AN ARITHMETIC SEQUENCE

When Carl Friederich Gauss (1777–1855) was at school he was always quick to answer mathematics questions. One day his teacher, hoping for half an hour of peace and quiet, told his class to add up all the whole numbers from 1 to 100. Almost at once the 10-year-old Gauss announced that he had done it and that the answer was 5050.

Gauss had not of course added the terms, one by one. Instead he wrote the series down twice, once in the given order and once backwards, and added the two together:

$$S = \quad 1 + \quad 2 + \quad 3 + \quad ... + \quad 98 + \quad 99 + \quad 100$$
$$S = \quad 100 + \quad 99 + \quad 98 + \quad ... + \quad 3 + \quad 2 + \quad 1$$

Adding, $2S = 101 + 101 + 101 + ... + 101 + 101 + 101$

Since there are 100 terms in the series:

$$2S = 101 \times 100$$
$$S = 5050$$

The numbers 1, 2, 3, ..., 100 form an arithmetic sequence with common difference 1. Gauss' method can be used for finding the sum of any arithmetic series.

It is common to use the letter S to denote the sum of a series. When there is any doubt as to the number of terms that are being summed, this is indicated by a subscript: S_5 indicates the sum of 5 terms, S_n indicates the sum of n terms.

EXAMPLE 3.7

Find the value of $8 + 6 + 4 + \ldots + (-32)$.

Solution This is an arithmetic series, with common difference -2. The number of terms, n, may be calculated using:

$$l = a + (n - 1)d$$
$$-32 = 8 + (n - 1) \times -2$$
$$2(n - 1) = 40$$
$$n = 21$$

The sum S of the series is then found as follows.

$$S = \quad 8 + \quad 6 + \ldots -30 -32$$
$$\underline{S = -32 \ -30 - \ldots + \ 6 + 8}$$
$$2S = -24 \ -24 - \ldots -24 -24$$

Since there are 21 terms, this gives $2S = -24 \times 21$, so $S = -12 \times 21 = -252$.

Generalising this method by writing the series in the conventional notation gives:

$$S_n = \quad [a] \quad + \quad [a + d] \quad + \ldots + \quad [a + (n - 2)d] \quad + \quad [a + (n - 1)d]$$
$$\underline{S_n = [a + (n - 1)d] \ + \ [a + (n - 2)d] \ + \ldots + \quad [a + d] \quad + \quad [a]}$$
$$2S_n = [2a + (n - 1)d] \ + \ [2a + (n - 1)d] \ + \ldots + \ [2a + (n - 1)d] \ + \ [2a + (n - 1)d]$$

Since there are n terms, it follows that:

$$S_n = \tfrac{1}{2}n[2a + (n - 1)d]$$ (You should know the proof of the derivation of this formula.)

This result may also be written:

$$S_n = \tfrac{1}{2}n(a + l)$$

EXAMPLE 3.8

Find the sum of the first 100 terms of this sequence.

$$1, 1\tfrac{1}{3}, 1\tfrac{2}{3}, 2, \ldots$$

Solution In this arithmetic sequence $a = 1$, $d = \tfrac{1}{3}$ and $n = 100$.

Using $S_n = \tfrac{1}{2}n[2a + (n - 1)d]$ gives:

$$S_{100} = \tfrac{1}{2} \times 100(2 + 99 \times \tfrac{1}{3})$$
$$= 1750$$

EXAMPLE 3.9

In an arithmetic progression the sum of the first six terms is 30 and the sum of the first ten terms is 10. Find the first term, the common difference and the sum of the first 20 terms.

Solution

$$S_n = \tfrac{1}{2}n[2a + (n-1)d]$$

$$S_6 = 3[2a + 5d] = 30 \qquad\qquad S_{10} = 5[2a + 9d] = 10$$
$$2a + 5d = 10 \quad ① \qquad\qquad\qquad 2a + 9d = 2 \quad ②$$

$$② - ① \qquad\qquad 4d = -8$$
$$d = -2 \ \text{ and } \ a = 10$$

$$S_{20} = \tfrac{1}{2} \times 20 \times [2 \times 10 + 19 \times -2]$$
$$= -180$$

EXAMPLE 3.10

Evaluate $\displaystyle\sum_{r=1}^{20}(2r + 3)$.

Solution   Write out a few of the terms.

$$\sum_{r=1}^{20}(2r + 3) = 5 + 7 + 9 + \ldots + 43$$

This is an arithmetic progression with $a = 5$, $d = 2$ and $n = 20$.

So $S_n \quad = \tfrac{1}{2}n[2a + (n-1)d]$

$$S_{20} \quad = \sum_{r=1}^{20}(2r + 3) = \tfrac{1}{2} \times 20 \times [2 \times 5 + 19 \times 2]$$

$$\therefore \sum_{r=1}^{20}(2r + 3) = 480$$

EXAMPLE 3.11

Express $1 + 3 + 5 + \ldots + 101$ in $\sum$ notation.

Solution   You will recognise that the terms form an arithmetic progression with $a = 1$ and $d = 2$.

The rth term is $a + (r-1)d \quad = 1 + (r-1) \times 2$
$$= 2r - 1$$

The number of terms n is given by:

$$l = a + (n-1)d$$
$$101 = 1 + (n-1) \times 2$$
$$n = 51$$

The series can be written as:

$$1 + 3 + 5 + \ldots 101 = \sum_{r=1}^{51}(2r - 1)$$

EXAMPLE 3.12

Jamila starts a job on a salary of £9000 per year, and this increases by an annual increment of £1000. Assuming that, apart from the increment, Jamila's salary does not increase, find:

(a) her salary in the 12th year
(b) the length of time she has been working when her total earnings are £100 000.

Solution   Jamila's annual salaries (in pounds) form the arithmetic sequence:

$$9000, 10\,000, 11\,000, ...$$

The first term, $a = 9000$, and the common difference, $d = 1000$.

(a) Her salary in the 12th year is calculated using:

$$u_n = a + (n - 1)d$$
$$\Rightarrow \quad u_{12} = 9000 + (12 - 1) \times 1000$$
$$= 20\,000$$

(b) The number of years that have elapsed when her total earnings are £100 000 is given by:

$$S_n = \tfrac{1}{2}n[2a + (n - 1)d]$$

where $S_n = 100\,000$, $a = 9000$ and $d = 1000$.

This gives $\quad 100\,000 = \tfrac{1}{2}n[2 \times 9000 + 1000\,(n - 1)]$

This simplifies to the quadratic equation:

$$n^2 + 17n - 200 = 0$$

Factorising:

$$(n - 8)(n + 25) = 0$$
$$\Rightarrow n = 8 \text{ or } -25$$

The root $n = -25$ is irrelevant, so the answer is $n = 8$.

Jamila has earned a total of £100 000 after 8 years.

EXERCISE 3C   **1** Are the following sequences arithmetic? If so, state the common difference and the 7th term.

(a) 27, 29, 31, 33, ...   (b) 1, 2, 3, 5, 8, ...   (c) 2, 4, 8, 16, ...
(d) 3, 7, 11, 15, ...   (e) 8, 6, 4, 2, ...

2 The first term of an arithmetic sequence is −8 and the common difference is 3.

(a) Find the 7th term of the sequence.
(b) The last term of the sequence is 100. How many terms are there in the sequence?

3 The first term of an arithmetic sequence is 12, the 7th term is 36 and the last term is 144.

(a) Find the common difference.
(b) Find how many terms there are in the sequence.

4 There are 20 terms in an arithmetic sequence; the first term is −5 and the last term is 90.

(a) Find the common difference.
(b) Find the sum of the terms in the sequence.

5 The nth term of an arithmetic sequence is given by this rule.

$$u_n = 14 + 2n$$

(a) Write down the first three terms of the sequence.
(b) Calculate the sum of the first 12 terms of this sequence.

6 Below is an arithmetic sequence.

120, 114, ..., 36

(a) How many terms are there in the sequence?
(b) What is the sum of the terms in the sequence?

7 The 5th term of an arithmetic sequence is 28 and the 10th term is 58.

(a) Find the first term and the common difference.
(b) The sum of all the terms in this sequence is 444. How many terms are there?

8 The 6th term of an arithmetic sequence is twice the 3rd term, and the first term is 3. The sequence has 10 terms.

(a) Find the common difference.
(b) Find the sum of all the terms in the sequence.

9 A set of 12 new stamps is to be issued; the denominations increase in steps of 2p, starting with 1p:

$$1p, 3p, 5p, 7p, \ldots$$

(a) What is the highest denomination of stamp in the set?
(b) What is the total cost of the complete set?

10 (a) Find the sum of all the odd numbers between 50 and 150.
   (b) Find the sum of all the even numbers from 50 to 150, inclusive.
   (c) Find the sum of the terms of the arithmetic sequence with first term 50, common difference 1 and 101 terms.
   (d) Explain the relationship between your answers to (a), (b) and (c).

11 The first term of an arithmetic sequence is 3000 and the 10th term is 1200.

   (a) Find the sum of the first 20 terms of the sequence.
   (b) Which is the first negative term of the sequence?

12 An arithmetic progression has first term 7 and common difference 3.

   (a) Write down a formula for the nth term of the progression. Which term of the progression equals 73?
   (b) Write down a formula for the sum of the first n terms of the progression. How many terms of the progression are required to give a sum equal to 6300?

                                                                          [MEI]

13 (a) The nth term of an arithmetic progression is given by $u_n = 3 + 7n$. Write down the first term and the common difference of this progression.

   Calculate also $\displaystyle\sum_{r=1}^{100} u_r$.

   (b) The sum S_n of the first n terms of an arithmetic progression is given by $S_n = 3n + 2n^2$. Write down the first and second terms of the progression and find a formula for the nth term.

                                                                          [MEI]

14 Paul's starting salary in a company is £14 000 and during the time he stays with the company it increases by £500 each year.

   (a) What is his salary in his 6th year?
   (b) How many years has Paul been working for the company when his total earnings for all his years there are £126 000?

EXERCISE 3D   **Examination-style questions**

1   A sequence of terms has sum $\sum\limits_{r=1}^{30}(4r+9)$.

   (a) Write down the first three terms and show that the last term is 129.

   (b) Find the sum of the series.

2   A sequence is defined by $u_n = 3n^2$. Write down the value of u_1, u_2, u_3, u_4 and u_5. Given that $t_n = u_{n+1} - u_n$ find $t_1, t_2, t_3,$ and t_4 and express t_n in terms of n.

   Find $\sum\limits_{r=1}^{50} t_n$

3   The sum of the first ten terms of an arithmetic series is 270 and the sum of the first 20 terms is 340.

   (a) Find the first term and the common difference.

   (b) Find the number of terms required for the sum of them to be 0.

4   During the school holidays Nina works in a factory. She is paid £25 on the first day and each following day she is given an increase of 20p. How much does she earn on her 30th day and what is the total sum of money she has earned for the 30 days?

5   The triangular numbers 1, 3, 6, 10, … may be generated using this rule.

   $$u_n = \frac{1}{2}n(n+1)$$

   (a) Find the tenth and the one hundredth triangular number.

   (b) Evaluate $u_1 + u_2, u_2 + u_3, u_3 + u_4$. What do you notice?

   (c) Find $u_n + u_{n+1}$ in terms of n and show that it is a perfect square.

6   Initially, the number of fish in a pond is 500. The population is then modelled by this recurrence relation.

   $$u_{n+1} = 1.1u_n - d, \ u_0 = 500$$

   In this relation, u_n is the number of fish in the pond after n years and d is the number of fish that are caught every year.

   Given that $d = 10$:

   (a) calculate u_1, u_2 and u_3 and comment briefly on your results.

   Given instead that $d = 150$:

   (b) show that the population of fish dies out in the fifth year.

   (c) Find the value of d that would leave the population each year unchanged.

   [Edexcel adapted]

7   The fifth term of an arithmetic progression is 4 and the tenth term is 24. For this progression find:

   (a) the first term and the common difference
   (b) the sum of the first 15 terms
   (c) how many terms are needed for the sum to exceed 1000.

8   The sum of the first ten terms of an arithmetic series with common difference 3 is S. The sum of the first 20 terms is $4S$.

   (a) Find the first term and the value of S.
   (b) Find the number of terms for which the sum is $9S$.

9   The sum of the first two terms of an arithmetic series is 47. The thirtieth term of the series is -62. Find:

   (a) the first term of the series and the common difference
   (b) the sum of the first 60 terms of the series.

[Edexcel]

10 (a) Find the sum of the integers which are divisible by 3 and lie between 1 and 400.
   (b) Hence, or otherwise, find the sum of the integers, from 1 to 400 inclusive, which are not divisible by 3.

[Edexcel]

11 A polygon has ten sides. The lengths of the sides, starting with the smallest, form an arithmetic series. The perimeter of the polygon is 675 cm and the length of the longest side is twice that of the shortest side. Find for this series:

   (a) the common difference
   (b) the first term.

[Edexcel]

12 Work out $\displaystyle\sum_{r=1}^{100} r - \sum_{r=1}^{33} 3r$

   Describe the set of numbers whose sum you have found.

[Edexcel]

13 (a) Find the sum of all the integers between 1 and 1000 that are divisible by 7.
   (b) Hence, or otherside, evaluate $\displaystyle\sum_{r=1}^{142}(7r + 2)$.

[Edexcel]

14 Each year, for 40 years, Anne will pay money into a savings scheme. In the first year she pays in £500. Her payments then increase by £50 each year, so that she pays in £550 in the second year, £600 in the third year, and so on.

(a) Find the amount that Anne will pay in the 40th year.
(b) Find the total amount that Anne will pay in over the 40 years.

Over the same 40 years, Brian will also pay money into the savings scheme. In the first year he pays £890 and his payments then increase by £d each year.

Given that Anne and Brian will pay in exactly the same amount over the 40 years:
(c) find the value of d.

[Edexcel]

15 In the first month after opening, a mobile phone shop sold 280 phones. A model for future trading assumes that sales will increase by x phones per month for the next 35 months, so that $(280 + x)$ phones will be sold in the second month, $(280 + 2x)$ in the third month, and so on.

Using this model, with $x = 5$, calculate:

(a) (i) the number of phones sold in the 36th month
    (ii) the total number of phones sold over the 36 months.

The shop sets a sales target of 17 000 phones to be sold over the 36 months. Using the same model:

(b) find the least value of x required to achieve this target.

[Edexcel]

KEY POINTS

1 Summations:
$$\sum_{r=1}^{n} f(r) = f(1) + f(2) + f(3) + \ldots + f(n).$$

2 Rules:

Term-to-term   e.g. $u_{n+1} = 2u_n + 1$, $u_1 = 1$
Position-to-term e.g. $u_n = 2n + 1$

3 An **arithmetic progression** (AP) is a series of terms in which each term can be found from the previous term by adding (or subtracting) a fixed number.

a = first term, d = common difference, l = last term

The **nth term** is:
$$u_n = a + (n - 1)d$$

The **sum of the first n terms** is:
$$S_n = \tfrac{1}{2}n[2a + (n-1)d] = \tfrac{1}{2}n(a + l)$$

Chapter four

DIFFERENTIATION

Hold infinity in the palm of your hand.

William Blake

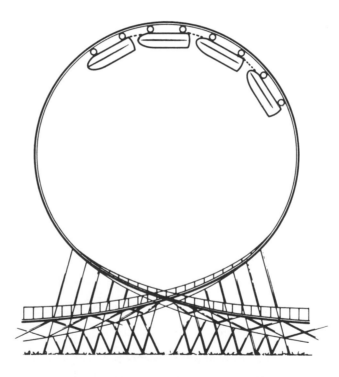

This picture illustrates one of the more frightening rides at an amusement park. To ensure that the ride is absolutely safe, its designers need to know the *gradient of the curve* at any point. What do we mean by the gradient of a curve?

THE GRADIENT OF A CURVE

To understand what this means, think of a log on a log-flume, as in figure 4.1. If you draw the straight line $y = mx + c$ passing along the bottom of the log, then this line is a tangent to the curve at the point of contact. The gradient m of the tangent is the gradient of the curve at the point of contact.

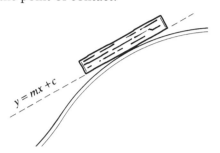

FIGURE 4.1

One method of finding the gradient of a curve is shown for point A in figure 4.2.

$$\text{Gradient} = \frac{y\text{-step}}{x\text{-step}}$$

$$= \frac{5.3}{1.5}$$

$$= 3.5$$

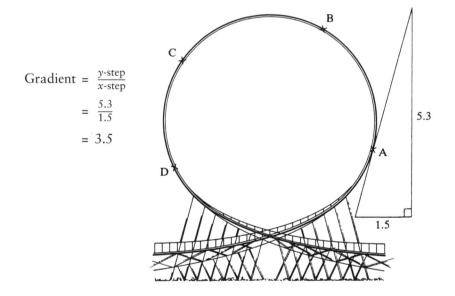

FIGURE 4.2

Find the gradient at the points B, C and D using the method shown in figure 4.2. (Use a piece of tracing paper to avoid drawing directly on the book!) Repeat the process for each point, using different triangles, and see whether you get the same answers.

You probably found that your answers were slightly different each time, because they depended on the accuracy of your drawing and measuring. Clearly you need a more accurate method of finding the gradient at a point. As you will see in this chapter, a method is available which can be used on many types of curve, and which does not involve any drawing at all.

FINDING THE GRADIENT OF A CURVE

Figure 4.3 shows the part of the graph $y = x^2$ which lies between $x = -1$ and $x = 3$. What is the value of the gradient at the point P(3, 9)?

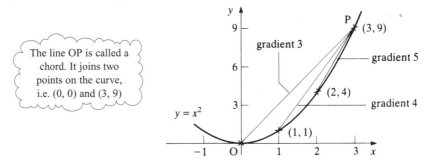

> The line OP is called a chord. It joins two points on the curve, i.e. (0, 0) and (3, 9)

FIGURE 4.3

Drawing the tangent at the point by hand provides only an approximate answer.

Another approach is to calculate the gradients of chords to the curve. Again, these give only approximate answers for the gradient of the curve, but based entirely on calculation and not drawing skill. Three chords are marked on figure 4.3.

Chord (0, 0) to (3, 9):   gradient $= \dfrac{9-0}{3-0} = 3$

Chord (1, 1) to (3, 9):   gradient $= \dfrac{9-1}{3-1} = 4$

Chord (2, 4) to (3, 9):   gradient $= \dfrac{9-4}{3-2} = 5$

Clearly none of these three answers is exact, but which of them is the most accurate?

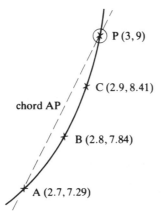

FIGURE 4.4

Of the three chords, the one closest to being a tangent is that joining (2, 4) to (3, 9), the two points that are nearest together.

You can take this process further by 'zooming in' on the point (3, 9) and using points which are much closer to it, as in figure 4.4.

The x-coordinate of point A is 2.7, the y-coordinate 2.7^2 or 7.29 (since the point lies on the curve $y = x^2$). Similarly B and C are (2.8, 7.84) and (2.9, 8.41). The gradients of the chords joining each point to (3, 9) are as follows.

Chord (2.7, 7.29) to (3, 9):   gradient $= \dfrac{9-7.29}{3-2.7} = 5.7$

Chord (2.8, 7.84) to (3, 9):   gradient $= \dfrac{9-7.84}{3-2.8} = 5.8$

Chord (2.9, 8.41) to (3, 9):   gradient $= \dfrac{9-8.41}{3-2.9} = 5.9$

These results are getting closer to the gradient of the tangent. What happens if you take points much closer to (3, 9), for example (2.99, 8.9401) and (2.999, 8.994 001)?

The gradients of the chords joining these to (3, 9) work out to be 5.99 and 5.999 respectively.

You might wish to take points X, Y, Z on the curve $y = x^2$ with x-coordinates 3.1, 3.01 and 3.001 respectively, and find the gradients of the chords joining each of these points to (3, 9).

It looks as if the gradients are approaching the value 6, and if so this is the gradient of the tangent at (3, 9).

Taking this method to its logical conclusion, you might try to calculate the gradient of the 'chord' from (3, 9) to (3, 9), but this is undefined because there is a 0 in the denominator. So although you can find the gradient of a chord which is as close as you like to the tangent, it can never be exactly that of the tangent. What we need is a way of making that final step from a chord to a tangent.

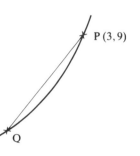

The concept of a *limit* enables us to do this, as you will see in the next section. It allows us to confirm that in the limit as point Q tends to point P(3, 9), the chord QP tends to the tangent of the curve at P, and the gradient of QP tends to 6 (figure 4.5).

FIGURE 4.5

The idea of a limit is central to calculus, which is sometimes described as the study of limits.

Historical note   This method of using chords approaching the tangent at P to calculate the gradient of the tangent was first described clearly by Pierre de Fermat (1601–65). He spent his working life as a civil servant in Toulouse and produced an astonishing amount of original mathematics in his spare time.

FINDING THE GRADIENT FROM FIRST PRINCIPLES

Although the work in the previous section was more formal than the method of drawing a tangent and measuring its gradient, it was still somewhat experimental. The result that the gradient of $y = x^2$ at (3, 9) is 6 was a sensible conclusion, rather than a proved fact.

In this section the method is formalised and extended.

Take the point P(3, 9) and another point Q on the curve $y = x^2$ close to (3, 9). Let the x-coordinate of Q be $3 + h$ where h is small. Since $y = x^2$ at Q, the y-coordinate of Q will be $(3 + h)^2$.

Note

Figure 4.6 shows Q in a position where h is positive, but negative values of h would put Q to the left of P.

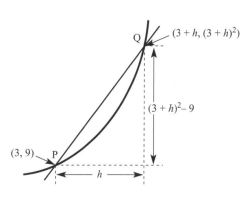

FIGURE 4.6

From figure 4.6, the gradient of PQ is:

$$\frac{(3+h)^2 - 9}{h} = \frac{9 + 6h + h^2 - 9}{h}$$

$$= \frac{6h + h^2}{h}$$

$$= \frac{h(6 + h)}{h}$$

$$= 6 + h$$

For example, when $h = 0.001$, the gradient of PQ is 6.001, and when $h = -0.001$, the gradient of PQ is 5.999. The gradient of the tangent at P is between these two values. Similarly the gradient of the tangent would be between $6 - h$ and $6 + h$ for all small non-zero values of h.

For this to be true the gradient of the tangent at (3, 9) must be *exactly* 6.

Using a similar method, show that the gradient of the tangent to the curve at:

(i) (1, 1) is 2   **(ii)** (−2, 4) is −4   **(iii)** (4, 16) is 8.

You will notice that the gradient is always twice the value of the x-coordinate.

THE GRADIENT FUNCTION

The work so far has involved finding the gradient of the curve $y = x^2$ at a particular point (3, 9), but this is not the way in which you would normally find the gradient at a point. Rather, you would consider the general point, (x, y), and then substitute the value(s) of x (and/or y) corresponding to the point of interest.

EXAMPLE 4.1

Find the gradient of the curve $y = x^3$ at the general point (x, y).

Solution

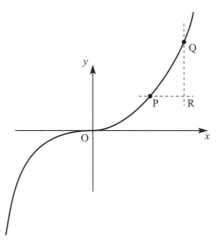

FIGURE 4.7

Let P have the general value x as its x-coordinate, so P is the point (x, x^3) (since it is on the curve $y = x^3$). Let the x-coordinate of Q be $(x + h)$ so Q is $((x + h), (x + h)^3)$. The gradient of the chord PQ is given by:

$$\frac{QR}{PR} = \frac{(x + h)^3 - x^3}{(x + h) - x}$$

$$= \frac{x^3 + 3x^2h + 3xh^2 + h^3 - x^3}{h}$$

$$= \frac{3x^2h + 3xh^2 + h^3}{h}$$

$$= \frac{h(3x^2 + 3xh + h^2)}{h}$$

$$= 3x^2 + 3xh + h^2$$

As Q takes values closer to P, h takes smaller and smaller values and the gradient approaches the value of $3x^2$ which is the gradient of the tangent at P. The gradient of the curve $y = x^3$ at the point (x, y) is equal to $3x^2$.

Note

If the equation of the curve is written as $y = f(x)$, then the *gradient function* (i.e. the gradient at the general point (x, y)) is written as $f'(x)$. Using this notation the result above can be written as $f(x) = x^3 \Rightarrow f'(x) = 3x^2$.

EXERCISE 4A

1 Use the method in Example 4.1 to prove that the gradient of the curve $y = x^2$ at the point (x, y) is equal to $2x$.

2 Copy the table below and suggest how the gradient pattern should continue when $f(x) = x^4$, $f(x) = x^5$, $f(x) = x^6$ and $f(x) = x^n$ (where n is a positive whole number).

$f(x)$	$f'(x)$ (gradient at (x, y))
x^2	$2x$
x^3	$3x^2$
x^4	
x^5	
x^6	
$\vdots$	
x^n	

3 Prove the result when $f(x) = x^5$.

....

Note   The result you should have obtained to the last part of question 2 can be used as a formula.
That is:

$$f(x) = x^n \Rightarrow f'(x) = nx^{n-1}$$
....

AN ALTERNATIVE NOTATION

So far h has been used to denote the difference between the x-coordinates of our points P and Q, where Q is close to P.

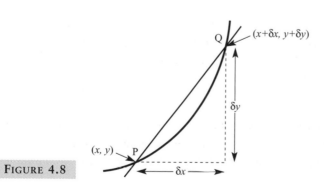

FIGURE 4.8

h is sometimes replaced by δx. The Greek letter δ is shorthand for 'a small change in' and so δx represents a small change in x and δy a corresponding small change in y.

In figure 4.8 the gradient of the chord PQ is $\frac{\delta y}{\delta x}$.

In the limit as $\delta x \to 0$, δx and δy both become infinitesimally small and the value obtained for $\frac{\delta y}{\delta x}$ approaches the gradient of the tangent at P.

So the gradient of the tangent is:

$$\underset{\delta x \to 0}{\text{Lim}} \frac{\delta y}{\delta x} \text{ which is written as } \frac{dy}{dx}.$$

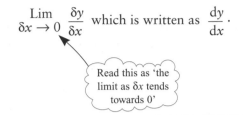

Read this as 'the limit as δx tends towards 0'

So we now have that: if $y = f(x)$ then the gradient function is $\frac{dy}{dx} = f'(x)$

Using this notation, the rule becomes: $y = x^n \Rightarrow \frac{dy}{dx} = nx^{n-1}$

The gradient function, $\frac{dy}{dx}$ or $f'(x)$ is sometimes called the *derivative* of y with respect to x, and when you find it you have *differentiated* y with respect to x.

Note

There is nothing special about the letters x, y and f.

If, for example, your curve represented time (t) on the horizontal axis and velocity (v) on the vertical axis, then the relationship may be referred to as $v = g(t)$, i.e. v is a function of t, and the gradient function is given by $\frac{dv}{dt} = g'(t)$.

RATES OF CHANGE

When finding the derivative $\frac{dy}{dx}$ you are not just finding the gradient of the function but you are also finding the rate at which the y-values are changing with respect to the x-values.

The value of $\frac{dv}{dt}$ above is the rate of change of v with respect to t. If, in this case v is the velocity of a moving particle and t is the time, then $\frac{dv}{dt}$ is the rate of change of velocity with respect to time: that is, the acceleration of the particle.

Historical note

The notation $\frac{dy}{dx}$ was first used by the German mathematician and philosopher Gottfried Leibniz (1646–1716) in 1675. Leibniz was a child prodigy and a self-taught mathematician. The terms 'function' and 'coordinates' are due to him and, because of his influence, the sign '=' is used for equality and '×' for multiplication. In 1684 he published his work on calculus (which deals with the way in which quantities change) in a six-page article in the periodical *Acta Eruditorum*.

Sir Isaac Newton (1642–1727) worked independently on calculus but Leibniz published his work first. Newton always hesitated to publish his discoveries. Newton used different notation (introducing 'fluxions' and 'moments of fluxions') and his expressions were thought to be rather vague. Over the years the best aspects of the two approaches have been combined, but at the time the dispute as to who 'discovered' calculus first was the subject of many articles and reports, and a cause of great controversy.

DIFFERENTIATING BY USING STANDARD RESULTS

The method of differentiation from first principles will always give the gradient function, but it is rather tedious and, in practice, it is hardly ever used. Its value is in establishing a formal basis for differentiation rather than as a working tool.

If you look at the results of differentiating $y = x^n$ for different values of n a pattern is immediately apparent, particularly when you include the result that the line $y = x$ has constant gradient 1.

y	$\dfrac{dy}{dx}$
x^1	1
x^2	$2x^1$
x^3	$3x^2$

This pattern continues and, in general:

$$y = x^n \quad \Longrightarrow \quad \frac{dy}{dx} = nx^{n-1}$$

This can be extended to functions of the type $y = kx^n$ for any constant k, to give:

$$y = kx^n \quad \Longrightarrow \quad \frac{dy}{dx} = nkx^{n-1}$$

Another important result is that:

$$y = c \quad \Longrightarrow \quad \frac{dy}{dx} = 0$$

where c is any constant.

This follows from the fact that the graph of $y = c$ is a horizontal line with gradient 0 (figure 4.9).

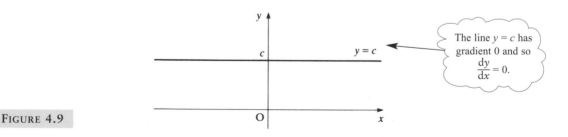

The line $y = c$ has gradient 0 and so $\dfrac{dy}{dx} = 0$.

FIGURE 4.9

EXAMPLE 4.2

For each of the given functions of x find the gradient function.

(a) $y = x^5$ (b) $z = 7x^6$ (c) $p = 11$

Solution   (a) $\dfrac{dy}{dx} = 5x^4$

(b) $\dfrac{dz}{dx} = 6 \times 7x^5 = 42x^5$

(c) $\dfrac{dp}{dx} = 0$

DIFFERENTIATING $y = x^n$ WHEN n IS NEGATIVE OR A FRACTION

The rule applies in exactly the same way as for positive n.

EXAMPLE 4.3

Differentiate each of the following with respect to x.

(a) x^{-3} (b) $\dfrac{3}{x^2}$ (c) $x^{\frac{5}{4}}$

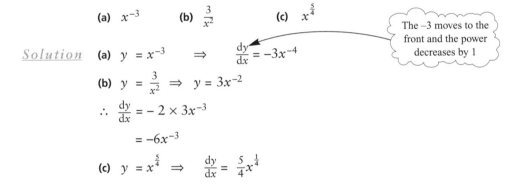

The −3 moves to the front and the power decreases by 1

Solution   (a) $y = x^{-3} \quad \Rightarrow \quad \dfrac{dy}{dx} = -3x^{-4}$

(b) $y = \dfrac{3}{x^2} \Rightarrow y = 3x^{-2}$

$\therefore \dfrac{dy}{dx} = -2 \times 3x^{-3}$

$= -6x^{-3}$

(c) $y = x^{\frac{5}{4}} \quad \Rightarrow \quad \dfrac{dy}{dx} = \dfrac{5}{4}x^{\frac{1}{4}}$

SUMS AND DIFFERENCES OF FUNCTIONS

Many of the functions you will meet are sums or differences of simpler ones. For example, the function $(3x^2 + 4x^3)$ is the sum of the functions $3x^2$ and $4x^3$.

To differentiate a function such as this you differentiate each part separately and then add the results together.

EXAMPLE 4.4

Differentiate $y = 3x^2 + 4x^3$.

Solution $\frac{dy}{dx} = 6x + 12x^2$

Note

This may be written in general form as:

$$y = f(x) + g(x) \quad\Rightarrow\quad \frac{dy}{dx} = f'(x) + g'(x).$$

EXAMPLE 4.5

You are given that $y = x^3 - 4x^2 + x - 1$.

(a) Find $\frac{dy}{dx}$.

(b) Find the gradient of the curve at the point $(3, -7)$.

Solution

(a) $\frac{dy}{dx} = 3x^2 - 8x + 1$

(b) At $(3, -7)$, $x = 3$.

Substituting $x = 3$ in the expression for $\frac{dy}{dx}$ gives:

$$\frac{dy}{dx} = 3 \times 3^2 - 8 \times 3 + 1$$
$$= 4$$

PRODUCTS AND QUOTIENTS OF FUNCTIONS

The products (in which two functions are multiplied) and the quotients (in which one function is divided by another) that you will meet in *Pure Mathematics: Core 1* must first be multiplied or divided before you differentiate.

EXAMPLE 4.6

Differentiate $(2x + 3)(3x - 1)$ with respect to x.

Solution $y = (2x + 3)(3x - 1)$

Multiply out the brackets.

$$y = 6x^2 + 7x - 3 \quad\Rightarrow\quad \frac{dy}{dx} = 12x + 7$$

EXAMPLE 4.7

Differentiate $y = \dfrac{x^3 + 2x^2 - 7}{x}$.

Solution   Divide through by x.

$$y = \frac{x^3}{x} + \frac{2x^2}{x} - \frac{7}{x} = x^2 + 2x - 7x^{-1}$$

$$\Rightarrow \frac{dy}{dx} = 2x + 2 + \frac{7}{x^2}$$

GRADIENTS

You have seen that the function $\dfrac{dy}{dx}$ or $f'(x)$ gives the gradient of a curve.

EXAMPLE 4.8

Given that $y = x^3 - 6x^2 + 15x - 4$:

(a) find the gradient of the curve at the point where $x = 2$

(b) find the coordinates of the points on the curve where the gradient is 6.

Solution   First find the derivative.

$$\frac{dy}{dx} = 3x^2 - 12x + 15$$

(a) When $x = 2$:

the gradient $= \dfrac{dy}{dx} = 3 \times 2^2 - 12 \times 2 + 15$

$$= 12 - 24 + 15$$

$$= 3$$

(b) When the gradient is 6:

$$\frac{dy}{dx} = 3x^2 - 12x + 15 = 6$$

$$\therefore 3x^2 - 12x + 9 = 0$$

$$x^2 - 4x + 3 = 0$$

$$(x - 1)(x - 3) = 0$$

$$x = 1 \text{ or } x = 3$$

When $x = 1$, $y = 1^3 - 6 \times 1^2 + 15 \times 1 - 4 = 6$

When $x = 3$, $y = 3^3 - 6 \times 3^2 + 15 \times 3 - 4 = 14$

So the points where the gradient is 6 are $(1, 6)$ and $(3, 14)$.

EXERCISE 4B

1 Differentiate the following functions using the rules:

$$y = kx^n \implies \frac{dy}{dx} = nkx^{n-1}$$

and $y = f(x) + g(x) \implies \frac{dy}{dx} = f'(x) + g'(x)$

(a) $y = x^5$ (b) $y = 4x^2$ (c) $y = 2x^3$

(d) $y = x^{11}$ (e) $y = 4x^{10}$ (f) $y = 3x^5$

(g) $y = 7$ (h) $y = 7x$ (i) $y = 2x^3 + 3x^5$

(j) $y = x^7 - x^4$ (k) $y = x^2 + 1$ (l) $y = x^3 + 3x^2 + 3x + 1$

(m) $y = x^3 - 9$ (n) $y = \frac{1}{2}x^2 + x + 1$ (o) $y = 3x^2 + 6x + 6$

2 Differentiate the following with respect to x.

(a) $y = x^{-2}$ (b) $y = x^{\frac{2}{3}}$ (c) $y = 3x^{-4}$

(d) $y = \frac{2}{x}$ (e) $y = \frac{4}{3}x^{\frac{3}{4}}$ (f) $y = \sqrt[3]{x}$

(g) $y = \sqrt{x} + \frac{1}{\sqrt{x}}$ (h) $y = \frac{x^2}{3} + \frac{3}{x^2}$ (i) $y = \frac{3}{x} + \frac{1}{\sqrt[3]{x}}$ (j) $y = 6\sqrt{x^3}$

3 Differentiate the following with respect to x.

(a) $y = 2x(3x + 1)$ (b) $y = (2x + 3)(3x + 1)$

(c) $y = (x + 2)(x^2 + 3x - 4)$ (d) $y = (x - 3)^2$

(e) $y = 2\sqrt{x}(1 + x)$ (f) $y = (1 + \sqrt{x})^2$

(g) $y = 3x(x^{\frac{1}{3}} + x^{-\frac{1}{3}})$ (h) $y = x^{-2}(x^2 + 4)$

4 Differentiate the following with respect to x.

(a) $y = \dfrac{3x^4 + 5x}{x^2}$ (b) $y = \dfrac{x + 1}{\sqrt{x}}$

(c) $y = \dfrac{x^2 - 3x + 4}{2x^3}$ (d) $y = \dfrac{2x^2 + 4x - 3}{3x^{\frac{1}{3}}}$

(e) $y = \dfrac{\sqrt{x} + 1}{x}$ (f) $y = \dfrac{2x^4 + 3}{x^3}$

(g) $y = \dfrac{(1 - \sqrt{x})^2}{\sqrt{x}}$ (h) $y = \dfrac{2 + x - x^2}{1 + x}$

5 Differentiate the following with respect to r.

(a) $A = 4\pi r^2$ (b) $V = \frac{4}{3}\pi r^3$ (c) $C = 2\pi r$

6 (a) Find the gradient function of $f(x) = \dfrac{(x + 3)(x - 2)}{2x}$.

(b) Show that $f'(2) = \frac{5}{4}$.

7 Find the gradient of each of the following curves at the given point.

(a) $y = x^2 + 2x - 3$ at $(1, 0)$ (b) $y = 4 + 3x - x^2$ at $(-1, -4)$

(c) $y = (x - 1)(x + 2)$ at $(3, 10)$ (d) $y = x^3 + 2x^2 - 3x - 4$ at $(2, 6)$

(e) $y = \frac{1}{6}x^3 - \frac{1}{4}x^2 + \frac{1}{12}x - 1$ at $(1, -1)$ (f) $y = (x - 1)(x - 2)(x - 3)$ at $(0, -6)$

(g) $y = \frac{2}{x}$ at $(4, \frac{1}{2})$ (h) $y = \sqrt{x} + 1$ at $(9, 4)$

(i) $y = \frac{3x^2 - 2}{x}$ at $(-2, -5)$ (j) $y = \frac{(\sqrt{x} + 1)^2}{\sqrt{x}}$ at $(4, 4\frac{1}{2})$

8 In each of the following find the coordinates of the point or points on the curve which have the given gradient.

(a) $y = x^2$ with gradient 4   (b) $y = 3x^2 + 2x - 1$ with gradient -4

(c) $y = (x + 1)(x - 1)$ with gradient 1   (d) $y = 2 + \sqrt{x}$ with gradient 2

(e) $y = \frac{3}{x}$ with gradient -12 (f) $y = \frac{2x^2 + 3}{x}$ with gradient -1

(g) $y = 2x^3 - 9x^2 + 18x - 24$ with gradient 6

(h) $y = x^3 - 3x^2 + 3x - 2$ with gradient 12

(i) $y = \sqrt{x}(1 + \sqrt{x})$ with gradient $\frac{1}{2}$ (j) $y = (x - 1)^3$ with gradient 12

SECOND DERIVATIVES

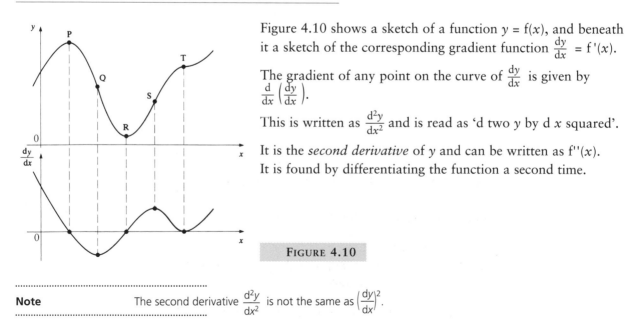

Figure 4.10 shows a sketch of a function $y = f(x)$, and beneath it a sketch of the corresponding gradient function $\frac{dy}{dx} = f'(x)$.

The gradient of any point on the curve of $\frac{dy}{dx}$ is given by $\frac{d}{dx}\left(\frac{dy}{dx}\right)$.

This is written as $\frac{d^2y}{dx^2}$ and is read as 'd two y by d x squared'.

It is the *second derivative* of y and can be written as $f''(x)$. It is found by differentiating the function a second time.

FIGURE 4.10

....
Note   The second derivative $\frac{d^2y}{dx^2}$ is not the same as $\left(\frac{dy}{dx}\right)^2$.
....

EXAMPLE 4.9

Given that $y = x^3 + 2x$, find $\dfrac{d^2y}{dx^2}$.

Solution $y = x^3 + 2x$

$\dfrac{dy}{dx} = 3x^2 + 2$

$\dfrac{d^2y}{dx^2} = 6x$

EXAMPLE 4.10

Given that $f(x) = \dfrac{2}{x} + 4\sqrt{x}$ show that $f''(4) = -\dfrac{1}{16}$.

Solution $f(x) = \dfrac{2}{x} + 4\sqrt{x} = 2x^{-1} + 4x^{\frac{1}{2}}$

$\qquad\quad f'(x) = -2x^{-2} + 2x^{-\frac{1}{2}}$

$\qquad\quad f''(x) = 4x^{-3} - x^{-\frac{3}{2}}$

$\Rightarrow f''(4) = \dfrac{4}{4^3} - \dfrac{1}{\sqrt{4^3}} = \dfrac{1}{16} - \dfrac{1}{8} = -\dfrac{1}{16}$

EXERCISE 4C

1 For each of the following, find $\dfrac{dy}{dx}$ and $\dfrac{d^2y}{dx^2}$.

(a) $y = x^3$ (b) $y = x^5$ (c) $y = 4x^2$

(d) $y = x^{-2}$ (e) $y = x^{\frac{3}{2}}$ (f) $y = x^4 - \dfrac{2}{x^3}$

(g) $y = 9\sqrt[3]{x}$ (h) $y = (4x + 3)^2$ (i) $y = \dfrac{x^2 - 1}{x}$

(j) $y = (2x - 1)(x^2 + 3x - 1)$

2 For each of the following find $f''(x)$ and the value of $f''(4)$.

(a) $f(x) = x^5 - x^4$ (b) $f(x) = 6\sqrt{x}$

(c) $f(x) = \dfrac{8}{x^2} - \dfrac{x^2}{8}$ (d) $f(x) = (x^3 + 1)(x^3 - 1)$

(e) $f(x) = 2x + 3$ (f) $f(x) = x^{\frac{1}{2}} + x^{\frac{3}{2}}$

(g) $f(x) = \dfrac{(2 + \sqrt{x})^2}{x}$ (h) $f(x) = x^{\frac{3}{2}} + \dfrac{1}{x^{\frac{3}{2}}}$

3 Given that $y = 4x - x^2$:

(a) find the values of x when $y = 0$

(b) find the values of x and y at the point where $\dfrac{dy}{dx} = 0$ and also the value of $\dfrac{d^2y}{dx^2}$ at this point

(c) find the values of x for which $\dfrac{dy}{dx} > 0$ and $\dfrac{dy}{dx} < 0$

(d) sketch the graph of $y = 4x - x^2$.

4 Given that $f(x) = \dfrac{x^{\frac{5}{2}}}{5} + \dfrac{64}{x}$:

    (a) find the values of x and y when $f'(x) = 0$.

    (b) find the value of $f''(x)$ at the point where $f'(x) = 0$.

5 Find the value of y at the point where $\dfrac{d^2y}{dx^2} = \dfrac{3}{16}$ given that

    $y = \dfrac{(x + 2)(x - 3)}{x}$.

GRAPHS AND GRADIENTS

EXAMPLE 4.11

Figure 4.11 shows the graph of $y = x^2(x - 6) = x^3 - 6x^2$.
Find the gradient of the curve at the points A and B where it meets the x-axis.

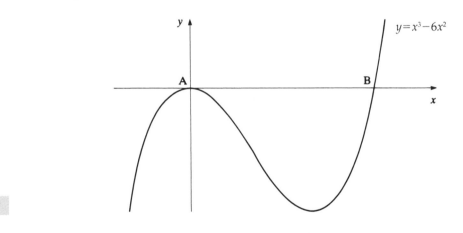

FIGURE 4.11

Solution   The curve cuts the x-axis when $y = 0$, and so at these points:

$$x^2(x - 6) = 0$$
$$\Rightarrow \quad x = 0 \text{ (twice) or } 6$$

Differentiating $y = x^3 - 6x^2$ gives:

$$\frac{dy}{dx} = 3x^2 - 12x$$

At the point $(0, 0)$: $\dfrac{dy}{dx} = 0$

and at $(6, 0)$: $\dfrac{dy}{dx} = 3 \times 6^2 - 12 \times 6 = 36$

At $A(0, 0)$ the gradient of the curve is 0 and at $B(6, 0)$ the gradient of the curve is 36.

Note

This curve goes through the origin. You can see from the graph and from the value of $\frac{dy}{dx}$ that the *x*-axis is a tangent to the curve at this point. You could also have deduced this from the fact that $x = 0$ is a repeated root of the equation $x^3 - 6x^2 = 0$.

EXAMPLE 4.12

Find the points on the curve for which the equation is $y = x^3 + 6x^2 + 5$ where the value of the gradient is –9.

Solution   The gradient at any point on the curve is given by:

$$\frac{dy}{dx} = 3x^2 + 12x$$

Therefore we need to find points at which $\frac{dy}{dx} = -9$, i.e.

$$3x^2 + 12x = -9$$
$$3x^2 + 12x + 9 = 0$$
$$3(x^2 + 4x + 3) = 0$$
$$3(x + 1)(x + 3) = 0$$
$$\Rightarrow \qquad x = -1 \text{ or } -3$$

When $x = -1$, $y = (-1)^3 + 6(-1)^2 + 5 = 10$.

When $x = -3$, $y = (-3)^3 + 6(-3)^2 + 5 = 32$.

Therefore the gradient is –9 at the points (–1, 10) and (–3, 32) (figure 4.12).

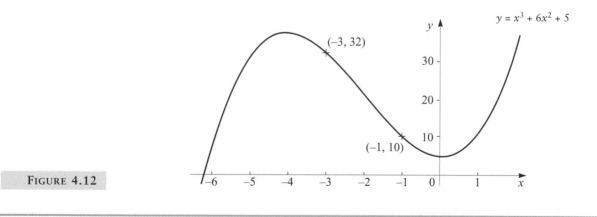

FIGURE 4.12

EXERCISE 4D

1 (a) Sketch the curve $y = x^2 - 4$.

(b) Write down the coordinates of the points where the curve crosses the x-axis.

(c) Differentiate $y = x^2 - 4$.

(d) Find the gradients of the curve at the points where it crosses the x-axis.

2 (a) Sketch the curve $y = x^2 - 6x$.

(b) Differentiate $y = x^2 - 6x$.

(c) Show that the point $(3, -9)$ lies on the curve $y = x^2 - 6x$ and find the gradient of the curve at this point.

(d) Relate your answer to the shape of the curve.

3 (a) Plot, on the same axes, the graphs for which the equations are

$$y = 2x + 5 \quad \text{and} \quad y = 4 - x^2 \quad \text{for } -3 \leqslant x \leqslant 3.$$

(b) Show that the point $(-1, 3)$ lies on both graphs.

(c) Differentiate $y = 4 - x^2$ and so find its gradient at $(-1, 3)$.

(d) Do you have sufficient evidence to decide whether the line $y = 2x + 5$ is a tangent to the curve $y = 4 - x^2$?

4 (a) Sketch the curve $y = (x - 1)(x - 2)(x - 3)$.

(b) Show that $y = x^3 - 6x^2 + 11x - 6$.

(c) Differentiate $y = x^3 - 6x^2 + 11x - 6$.

(d) Find the gradients of the curve at the points at which it cuts the x-axis.

5 (a) Sketch the curve $y = x^2 + 3x - 1$.

(b) Differentiate $y = x^2 + 3x - 1$.

(c) Find the coordinates of the point on the curve $y = x^2 + 3x - 1$ at which it is parallel to the line $y = 5x - 1$.

6 (a) Sketch, on the same axes, the curves for which the equations are

$$y = x^2 - 9 \quad \text{and} \quad y = 9 - x^2 \quad \text{for } -4 \leqslant x \leqslant 4.$$

(b) Differentiate $y = x^2 - 9$.

(c) Find the gradient of $y = x^2 - 9$ at the points $(2, -5)$ and $(-2, -5)$.

(d) Find the gradient of the curve $y = 9 - x^2$ at the points $(2, 5)$ and $(-2, 5)$.

7 (a) Sketch the graph of $y = \frac{1}{x} + 1$.

(b) Find the gradient of the curve at the point where $y = \frac{5}{4}$.

(c) Find the coordinates of the other point on the curve where the gradient is the same as that at $y = \frac{5}{4}$.

8 The function $f(x) = ax^3 + bx + 4$, where a and b are constants, goes through the point $(2, 14)$ with gradient 21.

(a) Using the fact that $(2, 14)$ lies on the curve, find an equation involving a and b.

(b) Differentiate $f(x)$ and, using the fact that the gradient is 21 when $x = 2$, form another equation involving a and b.

(c) By solving these two equations simultaneously find the values of a and b.

9 In his book *Mathematician's Delight*, W.W. Sawyer observes that the arch of Victoria Falls Bridge appears to agree with the curve:

$$y = \frac{(116 - 21x^2)}{120}$$

taking the origin as the point mid-way between the feet of the arch, and taking the distance between its feet as 4.7 units.

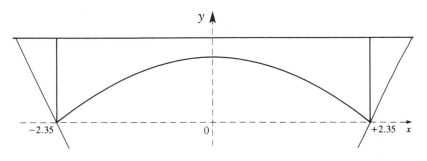

(a) Find $\frac{dy}{dx}$.

(b) Evaluate $\frac{dy}{dx}$ when $x = -2.35$ and when $x = 2.35$.

(c) Find the value of x for which $\frac{dy}{dx} = -0.5$.

10 The diagram shows part of the graph of $y = f(x)$. The coordinates of A and B are respectively $(a, f(a))$ and $(b, f(b))$.

(a) In the case when $f(x) = x^2$, $a = 4$, $b = 4.01$, calculate the exact value of the expression:

$$\frac{f(b) - f(a)}{b - a}$$

(b) State what is represented in the diagram by the expression:

$$\frac{f(b) - f(a)}{b - a}$$

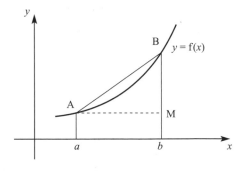

(c) Interpret the meaning of this expression in the limit as b tends to a.

[MEI]

11 The graph of $y = f(x) = 3x^2 + 1$ is shown below.

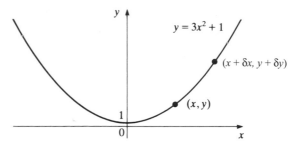

(a) Write down and expand the expression for the y-coordinate, $f(x + \delta x)$, when x is replaced by $x + \delta x$. Call this expression $y + \delta y$.

(b) Find $\frac{\delta y}{\delta x}$ by substituting your answer from (a) in

$$\frac{\delta y}{\delta x} = \frac{f(x + \delta x) - f(x)}{(x + \delta x) - x}$$

   i.e. find the gradient of the chord joining (x, y) to $(x + \delta x, y + \delta y)$.

(c) Now let δx tend to 0, and show that your answer from (b) gives $6x$ as the gradient function $\frac{dy}{dx}$.

12 A curve is given by $y = 2x - x^2$.

(a) Find an expression for the y-coordinate, $y + \delta y$, of the point on the curve with x-coordinate $x + \delta x$, in terms of x and δx.

(b) Hence write down an expression for $\frac{\delta y}{\delta x}$ in terms of x and δx, and simplify it.

(c) By considering the limit as δx tends to 0, obtain an expression for the gradient function $\frac{dy}{dx}$ for $y = 2x - x^2$.

TANGENTS AND NORMALS

A tangent is a straight line that touches a curve at a given point. The normal is the straight line that is at right angles to the tangent at that point (see figure 4.13).

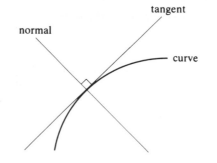

FIGURE 4.13

Now that you know how to find the gradient of a curve at any point you can use this value as the gradient, m, of the tangent. You will recall from Chapter 2 that the general equation of a straight line is $y - y_1 = m(x - x_1)$ and this can be used to find the equation of the tangent.

You will also recall that for perpendicular lines with gradients m_1 and m_2 the product $m_1m_2 = -1$. This enables you to find the gradient of the normal and, hence, the equation of the normal.

EXAMPLE 4.13

Find the equation of the tangent to the curve $y = x^2 + 3x + 2$ at the point $(2, 12)$.

Solution $y = x^2 + 3x + 2 \implies \dfrac{dy}{dx} = 2x + 3$

Substituting $x = 2$ into the expression $\dfrac{dy}{dx}$ to find the gradient m of the tangent at that point:

$$\begin{aligned} m &= 2 \times 2 + 3 \\ &= 7 \end{aligned}$$

The equation of the tangent is given by:

$$y - y_1 = m(x - x_1)$$

In this case $x_1 = 2$, $y_1 = 12$ so:

$$\begin{aligned} y - 12 &= 7(x - 2) \\ \implies \quad y &= 7x - 2 \end{aligned}$$

This is the equation of the tangent.

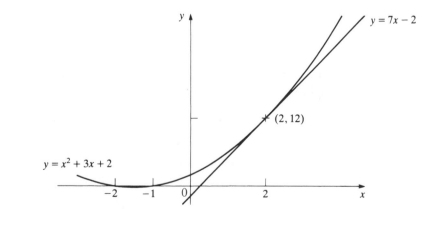

FIGURE 4.14

EXAMPLE 4.14

Find the equation of the normal to the curve $y = x^3 - 2x + 1$ at the point $(2, 5)$.

Solution $y = x^3 - 2x + 1 \implies \dfrac{dy}{dx} = 3x^2 - 2$

Substituting $x = 2$ to find the gradient m_1 of the tangent at the point $(2, 5)$:

$$m_1 = 3 \times 2^2 - 2 = 10$$

The gradient m_2 of the normal to the curve at this point is given by:

$$m_2 = -\frac{1}{m_1} = -\frac{1}{10}$$

The equation of the normal is given by:

$$y - y_1 = m_2(x - x_1)$$

and in this case $x_1 = 2$, $y_1 = 5$ so:

$$y - 5 = -\frac{1}{10}(x - 2)$$
$$y = -\frac{1}{10}x + \frac{26}{5}$$

Tidy this up by multiplying both sides by 10.

$$10y = -x + 52 \qquad \text{or} \qquad x + 10y = 52$$

EXERCISE 4E

1 The graph of $y = 6x - x^2$ is shown below.

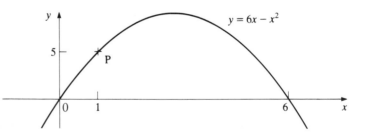

The marked point, P, is $(1, 5)$.

(a) Find the gradient function $\dfrac{dy}{dx}$.

(b) Find the gradient of the curve at P.

(c) Find the equation of the tangent at P.

2 (a) Sketch the curve $y = 4x - x^2$.

(b) Differentiate $y = 4x - x^2$.

(c) Find the gradient of $y = 4x - x^2$ at the point $(1, 3)$.

(d) Find the equation of the tangent to the curve $y = 4x - x^2$ at the point $(1, 3)$.

3 (a) Sketch the curve $y = x^3 - 4x^2$.

(b) Differentiate $y = x^3 - 4x^2$.

(c) Find the gradient of $y = x^3 - 4x^2$ at the point $(2, -8)$.

(d) Find the equation of the tangent to the curve $y = x^3 - 4x^2$ at the point $(2, -8)$.

(e) Find the coordinates of the other point at which this tangent meets the curve.

4 (a) Sketch the curve $y = 6 - x^2$.

(b) Find the gradients of the curve at the points $(-1, 5)$ and $(1, 5)$.

(c) Find the equations of the tangents to the curve at these points.

(d) Find the coordinates of the point of intersection of these two tangents.

5 (a) Sketch the curve $y = x^4 - 4x^2$.

(b) What are the coordinates of the three points where this curve crosses or touches the x-axis?

(c) Differentiate $y = x^4 - 4x^2$ and hence find the gradients at the three points you found in (b).

(d) Find the equations of the tangents to the curve at the three points you found in (b).

(e) The points of intersection of the three tangents are the vertices of a triangle. What is the area of the triangle?

6 (a) Sketch the curve $y = x^2 + 4$, and the straight line $y = 4x$, on the same axes.

(b) Show that both $y = x^2 + 4$ and $y = 4x$ pass through the point $(2, 8)$.

(c) Show that $y = x^2 + 4$ and $y = 4x$ have the same gradient at $(2, 8)$, and state what you conclude from this result and that in (b).

7 (a) Sketch the curve $y = x^2 - 5$.

(b) Find the gradient of the curve $y = x^2 - 5$ at the points $(-2, -1)$ and $(2, -1)$.

(c) Find the equations of the normals to the curve $y = x^2 - 5$ at the points $(-2, -1)$ and $(2, -1)$.

(d) Find the coordinates of the point of intersection of the two normals you found in (c).

8 Given that $f(x) = x^3 + 3x^2 + 2x$:

(a) factorise $f(x)$ and hence find where the curve of $f(x)$ cuts the x-axis

(b) find $\frac{dy}{dx}$

(c) find the equation of the normal to the curve at each of the three points found in (a)

(d) state whether the three normals all intersect in one point. If they do, state the coordinates of the point. If not, describe how they do intersect.

9 (a) Find the equation of the tangent to the curve $y = 2x^3 - 15x^2 + 42x$ at $(2, 40)$.

(b) Using your expression for $\frac{dy}{dx}$, find the coordinates of another point on the curve at which the tangent is parallel to the one at $(2, 40)$.

(c) Find the equation of the normal at this point.

10 Sketch the curve with equation $y = \frac{4}{x}$.

P is the point where $x = 1$.

    (a) Find the gradient of the curve at P.

    (b) Show that the equation of the tangent to the curve at P is
$$4x + y - 8 = 0.$$

    (c) Find the equation of the normal to the curve at P.

    (d) Show that the normal intersects the curve again at the point $(-16, -\frac{1}{4})$.

11 **(a)** Sketch the curve for which the equation is $y = x^2 - 3x + 2$ and state the coordinates of the points A and B where it crosses the x-axis.

    (b) Find the gradient of the curve at A and at B.

    (c) Find the equations of the tangent and normal to the curve at both A and B.

    (d) The tangent at A meets the tangent at B at the point P. The normal at A meets the normal at B at the point Q. What shape is the figure APBQ?

12 **(a)** Find the points of intersection of $y = 2x^2 - 9x$ and $y = x - 8$.

    (b) Find $\frac{dy}{dx}$ for the curve and hence find the equation of the tangent to the curve at each of the points in (a).

    (c) Find the point of intersection of the two tangents.

    (d) The two tangents from a point to a circle are always equal in length. Are the two tangents to the curve $y = 2x^2 - 9x$ (a parabola) from the point you found in (c) equal in length?

13 Given that $y = x^3 - x + 6$:

    (a) find $\frac{dy}{dx}$.

On the curve representing y, P is the point where $x = -1$.

    (b) Calculate the y-coordinate of the point P.

    (c) Calculate the value of $\frac{dy}{dx}$ at P.

    (d) Find the equation of the tangent at P.

The tangent at the point Q is parallel to the tangent at P.

    (e) Find the coordinates of Q.

    (f) Find the equation of the normal to the curve at Q.

[MEI]

EXERCISE 4F   **Examination-style questions**

1 Differentiate the following expressions with respect to x.

(a) $\sqrt[3]{x} + \dfrac{1}{\sqrt[3]{x}}$

(b) $\dfrac{3x^2 + 2}{x}$

2 Differentiate the following expressions with respect to x.

(a) $(x + 1)(x^2 - 1)$

(b) $\dfrac{(x - 2)^2}{2\sqrt{x}}$

3 $f(x) = 4 + \dfrac{4}{x^2} + \dfrac{\sqrt{x}}{2}$

(a) Find $f'(x)$.

(b) Find the value α such that $f'(\alpha) = 0$.

(c) Find the value of $f''(\alpha)$.

4 $f(x) = \dfrac{(4 + 3\sqrt{x})^2}{x}$

(a) Express $f(x)$ in the form $Ax^{-1} + Bx^{-\frac{1}{2}} + C$, giving the values of the constants A, B and C.

(b) Find $f'(x)$.

(c) Find $f''(4)$.

5 The curve C has equation $y = (x - 2)(x - 1)(x + 1)$.

(a) Sketch C showing the points where it cuts the axes.

(b) Show that the equation may be written as $y = x^3 - 2x^2 - x + 2$.

(c) Find the equation of the tangent to C at the point $P(1, 0)$.

(d) Find the equation of the normal at P.

(e) The tangent cuts the y-axis at T and the normal cuts the y-axis at N. Show that the area of the triangle NPT is $\dfrac{5}{4}$ units squared.

6 It is given that $y = \dfrac{x + 16}{\sqrt[3]{x}}$.

(a) Find the values of x and y when $\dfrac{dy}{dx} = 0$.

(b) Show that when $x = 1$ the value of $\dfrac{d^2y}{dx^2}$ is $\dfrac{62}{9}$.

7 The diagram shows part of the graph of $y = x + \dfrac{4}{x}$.

(a) Show that the coordinates of the points where the gradient is zero are $(2, 4)$ and $(-2, -4)$ respectively.

(b) Show that the tangent at the point where $x = 1$ meets the tangent at the point where $x = 3$ at the point $\left(\dfrac{3}{2}, \dfrac{25}{6}\right)$.

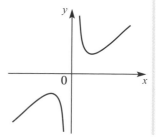

8 It is given that $y = x^{\frac{3}{2}} + \dfrac{48}{x}$, $x > 0$.

(a) Find: (i) $\dfrac{dy}{dx}$ (ii) $\dfrac{d^2y}{dx^2}$ in terms of x.

(b) Find the value of x and the value of y when $\dfrac{dy}{dx} = 0$.

(c) Find the equation of the tangent to the curve at the point where $x = 16$.

9 A curve has equation $y = 3 + 2x - x^2$.

(a) Express $3 + 2x - x^2$ in the form $a - (x + b)^2$ and state the coordinates of the vertex.

(b) Sketch the graph of $y = 3 + 2x - x^2$.

(c) A is the point on the curve where $x = -2$ and B is the point where $x = 2$. Find the gradient of the line through AB.

(d) Find the point P on the curve with gradient that is parallel to the line through AB.

(e) Find the equation of the normal through P.

(f) Show that this normal intersects the curve again at the point where $x = \dfrac{5}{2}$.

10 A curve C has equation $y = x^3 - 5x^2 + 5x + 2$.

(a) Find $\dfrac{dy}{dx}$ in terms of x.

The points P and Q lie on C. The gradient of C at both P and Q is 2. The x-coordinate of P is 3.

(b) Find the x-coordinate of Q.

(c) Find an equation for the tangent to C at P, giving your answer in the form $y = mx + c$, where m and c are constants.

The tangent intersects the coordinate axes at the points R and S.

(d) Find the length of RS, giving your answer as a surd.

[Edexcel]

KEY POINTS

1 Gradient

The gradient of a curve at a point:

= gradient of the tangent at that point

$$= \frac{dy}{dx}$$

$$= f'(x)$$

2 Derivative

- $y = x^n$ $\implies$ $\dfrac{dy}{dx} = nx^{n-1}$

- $y = c$ $\implies$ $\dfrac{dy}{dx} = 0$

- $y = f(x) + g(x)$ $\implies$ $\dfrac{dy}{dx} = f'(x) + g'(x).$

Where n is rational and c is a constant.

3 Second derivative

The derivative of $\dfrac{dy}{dx}$ is $\dfrac{d^2y}{dx^2}$

The derivative of $f'(x)$ is $f''(x)$

so $\dfrac{d^2y}{dx^2} = f''(x)$

4 Tangent and normal at (x_1, y_1)

A tangent is a line touching a curve.
The normal is at right angles to the tangent.

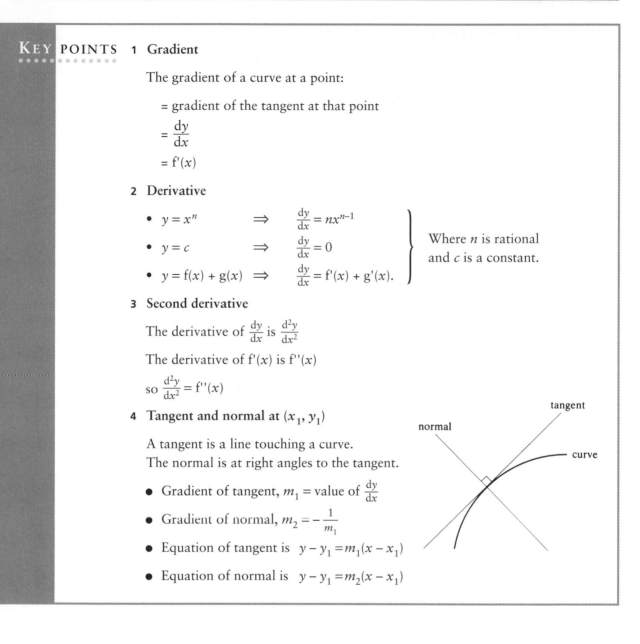

- Gradient of tangent, m_1 = value of $\dfrac{dy}{dx}$

- Gradient of normal, $m_2 = -\dfrac{1}{m_1}$

- Equation of tangent is $y - y_1 = m_1(x - x_1)$

- Equation of normal is $y - y_1 = m_2(x - x_1)$

Chapter five

INTEGRATION

Many small make a great.

Chaucer

REVERSING DIFFERENTIATION

In some situations you know the gradient function, $\frac{dy}{dx}$, and want to find the function itself, y. For example, you might know that $\frac{dy}{dx} = 2x$ and want to find y.

You know from the previous chapter that if $y = x^2$ then $\frac{dy}{dx} = 2x$, but $y = x^2 + 1$, $y = x^2 - 2$ and many other functions also give $\frac{dy}{dx} = 2x$.

Clearly if $y = x^2 + c$, where c is a number, then $\frac{dy}{dx} = 2x$.

So all that you can say at this point is that if $\frac{dy}{dx} = 2x$ then $y = x^2 + c$ where c is described as an *arbitrary constant*. An arbitrary constant may take any value.

The equation $\frac{dy}{dx} = 2x$ is an example of a *differential equation* and the process of solving this equation to find y is called *integration*.

So the solution of the differential equation $\frac{dy}{dx} = 2x$ is $y = x^2 + c$.

Such a solution is often referred to as the *general solution* of the differential equation. It may be drawn as a family of curves as in figure 5.1. Each curve corresponds to a particular value of c.

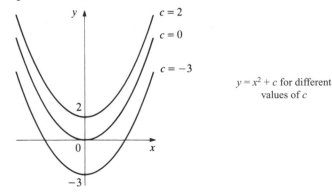

$y = x^2 + c$ for different values of c

FIGURE 5.1

142

PARTICULAR SOLUTIONS

Sometimes you are given more information about a problem and this enables you to find just one solution, called the *particular solution*.

Suppose that in the previous example, in which

$$\frac{dy}{dx} = 2x \qquad \Rightarrow \qquad y = x^2 + c$$

you were also told that when $x = 2$, $y = 1$.

Substituting these values in $y = x^2 + c$ gives:

$$1 = 2^2 + c$$
$$c = -3$$

and so the particular solution is:

$$y = x^2 - 3$$

This is one of the curves shown in figure 5.1.

THE RULE FOR INTEGRATING x^n $(n \neq -1)$

Recall the rule for differentiation:

$$y = x^n \qquad \Rightarrow \qquad \frac{dy}{dx} = nx^{n-1}$$

Similarly $y = x^{n+1} \qquad \Rightarrow \qquad \frac{dy}{dx} = (n+1)x^n$

or $y = \dfrac{1}{(n+1)} x^{n+1} \Rightarrow \dfrac{dy}{dx} = x^n$

Reversing this, integrating x^n gives $\dfrac{1}{(n+1)} x^{n+1} + c$, provided $n \neq -1$.

So:

$$\frac{dy}{dx} = x^n \quad \Rightarrow \quad y = \frac{1}{n+1}x^{n+1} + c, \, n \neq -1$$

Note   In words: to integrate a power of x, add 1 to the power and divide by the new power.

EXAMPLE 5.1

Find the general solution for each of the following differential equations.

(a) $\dfrac{dy}{dx} = 2x$ (b) $\dfrac{dy}{dx} = 2x^{\frac{1}{3}}$ (c) $\dfrac{dy}{dx} = \dfrac{2}{x^2}$ (d) $\dfrac{dy}{dx} = 2$

Solution (a) $\dfrac{dy}{dx} = 2 \times x^1 \Rightarrow y = 2 \times \dfrac{1}{2}x^{1+1} + c$

$y = x^2 + c$

(b) $\dfrac{dy}{dx} = 2 \times x^{\frac{1}{3}} \Rightarrow y = 2 \times \dfrac{1}{\frac{4}{3}}x^{1+\frac{1}{3}} + c$

$1 + \frac{1}{3} = \frac{4}{3}$

reciprocal of $\frac{4}{3}$ is $\frac{3}{4}$

$y = 2 \times \dfrac{3}{4}x^{\frac{4}{3}} + c$

Write as an improper fraction

$y = \dfrac{3}{2}x^{\frac{4}{3}} + c$

(c) $\dfrac{dy}{dx} = 2 \times x^{-2} \Rightarrow y = 2 \times \dfrac{1}{-1}x^{-2+1} + c$

$y = -2x^{-1} + c$

$y = -\dfrac{2}{x} + c$

(d) $\dfrac{dy}{dx} = 2 = 2 \times x^0 \Rightarrow y = 2 \times \dfrac{1}{1}x^1 + c$

$x^0 = 1$

$y = 2x + c$

Note In general, if $\dfrac{dy}{dx} = k$, then $y = kx + c$.

What happens if you try to integrate $\dfrac{dy}{dx} = \dfrac{1}{x}$?

Using the rule on $\dfrac{dy}{dx} = \dfrac{1}{x} = x^{-1}$ gives $y = \dfrac{1}{0}x^0 + c$, but $\dfrac{1}{0}$ is undefined and so the rule fails. You will learn how to integrate $\dfrac{1}{x}$ in the *Pure Mathematics: Core 4* book.

EXAMPLE 5.2

Given that $f'(x) = 4x + 3\sqrt{x} + \dfrac{1}{x^2} - 5$, find $f(x)$.

Solution $f'(x) = 4x + 3\sqrt{x} + \dfrac{1}{x^2} - 5$

$f'(x) = 4x + 3x^{\frac{1}{2}} + x^{-2} - 5$

Integrating:

$f(x) = 4 \times \dfrac{1}{2}x^2 + 3 \times \dfrac{2}{3}x^{\frac{3}{2}} + \dfrac{1}{-1} \times x^{-1} - 5x + c$

$f(x) = 2x^2 + 2x^{\frac{3}{2}} - \dfrac{1}{x} - 5x + c$

EXAMPLE 5.3

Given that $f'(x) = \dfrac{(x+2)^2}{x^{\frac{1}{2}}}$, find $f(x)$.

Solution Before you can integrate you must multiply out the equation and simplify.

$f'(x) = \dfrac{x^2 + 4x + 4}{x^{\frac{1}{2}}} = \dfrac{x^2}{x^{\frac{1}{2}}} + \dfrac{4x}{x^{\frac{1}{2}}} + \dfrac{4}{x^{\frac{1}{2}}} = x^{\frac{3}{2}} + 4x^{\frac{1}{2}} + 4x^{-\frac{1}{2}}$

Integrating:

$$f(x) = \frac{2}{5}x^{\frac{5}{2}} + 4 \times \frac{2}{3}x^{\frac{3}{2}} + 4 \times \frac{2}{1}x^{\frac{1}{2}} + c = \frac{2}{5}x^{\frac{5}{2}} + \frac{8}{3}x^{\frac{3}{2}} + 8x^{\frac{1}{2}} + c$$

EXERCISE 5A

1 Find the general solution for each of the following.

(a) $\frac{dy}{dx} = x^2$ (b) $\frac{dy}{dx} = x^3$ (c) $\frac{dy}{dx} = 5x^4$ (d) $\frac{dy}{dx} = 4x$

(e) $\frac{dy}{dx} = 7$ (f) $\frac{dy}{dx} = \frac{4}{x^2}$ (g) $\frac{dy}{dx} = \frac{5}{x^3}$ (h) $\frac{dy}{dx} = 4\sqrt{x}$

(i) $\frac{dy}{dx} = 5\sqrt[3]{x}$ (j) $\frac{dy}{dx} = \frac{3}{\sqrt{x}}$

2 Find $f(x)$, given that $f'(x)$ is:

(a) $\sqrt{x} + \frac{1}{\sqrt{x}}$ (b) $2x^3 + \frac{4}{x^3}$ (c) $3x^2 + 2x + 4$

(d) $6x^3 + 4x^2 + 8x - 10$

(e) $x(x + 1)$ (f) $\sqrt{x}(x - 1)$ (g) $\frac{3x^2 + 5}{x^2}$ (h) $\frac{x - 1}{\sqrt{x}}$

(i) $(x + 2)^2$ (j) $\frac{(x + 1)(x - 2)}{\sqrt{x}}$

INDEFINITE INTEGRALS

So far in this chapter you have started with a differential equation and integrated to obtain the function. So, for example, if $\frac{dy}{dx} = 2x + 3$ then $y = x^2 + 3x + c$.

If you start with the function $2x + 3$ and wish to integrate it then you use the integral sign $\int$, which is an elongated Old English letter S, and write

$$\int (2x + 3)dx = x^2 + 3x + c.$$

This is an example of an *indefinite integral* with respect to x.

Using this notation our general rule for the integral of powers of x becomes:

$$\int x^n dx = \frac{1}{n + 1}x^{n+1} + c \quad n \neq -1$$

EXAMPLE 5.5

Find $\int \left(\frac{3}{\sqrt{x}} + \frac{1}{x^3} \right) dx$.

Solution $\int \left(\frac{3}{\sqrt{x}} + \frac{1}{x^3} \right) dx = \int \left(3x^{-\frac{1}{2}} + x^{-3} \right) dx$

$$= 6x^{\frac{1}{2}} - \frac{1}{2}x^{-2} + c$$

1 Find the following indefinite integrals. Remember to include the constant of integration.

(a) $\int 3x^2\,dx$

(b) $\int (5x^4 + 7x^6)dx$

(c) $\int (6x^2 + 5)dx$

(d) $\int (x^3 + x^2 + x + 1)dx$

(e) $\int (11x^{10} + 10x^9)dx$

(f) $\int (3x^2 + 2x + 1)dx$

(g) $\int (x^2 + 5)dx$

(h) $\int 5\,dx$

(i) $\int (6x^2 + 4x)dx$

(j) $\int (x^4 + 3x^2 + 2x + 1)dx$

2 Integrate these functions with respect to x.

(a) $3x^{-2}$

(b) $\sqrt[3]{x}$

(c) $\dfrac{1}{x^5}$

(d) $\dfrac{6x}{\sqrt{x}}$

(e) $3x(x^2 + 1)$

(f) $(4x + 1)^2$

(g) $\frac{1}{3}x^{\frac{2}{3}} - \frac{1}{6}x^{-\frac{1}{3}}$

(h) $2\sqrt{x}(x - 4)$

(i) $(2 + \sqrt{x})(3 - \sqrt{x})$

(j) $\dfrac{5 + 3x^4}{x^2}$

3 In each of the following questions, find the indefinite integral.

(a) $\int 10x^{-4}dx$

(b) $\int (2x + 3x^{-4})dx$

(c) $\int (2 + x^3 + 5x^{-3})dx$

(d) $\int (6x^2 - 7x^{-2})dx$

(e) $\int 5x^{\frac{1}{4}}dx$

(f) $\int \dfrac{1}{x^4}dx$

(g) $\int \sqrt{x}\,dx$

(h) $\int \left(2x^4 - \dfrac{4}{x^2}\right)dx$

EQUATIONS OF CURVES

It is now appropriate to look at particular solutions of differential equations in which the value of the constant is found. This will give equations of the curves for which you know the gradient functions.

EXAMPLE 5.5

Given that $\dfrac{dy}{dx} = 3x^2 + 4x + 3$:

(a) find the general solution of this differential equation

(b) find the equation of the curve with this gradient function which passes through (1, 10).

Solution   **(a)** By integration the general solution is:

$$y = \frac{3x^3}{3} + \frac{4x^2}{2} + 3x + c$$
$$= x^3 + 2x^2 + 3x + c, \text{ where } c \text{ is a constant}$$

(b) Since the graph passes through $(1, 10)$:

$$10 = 1^3 + 2(1)^2 + 3(1) + c$$
$$c = 4$$
$$\Rightarrow \qquad y = x^3 + 2x^2 + 3x + 4$$

EXAMPLE 5.6

The gradient function of a curve is $\frac{dy}{dx} = 4x - 12$.

(a) The minimum y-value is 16 and this occurs when $\frac{dy}{dx} = 0$. By considering the gradient function, find the corresponding x-value.

(b) Use the gradient function and your answer from (a) to find the equation of the curve.

Solution   **(a)** When $\frac{dy}{dx} = 0$:

$$4x - 12 = 0 \qquad \Rightarrow \qquad x = 3$$

So the minimum point is $(3, 16)$.

(b)
$$\frac{dy}{dx} = 4x - 12$$
$$\Rightarrow \qquad y = 2x^2 - 12x + c$$

At the minimum point, $x = 3$ and $y = 16$:

$$\Rightarrow \qquad 16 = 2 \times 3^2 - 12 \times 3 + c$$
$$\Rightarrow \qquad c = 34$$
$$\Rightarrow \qquad y = 2x^2 - 12x + 34$$

EXAMPLE 5.7

The curve C has gradient at any point given by $6\sqrt{x} - \dfrac{4}{x^2}$, $x > 0$.

(a) Find the equation of C given that it passes through $(4, 30)$.

(b) Verify that C passes through the point $(1, 5)$.

Solution   (a) $\dfrac{dy}{dx} = 6\sqrt{x} - \dfrac{4}{x^2} = 6x^{\frac{1}{2}} - 4x^{-2}$

$y = 6 \times \dfrac{2}{3}x^{\frac{3}{2}} - 4 \times \dfrac{1}{-1}x^{-1} + c$ Note that $1 \div \frac{3}{2} = \frac{2}{3}$

$\therefore y = 4x^{\frac{3}{2}} + \dfrac{4}{x} + c$

When $x = 4$, $y = 30$ $\Rightarrow$ $30 = 4 \times 4^{\frac{3}{2}} + \dfrac{4}{4} + c = 33 + c$

$\Rightarrow$ $c = -3$

So $y = 4x^{\frac{3}{2}} + \dfrac{4}{x} - 3$

(b) To verify that $(1, 5)$ is on C, putting $x = 1$ gives $y = 4 + 4 - 3 = 5$ which is correct.

EXERCISE 5C

1 Given that $\dfrac{dy}{dx} = 6x^2 + 5$:

(a) find the general solution of the differential equation

(b) find the equation of the curve of which the gradient function is $\dfrac{dy}{dx}$ and which passes through $(1, 9)$

(c) hence show that $(-1, -5)$ also lies on the curve.

2 The gradient function for a curve is $\dfrac{dy}{dx} = 4x$ and the curve passes through the point $(1, 5)$.

(a) Find the equation of the curve.

(b) Find the value of y when $x = -1$.

3 The curve C passes through the point $(2, 10)$ and its gradient at any point is given by $\dfrac{dy}{dx} = 6x^2$.

(a) Find the equation of the curve C.

(b) Show that the point $(1, -4)$ lies on the curve.

4 A stone is thrown upwards out of a window, and the rate of change of its height (h metres) is given by:

$$\dfrac{dh}{dt} = 15 - 10t$$

where t is the time (in seconds). When $t = 0$, $h = 20$.

(a) Show that the solution of the differential equation, under the given conditions, is:

$$h = 20 + 15t - 5t^2$$

(b) For what value of t does $h = 0$? (Assume $t \geqslant 0$.)

5 (a) Find the general solution of the differential equation:

$$\frac{dy}{dx} = 5$$

(b) Find the particular solution which passes through the point (1, 8).

(c) Sketch the graph of this particular solution.

6 A curve passes through the point (4, 1) and its gradient at any point is given by:

$$\frac{dy}{dx} = 2x - 6$$

(a) Find the equation of the curve.

(b) Draw a sketch of the curve and state whether it passes under, over or through the point (1, 4).

7 You are given that $\frac{dy}{dx} = \frac{4}{x^2}$.

(a) Find the values of x when the gradient is 1.

(b) Find the general solution of the differential equation.

(c) Find the equation of the curve satisfying this differential equation that passes through the point (2, 1).

(d) Sketch the curve, showing any asymptotes.

8 The gradient of a curve is given by:

$$\frac{dy}{dx} = \sqrt{x} - \frac{1}{\sqrt{x}}, \; x > 0$$

(a) Find the general solution of the differential equation.

(b) Find the particular solution given that the curve passes through (9, 15).

(c) The curve through (9, 15) has a minimum turning point where $\frac{dy}{dx} = 0$. Show that the value of y at this point is $\frac{5}{3}$.

9 A curve C has gradient given by:

$$\frac{dy}{dx} = \left(x + \frac{1}{x}\right)^2, \quad x \neq 0$$

and it passes through the point (1, 3).

(a) Solve the differential equation to find the equation of the curve.

(b) Show that there are two points on the curve at which the gradient is 4 and that at these points the values of y are $\frac{1}{3}$ and 3.

10 (a) Find the general solution of

$$\frac{dy}{dx} = \frac{(x + 2)^2}{\sqrt{x}}, \; x > 0$$

(b) Find the particular solution for which $y = 11$ when $x = 1$.

(c) Verify that when $x = 4$ the value of y is $50\frac{1}{15}$.

EXERCISE 5D   **Examination-style questions**

1 Find:

(a) $\int (x^2 + 3x - 4)dx$

(b) $\int \left(\sqrt[3]{x} + \frac{1}{\sqrt[3]{x}} \right) dx$

2 Find:

(a) $\int (x^3 + 4x^{\frac{1}{3}} - 2x^{-3})dx$

(b) $\int 2x^3(x^2 - 3x)dx$

3 Find:

(a) $\int (2 - 5x + 3x^2 - 6x^3)dx$

(b) $\int \left(\frac{x^2 - 3}{x^4} \right) dx$

4 Given that $f(x) = 8\sqrt{x} - \frac{4}{\sqrt{x}}$ find $\int f(x)dx$.

5 Given that $f(x) = \left(2x^2 - \frac{1}{2x^2} \right)\left(2x^2 + \frac{1}{2x^2} \right)$ find $\int f(x)dx$.

6 You are given that $\frac{dy}{dx} = 2x + 4$.

(a) Use integration to find y in terms of x.

(b) Given that $y = 3$ when $x = 2$ find an equation for y.

(c) Find the value of y at the point where the gradient of the curve is zero.

7 You are given that:

$\frac{dy}{dx} = (x - \frac{1}{x})^2, \; x > 0$

(a) Use integration to find y in terms of x.

(b) Given that $y = 2$ at $x = 3$, find the value of y at $x = 2$.

8 The function $f(x)$ is such that:

$f'(x) = 3x^2 - 8x + 3$

(a) Given that $f(0) = 0$, find $f(x)$.

(b) Find the values of x for which $f(x) = 0$.

(c) Sketch the graph of $y = f(x)$.

9 The function $f(x)$, defined for $x \in \Re$, $x > 0$, is such that:

$f'(x) = x^2 - 2 + \frac{1}{x^2}$

Given that $f(3) = 0$, find $f(x)$.

[Edexcel]

10 $\frac{dy}{dx} = 5 + \frac{1}{x^2}$

(a) Use integration to find y in terms of x.

(b) Given that $y = 7$ at $x = 1$, find the value of y at $x = 2$.

[Edexcel]

KEY POINTS

1 **Integration** is the reverse process of differentiation.

2 **Differential equations**

- These include terms in $\dfrac{dy}{dx}$.

- If $\dfrac{dy}{dx} = x^n$ then $y = \dfrac{1}{n+1} x^{n+1} + c$, $n \neq -1$

- For **general solutions** the answer includes the general term c.

- For **particular solutions** the value of c is found by using given values of x and y.

3 **Indefinite integrals**

$$\int x^n \, dx = \frac{1}{n+1} x^{n+1} + c, \, n \neq -1$$

Core 2

C2

Chapter six

ALGEBRA AND FUNCTIONS

Gather up the fragments that remain, that nothing be lost.

Bible

• • • • • • • • • • • • • • • • •

You will recall from *Pure Mathematics: Core 1* that a polynomial expression is one that involves powers of x. For example, $2x^2 - 3x + 4$ (a *quadratic* expression) and $5x^3 - 3x^2 + 2x + 4$ (a *cubic* expression) are both *polynomials*. You have studied quadratics and some cubic functions. Now you will learn how to divide expressions and how to find factors and remainders.

DIVISION OF POLYNOMIALS

Division of polynomials can be set out like arithmetical long division. It can also be done by comparing coefficients, after expansion, or, indeed, it can be done by inspection. The next example illustrates these three methods.

EXAMPLE 6.1

Divide $2x^3 - 3x^2 + x - 6$ by $x - 2$.

Solution   **Method 1: Long division**

$$
\begin{array}{r}
2x^2 \\
x - 2 \overline{\smash{\big)}\ 2x^3 - 3x^2 + x - 6} \\
2x^3 - 4x^2
\end{array}
$$

> Found by dividing $2x^3$ (the first term in $2x^3 - 3x^2 + x - 6$) by x (the first term in $x - 2$).

> $2x^2(x - 2)$

Now subtract $2x^3 - 4x^2$ from $2x^3 - 3x^2$, bring down the next term (i.e. x) and repeat the method above.

$$
\begin{array}{r}
2x^2 + x \\
x - 2 \overline{\smash{\big)}\ 2x^3 - 3x^2 + x - 6} \\
\underline{2x^3 - 4x^2 } \\
x^2 + x \\
x^2 - 2x
\end{array}
$$

> $x^2 \div x$

> $x \times (x - 2)$

Continuing gives:

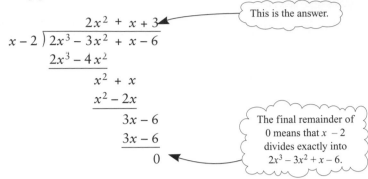

Thus $(2x^3 - 3x^2 + x - 6) \div (x - 2) = (2x^2 + x + 3)$.

Alternatively this may be set out as follows if you know that there is no remainder.

Solution   **Method 2: Expansion**

Let $(2x^3 - 3x^2 + x - 6) \div (x - 2) = ax^2 + bx + c$.

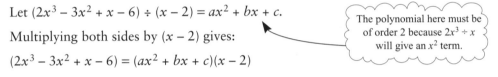

Multiplying both sides by $(x - 2)$ gives:

$(2x^3 - 3x^2 + x - 6) = (ax^2 + bx + c)(x - 2)$

Multiplying out the expression on the right:

$2x^3 - 3x^2 + x - 6 = ax^3 + (b - 2a)x^2 + (c - 2b)x - 2c$

Comparing coefficients of x^3:
$$2 = a$$

Comparing coefficients of x^2:
$$-3 = b - 2a$$
$$= b - 4$$
$$\Rightarrow b = 1$$

Comparing coefficients of x:
$$1 = c - 2b$$
$$= c - 2$$
$$\Rightarrow c = 3$$

Checking the constant term:
$$-6 = -2c \text{ (which agrees with } c = 3)$$

So $ax^2 + bx + c$ is $2x^2 + x + 3$

i.e. $(2x^3 - 3x^2 + x - 6) \div (x - 2) = 2x^2 + x + 3$

With practice you may be able to do this method 'by inspection'. The steps in this would be as shown in the next solution.

Solution   **Method 3: Inspection**

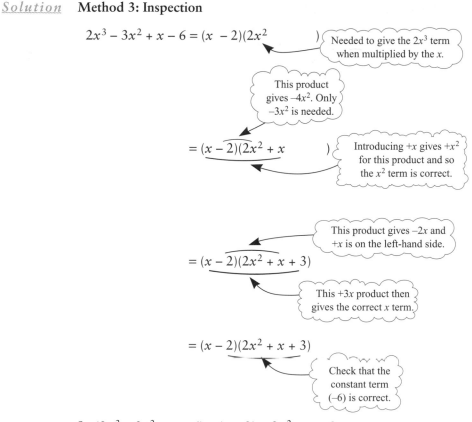

$$2x^3 - 3x^2 + x - 6 = (x - 2)(2x^2 \qquad)$$

Needed to give the $2x^3$ term when multiplied by the x.

This product gives $-4x^2$. Only $-3x^2$ is needed.

$$= (x - 2)(2x^2 + x \qquad)$$

Introducing $+x$ gives $+x^2$ for this product and so the x^2 term is correct.

This product gives $-2x$ and $+x$ is on the left-hand side.

$$= (x - 2)(2x^2 + x + 3)$$

This $+3x$ product then gives the correct x term.

$$= (x - 2)(2x^2 + x + 3)$$

Check that the constant term (-6) is correct.

So $(2x^3 - 3x^2 + x - 6) \div (x - 2) = 2x^2 + x + 3$.

EXERCISE 6A   Use whichever method you prefer.

1   Divide $x^2 + 7x + 10$ by $x + 2$

2   Divide $x^2 + x - 12$ by $x - 3$

3   Divide $x^2 - 11x + 28$ by $x - 4$

4   Divide $2x^2 + x - 3$ by $x - 1$

5   Divide $3x^2 + 8x - 35$ by $x + 5$

6   Divide $x^3 - 3x^2 - x + 3$ by $x - 1$

7   Divide $x^3 + x^2 - 6x - 8$ by $x + 2$

8   Divide $x^3 - x^2 - 7x - 20$ by $x - 4$

9   Divide $x^3 + x^2 - 6x$ by $x - 2$

10 Divide $2x^3 - x^2 - 5x + 10$ by $x + 2$

11 Divide $2x^3 - 6x^2 + 5x - 15$ by $x - 3$

12 Divide $3x^3 + 7x^2 - 4x + 6$ by $x + 3$

13 Divide $4x^3 + 18x^2 - 17x - 35$ by $x + 5$

14	Divide	$5x^3 - 38x^2 + 15x + 42$	by	$x - 7$
15	Divide	$x^3 + (3 - a)x^2 + (4 - 3a)x - 4a$	by	$x - a$
16	Divide	$x^4 + 2x^3 + 2x^2 + 2x + 1$	by	$x + 1$
17	Divide	$x^4 - 5x^3 + 8x^2 - 9x + 10$	by	$x - 2$
18	Divide	$2x^4 + x^3 - 8x^2 + 24x + 9$	by	$x + 3$
19	Divide	$3x^4 - 13x^3 - 14x^2 + 17x + 15$	by	$x - 5$
20	Divide	$x^4 + (2 - a)x^3 + (3 - 2a)x^2 + (4 - 3a)x - 4a$	by	$x - a$

THE FACTOR THEOREM

In *Pure Mathematics: Core 1* you learned how to factorise, where possible, quadratic expressions. For example:

$$x^2 - 5x + 6 = (x - 2)(x - 3)$$

That is, $x - 2$ and $x - 3$ are factors of $x^2 - 5x + 6$.

You also factorised some cubic expressions, for example:

$$x^3 - x^2 - 2x = x(x^2 - x - 2)$$
$$= x(x + 1)(x - 2)$$

In this case, x, $x + 1$ and $x - 2$ are factors of $x^3 - x^2 - 2x$.

The factor theorem helps you to find the factors of polynomial expressions.

Recalling that $x^2 - 5x + 6 = (x - 2)(x - 3)$ and writing this quadratic polynomial as p(x) then:

$$p(x) = x^2 - 5x + 6$$

and we note that:

$$p(2) = 2^2 - 5 \times 2 + 6 = 0 \quad \text{and } (x - 2) \text{ was a factor}$$
$$p(3) = 3^2 - 5 \times 3 + 6 = 0 \quad \text{and } (x - 3) \text{ was a factor.}$$

This leads to the factor theorem which states that:

> If p(α) = 0 then $(x - \alpha)$ is a factor of p(x)
> and, conversely:
> if $(x - \alpha)$ is a factor of p(x) then p(α) = 0.

To use the factor theorem substitute values into p(x) until 0 is obtained.

Note   These values must be factors of the number term in the polynomial.

EXAMPLE 6.2

Use the factor theorem to factorise $x^2 + 2x - 15$.

Solution   Let $\text{p}(x) = x^2 + 2x - 15$.

−15 has factors ±1, ±3, ±5, ±15, so try values in turn.

Then:

$$\text{p}(1) = 1^2 + 2 \times 1 - 15 = -12 \qquad \text{no good}$$
$$\text{p}(3) = 3^2 + 2 \times 3 - 15 = 0 \qquad \text{so } (x - 3) \text{ is a factor.}$$

Since $-15 = 3 \times -5$ the other factor should be found from $x = -5$.

$$\text{p}(-5) = (-5)^2 + 2 \times (-5) - 15 = 0 \quad \text{so } (x + 5) \text{ is the other factor.}$$

So:

$$x^2 + 2x - 15 = (x - 3)(x + 5)$$

EXAMPLE 6.3

Given that $x + 2$ is a factor of $\text{f}(x) = ax^2 + 5x - 2$ find the value of a.

Solution   If $x + 2$ is a factor then $\text{f}(-2) = 0$.

So:
$$a \times (-2)^2 + 5 \times (-2) - 2 = 0$$
$$4a - 12 = 0$$
$$a = 3$$

EXAMPLE 6.4

Factorise $\text{f}(x) = x^3 - 2x^2 - 5x + 6$. Sketch the graph of $y = \text{f}(x)$.

Solution $\text{f}(x) = x^3 - 2x^2 - 5x + 6$.

Note that 6 has factors ±1, ±2, ±3, ±6.

So:

$$\text{p}(1) = 1 - 2 - 5 + 6 = 0 \qquad \text{giving a factor } (x - 1).$$

You can then divide $(x - 1)$ into $\text{p}(x)$ or look for another factor.

$$\text{p}(2) = 8 - 8 - 10 + 6 \neq 0$$
$$\text{p}(3) = 27 - 18 - 15 + 6 = 0 \text{ giving a factor } (x - 3).$$

Hence:

$$x^3 - 2x^2 - 5x + 6 = (x - 1)(x - 3)(x + 2)$$

The third factor has to be $x + 2$ since $+6 = (-1) \times (-3) \times 2$.

So:

$$f(x) = (x - 1)(x - 3)(x + 2)$$

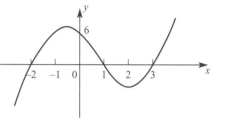

and when $f(x) = 0$, $x = 1, 3, -2$ which give the points where the curve cuts the x-axis.

When $x = 0$ then $f(x) = 6$, so the curve crosses the y-axis at $(0, 6)$.

Sketching gives the curve shown in figure 6.1.

FIGURE 6.1

Note

The curve in example 6.4 is dominated by the x^3 term when x is very large, and this term has a positive coefficient (of 1). This means that the curve must go up as x becomes large. If there had been a negative coefficient instead, then the curve would have gone down. This idea can help you to decide on the correct orientation of a curve sketch.

EXAMPLE 6.5

Express $f(x) = x^3 - x^2 - 4x - 6$ as the product of a linear and a quadratic factor.

Show that $f(x) = 0$ has only one real root.

Solution   Using the factor theorem:

$$f(1) = 1 - 1 - 4 - 6 \neq 0$$
$$f(2) = 8 - 4 - 8 - 6 \neq 0$$
$$f(3) = 27 - 9 - 12 - 6 = 0 \text{ so } (x - 3) \text{ is a factor.}$$

Dividing $x - 3$ into $f(x)$ gives:

$$x^3 - x^2 - 4x - 6 = (x - 3)(x^2 + 2x + 2)$$

If $f(x) = 0$ then $(x - 3)(x^2 + 2x + 2) = 0$ so that $x - 3 = 0$ or $x^2 + 2x + 2 = 0$.

$$x - 3 = 0 \text{ gives the solution } x = 3.$$

The discriminant of $x^2 + 2x + 2 = 0$ is -4, so it has no real roots.

Hence $f(x)$ has only one real root.

EXAMPLE 6.6

Factorise $6x^3 - 11x^2 - 4x + 4$.

Solution   Let $p(x) = 6x^3 - 11x^2 - 4x + 4$.

Using the factor theorem:

$p(1) = 6 - 11 - 4 + 4 \neq 0$

$p(2) = 48 - 44 - 8 + 4 = 0$ so $(x - 2)$ is a factor.

$p(4) = 384 - 176 - 16 + 4 \neq 0$

Other factors will come from the $6x^3$ term.

$6x^3$ has factors x, $2x$, $3x$, $6x$.

Try $2x - 1$ as a factor $\Rightarrow$ $2x - 1 = 0$

$$x = \tfrac{1}{2}$$

$p(\tfrac{1}{2}) = \tfrac{3}{4} - \tfrac{11}{4} - 2 + 4 = 0$ so $(2x - 1)$ is a factor.

You can find the third factor by writing it as $ax + b$.

$6x^3 - 11x^2 - 4x + 4 = (x - 2)(2x - 1)(ax + b)$

Looking at the coefficients:

x^3: $6 = 1 \times 2 \times a$ so $a = 3$
numbers: $4 = -2 \times - 1 \times b$ so $b = 2$

So the third factor is $3x + 2$.

Check: $p(-\tfrac{2}{3}) = -\tfrac{48}{27} - \tfrac{44}{9} + \tfrac{8}{3} + 4 = 0$ ✓

Hence: $p(x) = (x - 2)(2x - 1)(3x + 2)$

EXAMPLE 6.7

The polynomial $p(x) = 6x^3 + kx^2 - 19x + 6$. Show that if $x + 3$ is a factor of $p(x)$ then $k = 11$. Show that $2x - 1$ is a factor of $p(x)$. Factorise $p(x)$ completely.

Solution   If $x + 3$ is a factor then $p(-3) = 0$, so:

$$-162 + 9k + 57 + 6 = 0$$
$$9k = 99$$
$$k = 11$$

If $2x - 1$ is a factor then $p(\tfrac{1}{2}) = 0$.

$$p(\tfrac{1}{2}) = 6 \times \tfrac{1}{8} + 11 \times \tfrac{1}{4} - 19 \times \tfrac{1}{2} + 6$$
$$= \tfrac{3}{4} + 2\tfrac{3}{4} - 9\tfrac{1}{2} + 6$$
$$= 0$$

So $2x - 1$ is a factor.

Factorising completely:

$$p(x) = (x + 3)(2x - 1)(ax + b)$$
$$= (2x^2 + 5x - 3)(ax + b)$$

Since the coefficient of x^3 is 6 then $a = 3$ and since the constant term is 6 then $b = -2$, so:

$$p(x) = (x + 3)(2x - 1)(3x - 2)$$

GRAPHICAL SOLUTIONS

You know that it is possible to solve equations by sketching the curve and finding where it crosses the x-axis The factor theorem will enable you to check whether or not your answers are close to the exact values.

EXAMPLE 6.8

Solve the equation $4x^3 - 8x^2 - x + 2 = 0$.

Solution   Start by plotting the curve with equation $y = 4x^3 - 8x^2 - x + 2$. (You will also find it helpful at this stage to display it on a graphics calculator.)

x	−1	0	1	2	3
y	−9	+2	−3	0	35

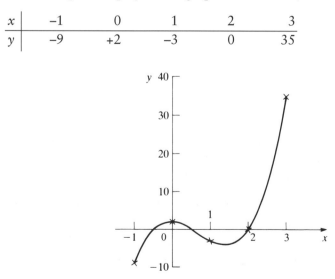

FIGURE 6.2

Figure 6.2 shows that one root is $x = 2$ and that there are two others. One is between $x = -1$ and $x = 0$ and the other between $x = 0$ and $x = 1$.

Using the factor theorem we see that:

when $\quad x = \frac{1}{2}, y = 0$

and when $\quad x = -\frac{1}{2}, y = 0$

So the solutions are $x = -\frac{1}{2}, \frac{1}{2}$ and 2.

---

1   Factorise the following expressions.
   (a) $x^3 - 2x^2 - 9x + 18$
   (b) $x^3 - 5x^2 - 2x + 24$
   (c) $x^3 + 3x^2 - 13x - 15$
   (d) $x^3 - 6x^2 - 16x + 96$
   (e) $60 + 7x - 6x^2 - x^3$
   (f) $2x^3 + 3x^2 - 8x + 3$
   (g) $3x^3 + 7x^2 - 22x - 8$
   (h) $6x^3 - 31x^2 + 4x + 5$
   (i) $27 + 36x - 3x^2 - 4x^3$
   (j) $24x^3 - 26x^2 + 9x - 1$

2   Given that $f(x) - x^3 + 2x^2 - 9x - 18$:

   (a) find $f(-3)$, $f(-2)$, $f(-1)$, $f(0)$, $f(1)$, $f(2)$ and $f(3)$
   (b) factorise $f(x)$
   (c) solve the equation $f(x) = 0$
   (d) sketch the curve for which the equation is $y = f(x)$.

3   The polynomial $p(x)$ is given by $p(x) = x^3 - 4x$.

   (a) Find the values of $p(-3)$, $p(-2)$, $p(-1)$, $p(0)$, $p(1)$, $p(2)$, $p(3)$.
   (b) Factorise $p(x)$.
   (c) Solve the equation $p(x) = 0$.
   (d) Sketch the curve for which the equation is $y = p(x)$.

4 (a) Show that $x - 3$ is a factor of $x^3 - 5x^2 - 2x + 24$.
   (b) Solve the equation $x^3 - 5x^2 - 2x + 24 = 0$.
   (c) Sketch the curve for which the equation is $y = x^3 - 5x^2 - 2x + 24$.

5 (a) The polynomial $p(x) = x^3 - 6x^2 + 9x + k$ has a factor $x - 4$. Find the value of k.
   (b) Find the other factors of the polynomial.
   (c) Sketch the curve for which the equation is $y = p(x)$.

6 The diagram shows the curve for which the equation is

$$y = (x + a)(x - b)^2$$

where a and b are positive integers.

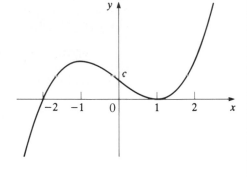

(a) Write down the values of a and b, and also of c, given that the curve crosses the y-axis at $(0, c)$.

(b) Solve the equation

$$(x + a)(x - b)^2 = c$$

(using the values of a, b and c you found in part (a)).

7 The function $f(x)$ is given by $f(x) = x^4 - 3x^2 - 4$.

(a) By treating $f(x)$ as a quadratic in x^2, factorise it in the form $(x^2 \ldots)(x^2 \ldots)$.

(b) Complete the factorisation to the extent that is possible.

(c) How many real roots has the equation $f(x) = 0$? What are they?

8 The equation $f(x) = x^3 - 4x^2 + x + 6 = 0$ has three integer roots.

(a) List the eight values of a for which it is sensible to check whether $f(a) = 0$, and check each of them.

(b) Solve $f(x) = 0$.

9 Factorise, as far as possible, the following expressions.

(a) $x^3 - x^2 - 4x + 4$ given that $(x - 1)$ is a factor

(b) $x^3 + 1$ given that $(x + 1)$ is a factor

(c) $x^3 + x - 10$ given that $(x - 2)$ is a factor

(d) $x^3 + x^2 + x + 6$ given that $(x + 2)$ is a factor.

10 The diagram shows an open rectangular tank of which the base is a square of side x metres and with volume $8\,\text{m}^3$.

(a) Write down an expression in terms of x for the height of the tank.

(b) Show that the surface area of the tank is

$$\left(x^2 + \frac{32}{x}\right)\,\text{m}^2.$$

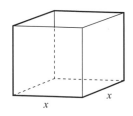

(c) Given that the surface area is $24\,\text{m}^2$ show that $x^3 - 24x + 32 = 0$.

(d) Solve $x^3 - 24x + 32 = 0$ to find the possible values of x.

THE REMAINDER THEOREM

You will recall from the previous section that for a polynomial p(x) in x the factor theorem is:

if p(α) = 0, then ($x - \alpha$) is a factor of p(x)

and, conversely:

if ($x - \alpha$) is a factor of p(x) then p(α) = 0

Clearly ($x - \alpha$) is not always a factor of p(x) and in these cases there will be a remainder R.

Let us look at a simple numerical case. If you divide 45 by 7 it goes 6 times with a remainder R of 3, i.e.

$$45 = 7 \times 6 + 3$$

and the number 6 is the quotient Q.

Dividing a polynomial p(x) by a linear expression ($x - \alpha$) gives a quotient Q(x) with a remainder R.

$$p(x) = (x - \alpha)Q(x) + R$$

and putting $x = \alpha$ gives:

$$p(\alpha) = (\alpha - \alpha)Q(\alpha) + R \implies R = p(\alpha)$$

This gives us the *remainder theorem*:

$$p(x) = (x - \alpha)Q(x) + p(\alpha)$$

EXAMPLE 6.9

Find the remainder when p(x) = $3x^4 - 5x^3 + 2x^2 - 7x + 2$ is divided by $x + 2$.

Solution Dividing p(x) by $x + 2$ gives:

$$p(x) = (x + 2)Q(x) + R$$

and putting $x = -2$ means that:

$$
\begin{aligned}
p(-2) &= R \\
&= 3 \times (-2)^4 - 5(-2)^3 + 2(-2)^2 - 7(-2) + 2 \\
&= 48 + 40 + 8 + 14 + 2 \\
&= 112
\end{aligned}
$$

So the remainder when $3x^4 - 5x^3 + 2x^2 - 7x + 2$ is divided by $x + 2$ is 112.

EXAMPLE 6.10

When $2x^3 + x^2 - 4x + k$ is divided by $2x - 3$ the remainder is 5. Find k.

Solution   Let $p(x) = 2x^3 + x^2 - 4x + k$.

Putting $2x - 3 = 0 \Rightarrow x = \frac{3}{2}$ gives:

$$p(\tfrac{3}{2}) = R$$

So $2 \times (\tfrac{3}{2})^3 + (\tfrac{3}{2})^2 - 4 \times (\tfrac{3}{2}) + k = 5$

$$\tfrac{27}{4} + \tfrac{9}{4} - 6 + k = 5$$

$$k = 2$$

EXAMPLE 6.11

When $p(x) = x^3 + bx^2 + cx + 5$ is divided by $x - 2$ the remainder is 3. When $p(x)$ is divided by $x + 3$ the remainder is −67. Find b and c.

Solution   By the remainder theorem $R = p(\alpha)$.

$$\begin{aligned}
\text{Dividing by } x - 2: \quad & 3 = p(2) \\
& 3 = 8 + 4b + 2c + 5 \\
\Rightarrow \quad & 4b + 2c = -10 \\
\Rightarrow \quad & 2b + c = -5 \qquad\qquad\qquad ① \\
\text{Dividing by } x + 3: \quad & -67 = p(-3) \\
& -67 = -27 + 9b - 3c + 5 \\
\Rightarrow \quad & 9b - 3c = -45 \\
\Rightarrow \quad & 3b - c = -15 \qquad\qquad\qquad ②
\end{aligned}$$

Solving ① and ② simultaneously:

$$① + ②: 5b = -20, \text{ so } b = -4$$

Substituting in ①: $-8 + c = -5$, so $c = 3$

The solution is $b = -4$ and $c = 3$.

You will note that the *factor theorem* is a special case of the *remainder theorem* when you have $R = 0$.

Quotient

To find the quotient you need to recall the method of dividing polynomials that you met in the first section of this chapter.

EXAMPLE 6.12

Find the quotient and the remainder when $2x^3 - 3x^2 + 4x - 5$ is divided by $x - 3$.

Solution **The long division method**

$$
\begin{array}{r}
2x^2 + 3x + 13 \\
x - 3 \overline{\smash{\big)}\,2x^3 - 3x^2 + 4x - 5} \\
\underline{2x^3 - 6x^2\phantom{{}+4x-5}} \\
3x^2 + 4x \\
\underline{3x^2 - 9x} \\
13x - 5 \\
\underline{13x - 39} \\
34
\end{array}
$$

Hence:

$$2x^3 - 3x^2 + 4x - 5 = (x - 3)(2x^2 + 3x + 13) + 34$$

So the quotient is $2x^2 + 3x + 13$ and the remainder is 34.

You can check that the remainder is correct using the remainder theorem, which would give:

$$
\begin{aligned}
R &= p(\alpha) \\
&= 2 \times 3^3 - 3 \times 3^2 + 4 \times 3 - 5 \\
&= 54 - 27 + 12 - 5 \\
&= 34
\end{aligned}
$$

Solution **The expansion method**

Let $2x^3 - 3x^2 + 4x - 5 = (x - 3)(ax^2 + bx + c) + R$
but $R = p(3) = 34$
and multiplying out gives:

$$2x^3 - 3x^2 + 4x - 5 = ax^3 + (b - 3a)x^2 + (c - 3b)x - 3c + 34$$

Comparing coefficients:
x^3: $a = 2$
x^2: $b - 3a = -3$ so $b = 3$
x: $c - 3b = 4$ so $c = 13$

Check with the number: $-3c + 34 = -5$ ✓

So the quotient is $2x^2 + 3x + 13$ and the remainder is 34.

EXERCISE 6C

1 Find the remainder when:

(a) $x^2 + 4x - 5$ is divided by $x - 2$

(b) $3x^2 + 6x + 17$ is divided by $x + 3$

(c) $2x^2 + 4x - 5$ is divided by $2x - 1$

(d) $3x^3 - x^2 + x - 4$ is divided by $x + 1$

(e) $3x^3 + 2x^2 + 2x + 1$ is divided by $x - 2$

(f) $4x^4 + 2x^3 - x^2 + x + 1$ is divided by $2x + 1$

(g) $3x^3 + 12x^2 - x - 1$ is divided by $x + 4$

(h) $16x^4 + 4x^2 + 3$ is divided by $4x - 1$

(i) $x + 3x^2 - 5x^3$ is divided by $2 + x$

(j) $9x^3 - 6x^2 + 12x - 8$ is divided by $2 - 3x$

2 When $p(x) = 2x^3 - x^2 - 13x + k$ is divided by $x - 2$ the remainder is -20.

(a) Find the value of k.   (b) Show that $x + 2$ is a factor of $p(x)$.

(c) Factorise $p(x)$.   (d) Sketch the graph of $y = p(x)$.

3 $p(x) = x^3 + ax^2 + bx + 8$. When $p(x)$ is divided by $x - 4$ the remainder is 88 and when $p(x)$ is divided by $x + 3$ the remainder is 11. Find the values of a and b.

4 The polynomial $p(x) = x^3 + ax^2 + bx + c$ leaves remainders -36, -20 and 0 on division by $x + 1$, $x + 2$ and $x + 3$.

(a) Find the values of a, b and c.   (b) Factorise $p(x)$ completely.

(c) Solve $p(x) = 0$.

5 Find the quotient and the remainder when $2x^3 - x^2 + 3x - 5$ is divided by $x - 2$.

6 Show that the quotient when $3x^3 + 8x^2 + 2x - 1$ is divided by $x + 3$ is $3x^2 - x + 5$, and find the remainder.

7 When $x^3 + 3x^2 + ax + b$ is divided by $x - 2$ the remainder is 15 and when it is divided by $x - 3$ the remainder is 42. Find the remainder when the cubic polynomial is divided by $x - 4$.

8 Given that $p(x) = ax^3 + bx^2 + x - 10$, that $p(2) = 36$ and that $x + 2$ is a factor of $p(x)$, show that $a = 2$ and find b.

9 When $4x^3 - 6x^2 + 2x + 1$ is divided by $ax + b$ the quotient is $2x^2 + 1$ and the remainder is 4. Find a and b.

10 (a) Write down an expression for the remainder when a polynomial $P(x)$ is divided by $(x - a)$.

When $f(x) = 2x^6 + kx^5 + 32x^2 - 26$ is divided by $(x + 1)$ the remainder is 15.

(b) Calculate the value of the constant k.

Now $f(x)$ is divided by $(x - 2)$, giving the quotient $g(x)$ and the remainder R, so that $f(x) = (x - 2)g(x) + R$.

(c) Calculate the remainder R.   (d) Calculate $g(-1)$.

[MEI (part)]

EXERCISE 6D   **Examination-style questions**

1 Divide $x^3 + 5x^2 + 5x - 2$ by $x + 2$.

2 Divide $2x^3 - 11x^2 + 13x - 4$ by $x - 4$.

3 Divide $3x^4 + 11x^3 + 2x^2 - 7x + 15$ by $x + 3$.

4 Find the quotient when $x^3 - 8x + 3$ is divided by $x - 2$ and show that the remainder is -5.

5 You are given that $p(x) = x^3 + 7x^2 + 14x + 8$.
 (a) Divide $p(x)$ by $x + 1$.
 (b) Divide $p(x)$ by $x + 2$.
 (c) State the third factor of $p(x)$ and show that this factor divides into $p(x)$ without leaving a remainder.

6 The cubic polynomial $2x^3 + x^2 + kx - 10$ has a factor $x - 2$. Find the value of k.

7 The cubic polynomial $x^3 + 5x^2 - 4x - 20$ is denoted by $f(x)$.
 (a) Show that $x - 2$ is a factor of $f(x)$.
 (b) Factorise $f(x)$ completely.
 (c) Sketch the graph of $y = f(x)$.
 (d) Solve $f(x) \geqslant 0$.

8 Given that $f(x) = 2x^3 - x^2 - 10x + 8$ find a linear factor and express $f(x)$ as the product of a linear factor and a quadratic factor.

 Hence find the two positive roots of $f(x) = 0$, expressing one of your answers in the form $\frac{a + \sqrt{b}}{4}$ where a and b are integers.

9 (a) Factorise $(x - 2)^2 - 25$.
 (b) Solve $(x - 2)^2 - 25 = 0$.

10 (a) Factorise $25x^2 - 20x + 3$.
 (b) By making a suitable substitution solve

 $$25y - 20y^{\frac{1}{2}} + 3 = 0$$

 leaving your answers as fractions.

11 It is given that $p(x) = x^4 + x^2$.
 (a) Find the quotient and the remainder when $p(x)$ is divided by $(x - 1)(x - 3)$.
 (b) Use the remainder theorem and your answer to part **(a)** to find the remainders **(i)** when $p(x)$ is divided by 1 and **(ii)** when $p(x)$ is divided by 3.

12 The polynomial $p(x) = 3x^3 + ax^2 + bx - 7$. The remainder when $p(x)$ is divided by $x + 2$ is -13 and when it is divided by $x - 3$ the remainder is 77.
 (a) Show that the value of a is 2 and find the value of b.
 (b) Find the remainder when $p(x)$ is divided by $3x - 1$.

13 (a) If f(x) is a polynomial in x show that when you divide f(x) by $x - \alpha$ the remainder is f(α).

(b) You are given that:

$$f(x) = x^3 + ax^2 + bx + c$$

When f(x) is divided by $x + 3$ the remainder is -63, when f(x) is divided by $x + 1$ the remainder is -13 and when f(x) is divided by $x - 2$ the remainder is 2. Show that $a = -2$ and find the values of b and c.

14 A polynomial in x is defined by:

$$f(x) = x^3 + ax^2 + bx + c$$

If f(x) is divided by $x + 2$ the remainder is -16. If f(x) is divided by $x^2 - 2x - 3$ the remainder is 4. Find the values of a, b and c.

15 A polynomial in x has equation:

$$f(x) = ax^3 - x^2 + bx - 6$$

It is given that $x - 3$ is a factor of f(x). When f(x) is divided by $2x - 1$ the remainder is $-12\frac{1}{2}$.

(a) Find the values of a and b.

(b) Factorise f(x) completely.

(c) Find the quotient when f(x) is divided by $x - 2$ and state the remainder.

KEY POINTS

1 Algebraic division

(a) Long division method

(b) Expansion method

(c) Inspection

2 Factor theorem

For a polynomial function p(x) divided by $(x - \alpha)$ then if p(α) = 0, $(x - \alpha)$ is a factor of p(x) and, conversely, if $(x - \alpha)$ is a factor of p(x) then p(α) = 0.

3 Remainder theorem

If a polynomial function p(x) is divided by $(x - \alpha)$ then the quotient is Q(x) and the remainder is R
where $p(x) = (x - \alpha)Q(x) + R$
and $R = p(\alpha)$.

Chapter seven

COORDINATE GEOMETRY IN THE (x, y) PLANE

The nature of God is a circle of which the centre is everywhere and the circumference is nowhere.

Attributed to Empedocles

• • • • • • • • • • • • • • •

THE CIRCLE

You are of course familiar with the circle, and have probably done calculations involving its area and circumference. In this section you are introduced to the *equation* of a circle.

The circle is defined as the *locus* of all the points in a plane which are at a fixed distance (the radius) from a given point (the centre). ('Locus' comes from the Latin, meaning 'place'.)

As you have seen in *Pure Mathematics: Core 1*, Chapter 2, the length of a line joining (x_1, y_1) to (x_2, y_2) is given by:

$$\text{length} = \sqrt{(x_2 - x_1)^2 + (y_2 - y_1)^2}$$

This is used to derive the equation of a circle.

In the case of a circle of radius 3, with its centre at the origin, as shown in figure 7.1, any point (x, y) on the circumference is distance 3 from the origin. Since the distance of (x, y) from $(0, 0)$ is given by $\sqrt{(x - 0)^2 + (y - 0)^2}$, this means that:

$$\sqrt{(x - 0)^2 + (y - 0)^2} = 3$$

or $x^2 + y^2 = 9$, and this is the equation of the circle.

Similarly a point (x, y) on the circumference of the circle centre $(9, 5)$, radius 4, is such that:

$$\sqrt{(x - 9)^2 + (y - 5)^2} = 4$$

or $(x - 9)^2 + (y - 5)^2 = 16$, and this is the equation of this circle.

This circle is also shown in figure 7.1.

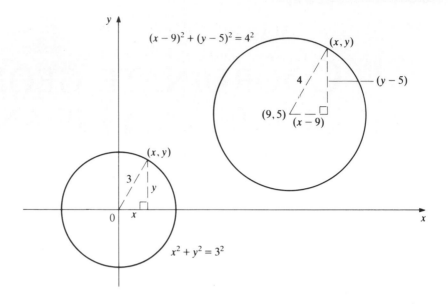

FIGURE 7.1

These results can be generalised to give equations for all circles.

1 Circle with centre (0, 0) and radius *r*

Using the equation for the distance between two points and taking the two points to be $(0, 0)$ and (x, y) then the equation of this circle is:

$$x^2 + y^2 = r^2$$

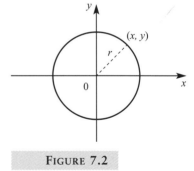

FIGURE 7.2

2 Circle with centre (*a*, *b*) and radius *r*

This time the equation of the circle is:

$$(x - a)^2 + (y - b)^2 = r^2$$

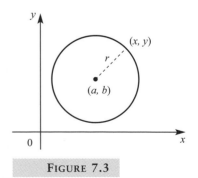

FIGURE 7.3

EXAMPLE 7.1

Write down the equation of the circle with centre $(2, -3)$ and radius 4.

Solution   Using the general equation $(x - a)^2 + (y - b)^2 = r^2$:

$$(x - 2)^2 + (y + 3)^2 = 4^2$$
$$(x - 2)^2 + (y + 3)^2 = 16$$

Expanding gives $x^2 - 4x + 4 + y^2 + 6y + 9 = 16$

$$\Rightarrow x^2 + y^2 - 4x + 6y - 3 = 0$$

EXAMPLE 7.2

Find the centre and radius of the circle $(x - 3)^2 + (y + 5)^2 = 49$.

Solution   Comparing with $(x - a)^2 + (y - b)^2 = r^2$, the centre is $(3, -5)$ and the radius is 7.

EXAMPLE 7.3

Find the centre and radius of the circle with equation $x^2 + y^2 - 10x + 4y + 13 = 0$.

Solution   Rewrite the equation as $x^2 - 10x + y^2 + 4y = -13$.

Completing the square gives $(x - 5)^2 - 25 + (y + 2)^2 - 4 = -13$

$$\therefore (x - 5)^2 + (y + 2)^2 = 16$$

Comparing with $(x - a)^2 + (y - b)^2 = r^2$, the centre is $(5, -2)$ and the radius is 4.

EXAMPLE 7.4

A circle has a radius of 5 units, and passes through the points $(0, 0)$ and $(0, 8)$. Sketch the two possible positions of the circle, and find their equations.

Solution   The line joining $(0, 0)$ to $(0, 8)$ is a chord of the circle, and the midpoint of the chord is $(0, 4)$. From symmetry (see Figure 7.4), the centre must lie on the line $y = 4$.

FIGURE 7.4

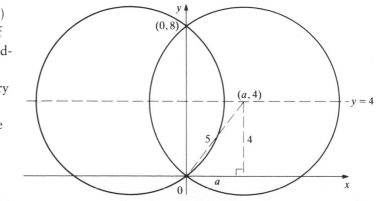

Let the centre be the point $(a, 4)$.

Using Pythagoras' theorem $a^2 + 16 = 25$,

$$\Rightarrow \qquad a^2 = 9$$
$$\Rightarrow \quad a = 3 \text{ or } a = -3$$

Possible equations are therefore:

$$(x - 3)^2 + (y - 4)^2 = 25 \quad \text{and} \quad (x - (-3))^2 + (y - 4)^2 = 25$$
$$(x + 3)^2 + (y - 4)^2 = 25$$

Expanding the equation $(x - a)^2 + (y - b)^2 = r^2$ gives:

$$x^2 - 2ax + a^2 + y^2 - 2by + b^2 = r^2$$
$$\Rightarrow \quad x^2 + y^2 - 2ax - 2by + a^2 + b^2 - r^2 = 0$$

This tells us two important features that show that an equation relates to a circle:

- the coefficients of x^2 and y^2 are equal
- there is no term in xy.

3 The general equation of a circle

This has the form: $x^2 + y^2 + 2gx + 2fy + c = 0$

Rewriting this gives: $x^2 + 2gx + y^2 + 2fy + c = 0$
$$(x + g)^2 + (y + f)^2 - g^2 - f^2 + c = 0$$
$$(x + g)^2 + (y + f)^2 = g^2 + f^2 - c$$

Comparing this with $(x - a)^2 + (y - b)^2 = r^2$ implies a circle with centre $(-g, -f)$ and radius $r = \sqrt{g^2 + f^2 - c}$ and requires that $g^2 + f^2 > c$.

EXAMPLE 7.5

Find the centre and radius of the circle with equation $x^2 + y^2 - 6x + 10y + 18 = 0$.

Solution By comparison with $x^2 + y^2 + 2gx + 2fy + c = 0$, the centre $(-g, -f)$ is $(3, -5)$ and the radius is given by:

$$r^2 = g^2 + f^2 - c$$

so $r^2 = 9 + 25 - 18$
$$r^2 = 16$$

That is, the radius is 4.

EXAMPLE 7.6

Find the centre of the circle with equation:

$$2x^2 + 2y^2 + 5x - 12y + 6 = 0$$

and show that its radius is $\frac{11}{4}$.

Solution   Dividing the equation of the circle by 2 gives:

$$x^2 + y^2 + \tfrac{5}{2}x - 6y + 3 = 0$$

By comparison with $x^2 + y^2 + 2gx + 2fy + c = 0$ the centre is $(-\tfrac{5}{4}, 3)$.

The radius is given by $r^2 = g^2 + f^2 - c$.

So $r^2 = \tfrac{25}{16} + 9 - 3 = \tfrac{121}{16}$

$\Rightarrow \quad r = \tfrac{11}{4}$

EXERCISE 7A

1 Find the equations of the following circles.
   (a) centre (1, 0), radius 4   **(b)** centre (2, −1), radius 3
   (c) centre (1, −3), radius 5   **(d)** centre (−2, −5), radius 1

2 State **(i)** the coordinates of the centre and **(ii)** the radius of the following circles.
   (a) $x^2 + y^2 = 9$
   (b) $x^2 + (y - 2)^2 = 25$
   (c) $(x - 3)^2 + (y + 1)^2 = 16$
   (d) $(x + 2)^2 + (y + 2)^2 = 4$
   (e) $(x + 4)^2 + y^2 = 8$
   (f) $x^2 + y^2 + 8x - 16y + 31 = 0$
   (g) $x^2 + y^2 - x + 2y + \tfrac{1}{4} = 0$
   (h) $2x^2 + 2y^2 - 18x - 10y + 3 = 0$
   (i) $3x^2 + 3y^2 - 18x - 24y + 39 = 0$
   (j) $5x^2 + 5y^2 + 20x + 24y - 9 = 0$

3 Find the equation of the circle with centre (1, 7) passing through the point (−4, −5). (Hint: Use the coordinates of these two points to find the radius.)

4 Sketch the circle $(x - 4)^2 + (y - 5)^2 = 16$.

5 Show that the equation $x^2 + y^2 + 2x - 4y + 1 = 0$ can be written in the form $(x + 1)^2 + (y - 2)^2 = r^2$, where the value of r is to be found. Hence give the coordinates of the centre and the radius of the circle.

6 (a) Find the mid-point C of AB where A and B are (1, 8) and (3, 14) respectively. Find also the distance AC.
   (b) Hence find the equation of the circle which has AB as diameter.

7 Sketch the circle of radius 4 units which touches the positive x- and y-axes, and find its equation.

8 A circle passes through the points (2, 0) and (8, 0) and has the y-axis as a tangent. Find the two possible equations.

9 Using the fact that a tangent to a circle is perpendicular to the radius passing through the point of contact, find the equation of the tangent to the circle $x^2 + (y + 4)^2 = 25$ at the point (−4, −1).

10 Sketch the circle with centre (2, 3) and radius 4.

    (a) Show that the distance between the points where the circle cuts the y-axis is $4\sqrt{3}$.

    (b) Find the distance between the two points where the circle cuts the x-axis.

PROPERTIES OF CIRCLES

You should be able to recall these first properties of circles from your study of GCSE mathematics.

1 The angle at the centre of a circle is twice the angle at the circumference.

O is the centre of the circle and A, B and C are points on the circle. Angle AOB is the angle y at the centre of the circle and angle ACB is the angle x at the circumference.

$$y = 2x$$

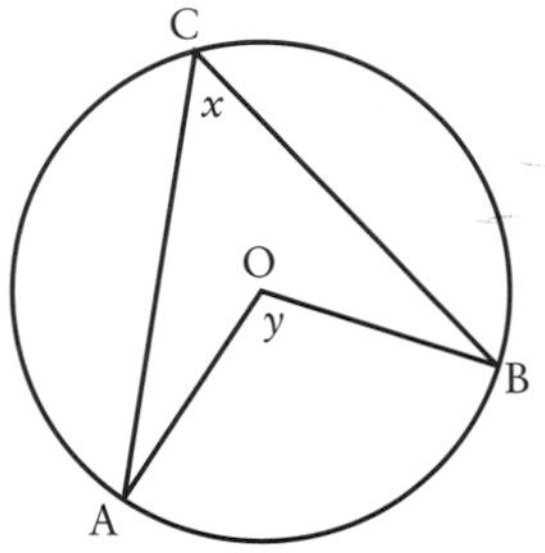

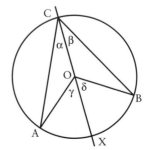

FIGURE 7.5

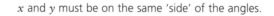

Note x and y must be on the same 'side' of the angles.

Proof:

Let angle ACO = α and angle AOX = γ.

Angle CAO = angle ACO (isosceles triangle)

∴ angle CAO = α

Hence $\gamma = 2\alpha$ (exterior angle of triangle AOC)

Similarly $\delta = 2\beta$

So, if $\alpha + \beta = x$ and $\gamma + \delta = 2(\alpha + \beta) = y$ then $y = 2x$.

FIGURE 7.6

2 Angles at the circumference, subtended from the same chord, are equal.

The angles ADB ($= x$) and ACB ($= y$) are both subtended from the chord AB (not drawn). Thus:

$$x = y$$

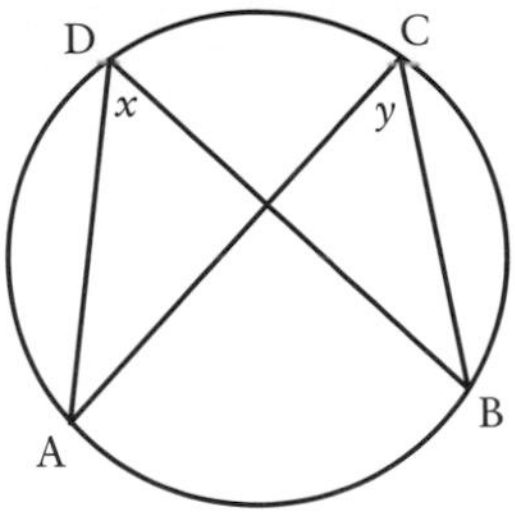

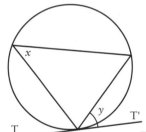

Note

C and D must be on the same side of the chord AB.

FIGURE 7.7

The proof follows from the proof of property 1, where the point C could be at any point on the major arc AB.

3 Opposite angles of a cyclic quadrilateral add up to 180°.

$$x + y = 180°$$

Proof:
If O is the centre of the circle, then from property 1:

$$X = 2x \text{ and } Y = 2y$$

but $X + Y = 360°$, hence:

$$x + y = 180°$$

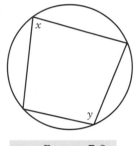

FIGURE 7.8

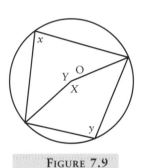

FIGURE 7.9

4 Alternate segment theorem

If TT' is a tangent to the circle then:

$$x = y$$

Proof:
Draw AD such that it is a diameter.

Let angle ADB $= x$ and angle T'AB $= y$.

Then angle ABD is a right angle.

Also, angle DAT' is a right angle.

Thus angle BAD $= 90° - y$ and so:

$$x = y.$$

Since the point D could be at any point on the major arc ADB the alternate segment theorem is proven.

FIGURE 7.10

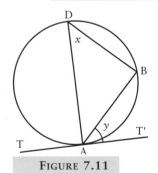

FIGURE 7.11

The following properties are specifically stated in the specification for Pure Mathematics: Core 2.

5 The angle in a semicircle is a right angle.

If O is the centre of the circle and AOB is a diameter then angle ACB = 90°.

Proof:
The proof follows from property 1.

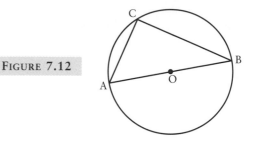

FIGURE 7.12

$$\text{Angle ACB} = \tfrac{1}{2} \text{ angle AOB}$$
$$= \tfrac{1}{2} \times 180°$$
$$= 90°$$

EXAMPLE 7.7

Show that the triangle with vertices A(6, 4), B(5, −3) C(−2, −2) is right-angled. Hence state the coordinates of the centre of the circle that passes through A, B and C. Find the equation of the circle.

Solution Gradients:

$$\text{A to B} \Rightarrow \frac{4 - {}^-3}{6 - 5} = 7 \qquad \text{C to B} \Rightarrow \frac{{}^-2 - {}^-3}{{}^-2 - 5} = -\frac{1}{7}$$

Since the product of these gradients is −1 the lines AB and BC are perpendicular.

The triangle ABC is right-angled.

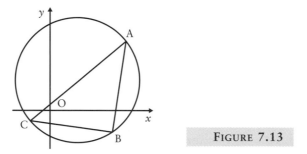

FIGURE 7.13

If A, B and C lie on a circle then AC will be a diameter (from property 5). The centre of the circle is thus the mid-point of AC, which is the point (2, 1).

The radius of the circle is the distance from (2, 1) to A(6, 4).

$$\text{Radius} = \sqrt{4^2 + 3^2} = 5$$

Hence the equation of the circle through ABC is:

$$(x - 2)^2 + (y - 1)^2 = 25$$

6 The perpendicular from the centre to a chord bisects the chord.

If O is the centre of the circle and AB is a chord of the circle then the perpendicular XY to the chord form the centre bisects the chord.

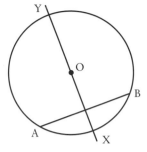

FIGURE 7.14

Proof:

Let XY be the line perpendicular to the chord AB which passes through the centre O of the circle. XY cuts AB at M.

Consider the triangles AOM and BOM.

OM is common to both triangles.

AO = BO (both radii).

Angle AMO = angle BMO (90°).

Hence triangles AOM and BOM are congruent by the RHS rule (right angle, hypotenuse, side).

∴ AM = BM

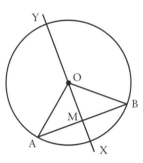

The chord AB is bisected by the perpendicular through O.

FIGURE 7.15

Note

If you have three points on a circle and you find the perpendicular bisectors of any two pairs of those points then the point of intersection of these bisectors will give you the centre of the circle.

EXAMPLE 7.8

X, Y and Z have coordinates (–1, 4), (6, 5) and (7, 4).

(a) Find the equations of the perpendicular bisectors of XY and YZ.

(b) Find the coordinates of the centre of the circle that passes through XYZ.

(c) Find the equation of the circle in the form $x^2 + y^2 + px + qy + r = 0$.

Solution   (a) The gradient of XY is $\frac{1}{7}$ and the mid-point is $\left(\frac{5}{2}, \frac{9}{2}\right)$.

Hence the equation of the perpendicular bisector is:

$$y - \frac{9}{2} = -7\left(x - \frac{5}{2}\right)$$

$$y = -7x + 22$$

The gradient of YZ is –1 and the mid-point is $\left(\frac{13}{2}, \frac{9}{2}\right)$.

Hence the equation of the perpendicular bisector is:

$$y - \frac{9}{2} = 1\left(x - \frac{13}{2}\right)$$

$$y = x - 2$$

(b) XY and YZ are chords to the circle that passes through X, Y and Z.

The perpendicular bisectors both pass through the centre of the circle, by property 6 above. Hence their point of intersection is the centre of the circle.

Solving the simultaneous equations:

$$y = -7x + 22 \qquad \qquad \text{①}$$

$$y = x - 2 \qquad \qquad \text{②}$$

gives ① + 7 × ②

$$8y = 8$$

$$y = 1 \quad \Rightarrow \quad x = 3$$

The centre of the circle is at (3, 1).

(c) The radius of the circle is the distance of (3, 1) from any of the points X, Y or Z. Taking distances from point Y gives:

$$\text{radius} = \sqrt{3^2 + 4^2} = 5$$

The equation of the circle is:

$$(x - 3)^2 + (y - 1)^2 = 25$$

$$x^2 + y^2 - 6x - 2y - 15 = 0$$

---

7 The tangent to a circle is perpendicular to the radius.

If O is the centre of the circle and OA is a radius then the tangent TT' at A meets OA at 90°.

Then angle OT = angle OT' = 90°.

This follows from the symmetry of the tangent at its point of contact with the circle.

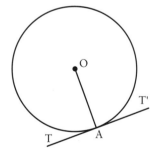

FIGURE 7.16

EXAMPLE 7.9

(a) Find the equation of the line through the centre of the circle with equation

$$x^2 + y^2 - 2x - 4y - 20 = 0$$

that passes through the point (4, 6).

(b) Find the equation of the tangent to the circle at the point (4, 6).

Solution   **(a)** Rewrite $x^2 + y^2 - 2x - 4y - 20 = 0$ to give:

$$(x - 1)^2 + (y - 2)^2 = 25$$

So the centre of the circle is at (1, 2).

The line through (1, 2) and (4, 6) has gradient $\dfrac{6-2}{4-1} = \dfrac{4}{3}$ so the equation is:

$$y - 6 = \frac{4}{3}(x - 4)$$
$$3y = 4x + 2$$

Note

This is the equation of the diameter through the point (4, 6).

(b) The tangent will have gradient $-\dfrac{3}{4}$ so the equation is:

$$y - 6 = -\frac{3}{4}(x - 4)$$
$$3x + 4y = 36$$

EXAMPLE 7.10

Sketch the graph of $x^2 + y^2 - 6x - 2y - 15 = 0$. From the point X(9, 9) draw a straight line which is a tangent to the circle.

Calculate the distance of X from the point where the tangent touches the circle.

Solution   Rewriting $x^2 + y^2 - 6x - 2y - 15 = 0$ gives:

$$(x - 3)^2 + (y - 1)^2 = 25$$

so the centre of the circle is at (3, 1).

You could also draw the tangent to touch the other side of the circle from C but, by symmetry, the distance to the point of contact would be the same as XT.

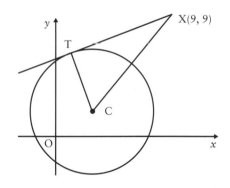

FIGURE 7.17

If the centre is C and the point of contact between the tangent and the circle is T then:

$$CX = \sqrt{(9 - 3)^2 + (9 - 1)^2}$$
$$= \sqrt{6^2 + 8^2}$$
$$= 10$$

The radius of the circle is $CT = \sqrt{25} = 5$.

Since CTX is a right angle:

$$XT^2 = 10^2 - 5^2 = 75$$
$$\therefore \quad XT = 5\sqrt{3}$$

EXAMPLE 7.11

A circle with equation $x^2 + y^2 - 6y = 0$ has a tangent with equation $x + y = k$. Find the exact values of k.

Solution   Solving the simultaneous equations:

$$x^2 + y^2 - 6y = 0$$

$$x + y = k$$

will give the points of intersection. As $x + y = k$ is a tangent there is only one point of intersection. So the quadratic equation obtained should have only one real root. Thus:

$$(k - y)^2 + y^2 - 6y = 0$$

$$\therefore\ 2y^2 - (2k + 6)y + k^2 = 0$$

For equal roots:

$$(2k + 6)^2 - 4 \times 2 \times k^2 = 0$$

$$\therefore\ 4k^2 - 24k - 36 = 0$$

$$\therefore\ k = 3 \pm 3\sqrt{2}$$

The two values of k correspond to the two possible tangents to the circle.

EXERCISE 7B

1   Find the equations of the lines going through the centres of the following circles and passing through the given points.
   (a) $(x - 3)^2 + (y - 5)^2 = 20$ $(1, 1)$
   (b) $(x + 2)^2 + (y + 1)^2 = 18$ $(-5, 2)$
   (c) $(x - 2)^2 + (y + 3)^2 = 25$ $(5, 1)$
   (d) $x^2 + y^2 - 2x - 4y - 15 = 0$ $(5, 4)$
   (e) $x^2 + y^2 + x - 5y - 16 = 0$ $(4, 4)$

2   Find the equations of the tangents to the circles in question 1 at the given points.

3   The equation of a tangent to a circle at a point $(3, 2)$ is $x + y = 5$.
   (a) Find the equation of the diameter passing through this point.
   (b) Given that the radius of the circle is $2\sqrt{2}$ find the two possible equations of the circle.

4   A circle has equation $x^2 + y^2 - 4x - 8y - 29 = 0$. Find its centre and radius. A line is drawn from the point X(7, -8) which is tangential to the circle. Show that the distance from X to the point where the tangent touches the circle is $2\sqrt{30}$.

5 A tangent to the circle $x^2 + y^2 + 6x - 4y - 3 = 0$ passes through the point (5, 8). Find the distance from the point of contact between the tangent and the circle and the point (5, 8).

6 The points X, Y and Z have coordinates (4, 16), (21, 9) and (14, −8). Show that XY is perpendicular to YZ.
Hence find the equation of the circle passing through points X, Y and Z.

7 A triangle has vertices A(1, 8), B(9, 2) and C(5, 10).
   (a) Prove that the triangle is right-angled.
   (b) If A and B are the ends of a diameter of a circle explain why C is on the circle.
   (c) Show that the equation of the circle is $x^2 + y^2 - 10(x + y) + 25 = 0$.

8 The points A, B and C have coordinates (15, 9), (9, −3) and (3, 5).
   (a) Find the equations of the perpendicular bisectors of AB and AC.
   (b) Show that the perpendicular bisectors intersect at the point (10, 4).
   (c) With the aid of a diagram explain why (10, 4) is the centre of the circle through A, B and C.
   (d) Find the equation of the circle in the form $x^2 + y^2 + px + qy + r = 0$.

9 The points A(1, 6), B(8, 5) and C(0, −1) lie on a circle. By finding the equations of the perpendicular bisectors of AB and AC, find the coordinates of the centre of the circle.
Obtain the equation of the circle.

10 A circle has equation $x^2 + y^2 - 4x = 0$. The circle has a tangent with equation $y = x + k$. Find the exact values of k.

EXERCISE 7C   **Examination-style questions**

1 A circle C has centre (3, −4) and radius $2\sqrt{3}$. Write the equation of C in the form $x^2 + y^2 + px + qy + r = 0$.

2 The points (-7, −2) and (−1, 6) are at the opposite ends of the diameter of a circle C.
   (a) Find the centre and radius of the circle C.
   (b) Find the equation of the circle in the form $x^2 + y^2 + px + qy + r = 0$.

3 A circle C has equation $x^2 + y^2 - 10x + 6y - 15 = 0$.
   (a) Find the coordinates of the centre of C.
   (b) Find the radius of C.

[Edexcel]

4   A circle C has equation $x^2 + y^2 - 2x - 6y - 16 = 0$.
   (a) Find the centre and the radius of the circle C.
   (b) Find the coordinates of the points where the line $y = x - 2$ cuts the circle.

5   A circle C has centre $(3, 4)$ and radius $3\sqrt{2}$. A straight line l has equation
   $y = x + 3$.
   (a) Write down the equation of the circle C.
   (b) Calculate the coordinates of the two points where the line l intersects C,
        giving your results in surds.
   (c) Find the distance between the two points.

                                                                                    [Edexcel]

6   A circle C has equation $x^2 + y^2 - 8x + 2y - 12 = 0$.
   (a) Show that the centre of C is $(4, -1)$ and find the radius of C.
   (b) Verify that the point P$(6, 4)$ lies on the circle C.
   (c) Find the equation of the tangent to the circle C at the point P.

7   The circle C has equation $x^2 + y^2 - 8x - 16y - 209 = 0$.
   (a) Find the coordinates of the centre of C and the radius of C.
   The point P(x, y) lies on C.
   (b) Find, in terms of x and y, the gradient of the tangent to C at P.
   (c) Hence, or otherwise, find an equation of the tangent to C at the
        point $(21, 8)$.

                                                                                    [Edexcel]

8   A circle C has equation $x^2 + y^2 - 6x + 8y - 75 = 0$.
   (a) Write down the coordinates of the centre of C, and calculate the radius of C.
        A second circle has centre at the point $(15, 12)$ and radius 10.
   (b) Sketch both circles in a single diagram and find the coordinates of the
        point where they touch.

                                                                                    [Edexcel]

9   A circle C has equation $x^2 + y^2 + 2x - 4y = 0$.
   (a) Find the coordinates of the centre of C and show that the radius of C is $\sqrt{5}$.
   (b) A line is drawn through the point S$(4, 6)$ and touches the circle C at the
        point T. Calculate the length ST.
   (c) What is the equation of the circle which has centre S and passes through
        the point T? Express your answer in the form $x^2 + y^2 + px + qy + r = 0$.

10 The points A, B and C have coordinates $(6, 2)$, $(7, 1)$ and $(-2, -8)$ respectively.
   (a) Find the equation of the perpendicular bisector of AB.
   (b) Find the equation of the perpendicular bisector of AC.
   (c) Find the coordinates of P which is the point of intersection of the
        perpendicular bisectors of AB and AC.
   (d) Hence, or otherwise, find the equation of the circle passing through A, B
        and C, expressing your answer in the form $x^2 + y^2 + px + qy + r = 0$.

11 P is the point with coordinates $(-10, 5)$, Q is the point $(4, 7)$ and R is the point $(6, 5)$.

   (a) Show that the perpendicular bisectors of PQ and QR intersect at the point $(-2, -1)$.

   (b) Show that the equation of the circle passing through P, Q and R is $x^2 + y^2 + 4x + 2y - 95 = 0$.

12 (a) On the same diagram plot the points $X(2, -1)$, $Y(1, 4)$ and $Z(6, 5)$.

   (b) Find the lengths of XY, YZ and XZ.

   (c) Find the equation of the circle through X, Y and Z in the form $(x - a)^2 + (y - b)^2 = c$.

13 A circle passes through the points $A(8 - 4\sqrt{3}, 0)$, $B(16, 4)$ and $C(8 + 4\sqrt{3}, 8)$.

   (a) Prove that AB is perpendicular to BC.

   (b) Find the equation of the circle.

14 A circle C has equation $x^2 + y^2 - 12x - 20y + 111 = 0$.

   (a) Find the centre of C and the radius of C.

   (b) Verify that $(3, 14)$ and $(10, 7)$ lie on C.

   (c) Show that the tangents to the circle C at the points $(3, 14)$ and $(10, 7)$ meet at the point $(31, 35)$.

15 Show that the circle C with centre $(7, 3)$ and passing through the point $(12, -5)$ has equation $x^2 + y^2 - 14x - 6y - 31 = 0$.

   (a) The line $y = x - 7$ intersects the circle at the points A and B. Find the coordinates of A and B.

   (b) Find the equation of the circle with diameter AB.

KEY POINTS

1 The **equation of a circle** with radius r and centre the origin is:

$x^2 + y^2 = r^2$

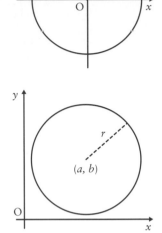

2 The **equation of a circle** with radius r and centre at the point (a, b) is:

$(x - a)^2 + (y - b)^2 = r^2$

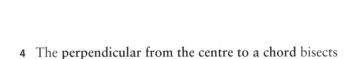

3 The **angle in a semicircle** is a right angle.

4 The **perpendicular from the centre to a chord** bisects the chord.

5 The **tangent to a circle** is perpendicular to the radius.

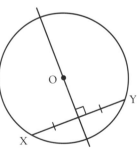

Chapter eight

SEQUENCES AND SERIES

For he, by geometric scale, could take the size of pots of ale,

And wisely, tell what hour o' th' day the clock doth strike,

By algebra!

Samuel Butler

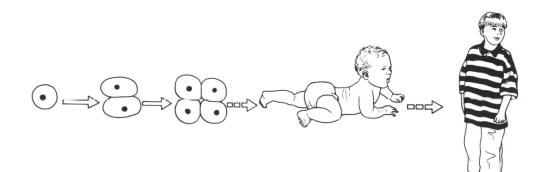

FIGURE 8.1

GEOMETRIC SEQUENCES AND SERIES

A human being begins life as one cell, which divides into two, then four... .

The terms of a geometric sequence are formed by multiplying one term by a fixed number, the common ratio, to obtain the next. This relationship between two consecutive terms can be written as:

$$u_{n+1} = ru_n \text{ with first term } u_1$$

The sum of the terms of a geometric sequence is called a *geometric series*. An alternative name is a *geometric progression*, shortened to GP.

NOTATION

When describing geometric sequences in this book, the following conventions are used:

- first term $u_1 = a$

- common ratio $= r$

- number of terms $= n$

- the general term u_n is that in position n (i.e. the nth term).

In the geometric sequence   3, 6, 12, 24, 48, $a = 3$, $r = 2$ and $n = 5$.

The terms of this sequence are formed as follows:

$$u_1 = a \qquad = 3$$
$$u_2 = a \times r = 3 \times 2 = 6$$
$$u_3 = a \times r^2 = 3 \times 2^2 = 12$$
$$u_4 = a \times r^3 = 3 \times 2^3 = 24$$
$$u_5 = a \times r^4 = 3 \times 2^4 = 48$$

You will see that in each case the power of r is one less than the number of the term: $u_5 = ar^4$ and 4 is one less than 5. The general term, u_k, can be written deductively as:

$$u_k = ar^{k-1}$$

and the last term is:

$$u_n = ar^{n-1}$$

These are both general formulae which apply to any geometric sequence.

Given two consecutive terms of a geometric sequence, you can always find the common ratio by dividing the later term by the earlier. For example, the geometric sequence ... 5, 8, ... has common ratio $r = \frac{8}{5}$.

EXAMPLE 8.1

Find the 7th term in the geometric sequence 8, 24, 72, 216, ...

Solution In the sequence, the first term, $a = 8$ and the common ratio, $r = 3$.

The nth term of a geometric sequence is given by $u_n = ar^{n-1}$,

and so $u_7 = 8 \times 3^6$
$$= 5832$$

EXAMPLE 8.2

How many terms are there in the geometric sequence 0.2, 1, 5, ..., 390 625?

Solution   The last term of the geometric sequence is $u_n = ar^{n-1}$.

In this example, $a = 0.2$ and $r = 5$, so:

$$0.2 \times 5^{n-1} = 390\,625$$
$$5^{n-1} = 1\,953\,125$$

Using trial and improvement:

$$\text{when } n = 10, \; 5^{n-1} = 5^9 = 1\,953\,125$$

So there are 10 terms in the geometric sequence.

EXERCISE 8A

1   Are the following sequences geometric? If so, state the common ratio and calculate the 7th term.

    (a) 5, 10, 20, 40, ...   (b) 2, 4, 6, 8, ...   (c) 1, −1, 1, −1, ...

    (d) 5, 5, 5, 5, ...   (e) 6, 3, 0, −3, ...   (f) $6, 3, 1\frac{1}{2}, \frac{3}{4}, ...$

    (g) 1, 1.1, 1.11, 1.111, ...

2   For each of the following geometric sequences, state the value of the common ratio and the value of the tenth term.

    (a) 2, 4, 8, 16, ...   (b) $1, \frac{1}{2}, \frac{1}{4}, \frac{1}{8}, ...$

    (c) 5, −10, 20, −40, ...   (d) $3, -1, \frac{1}{3}, -\frac{1}{9}, ...$

    (e) 11, 121, 1331, 14 641, ...   (f) 5, 2, 0.8, 0.32, ...

    (g) 60, 90, 135, 202.5, ...

3   A geometric sequence has first term $\frac{1}{9}$ and common ratio 3.

    (a) Find the fifth term.

    (b) Which is the first term of the sequence which exceeds 1000?

4   How many terms are there in each of the following sequences?

    (a) 2, 4, 8, ... , 4096   (b) $1, \frac{1}{2}, \frac{1}{4}, ... , \frac{1}{1024}$

    (c) 5, −10, 20, ... , −2560   (d) 5, 2, 0.8, ... , 0.020 48

    (e) $\frac{8}{3}, \frac{4}{3}, \frac{2}{3}, ... , \frac{1}{48}$

5   The third term of a geometric sequence is 150 and the fifth term is 3750. Find the first term, the common ratio and the eighth term.

6   The fourth term of a geometric sequence is −40 and the seventh term is 320. Find the tenth term.

7 The first term of a geometric sequence of positive terms is 5 and the 5th term is 1280.

   (a) Find the common ratio of the sequence.
   (b) Find the 8th term of the sequence.

8 Find, in terms of n, the nth term of a geometric sequence, given that the fifth term is $\frac{1}{2}$ and the eighth term is $\frac{1}{16}$.
   Which is the first term to have a value less than $\frac{1}{1000}$?

9 I invest £500 at a fixed annual rate of interest of 5%. At the end of the first year I shall have £525 in my account. I leave the money in the account for six years.

   (a) After how many full years will my account first have more than £600?
   (b) How much money shall I have in the account at the end of the sixth year?

10 I knock a nail into a piece of wood. On the first knock the nail goes in by 1 cm. Each time I hit the nail after that it goes in half as far as it did the time before.

   (a) How far will it go in on the fifth hit?
   (b) What is the total distance gone by the nail after five hits?
   (c) Will the nail ever go 2 cm into the wood?

THE SUM OF THE TERMS OF A GEOMETRIC SEQUENCE

If a geometric sequence has first term a and common ratio r then the sum of the first n terms is S_n where:

$$S_n = a + ar + ar^2 + \dots + ar^{n-1} \qquad ①$$

Multiplying by the common ratio r gives:

$$rS_n = ar + ar^2 + ar^3 + \dots + ar^n \qquad ②$$

Subtracting ② from ① gives:

$$(1 - r)S_n = a - ar^n$$
$$= a(1 - r^n)$$

This gives the general equation for the sum to n terms of a geometric sequence:

$$S_n = \frac{a(1 - r^n)}{1 - r}$$

The proof of the derivation of this formula should be known.

EXAMPLE 8.3

Find the number of terms in the progression

$$243, 81, 27, \dots, \tfrac{1}{27}$$

and find the sum of these terms.

Solution $a = 243$ and $r = \frac{1}{3}$

The last term ar^{n-1} is $\frac{1}{27}$ so:

$$243 \times \left(\tfrac{1}{3}\right)^{n-1} = \tfrac{1}{27}$$

$$\left(\tfrac{1}{3}\right)^{n-1} = \frac{1}{27 \times 243} = \left(\tfrac{1}{3}\right)^8$$

$$\Rightarrow n = 9 \qquad\qquad \text{so the progression has 9 terms.}$$

The sum of the 9 terms is:

$$S_9 = \frac{a(1 - r^9)}{1 - r} = \frac{243 \times \left(1 - \left(\tfrac{1}{3}\right)^9\right)}{1 - \tfrac{1}{3}} = 364.48 \text{ (to 5 significant figures).}$$

EXAMPLE 8.4

Find the sum of the first ten terms of the geometric series with fifth term 48 and ninth term 768 given that all the terms are positive.

Solution   Fifth term $ar^4 = 48$ ①

Ninth term $ar^8 = 768$ ②

② ÷ ① $r^4 = 16$

so $r = 2$ (Discount the solution $r = -2$ since all the terms are positive.)

and $a = 3$.

The sum of the first ten terms is:

$$S_{10} = \frac{a(1 - r^{10})}{1 - r} = \frac{3(1 - 2^{10})}{1 - 2} = 3069$$

EXAMPLE 8.5

How many terms of $4 + 12 + 36 + \ldots$ must be taken for the sum to exceed one million?

$a = 4$ and $r = 3$ and the sum to n terms gives:

$$S_n = \frac{a(1 - r^n)}{1 - r}$$

$$\frac{4(1 - 3^n)}{1 - 3} = 1\,000\,000$$

$$3^n - 1 = 500\,000$$

$$3^n = 500\,001$$

By trial and improvement:

when $n = 11$, $3^{11} = 177\,147$ too small

when $n = 12$, $3^{12} = 531\,441$

So 12 terms must be taken for the sum to exceed one million.

EXERCISE 8B

1 Find the sum of the first ten terms of each of the following geometric sequences.

    (a) $2, 4, 8, 16, \ldots$ **(b)** $2, 1, \frac{1}{2}, \frac{1}{4}, \ldots$ **(c)** $\frac{1}{2}, 2, 8, 32, \ldots$

    (d) $3, 2, 1\frac{1}{3}, \frac{8}{9}, \ldots$ **(e)** $2, -6, 18, -54, \ldots$ **(f)** $1, -\frac{1}{4}, \frac{1}{16}, -\frac{1}{64}, \ldots$

    (g) $128, -64, 32, -16, \ldots$

2 The first term of a geometric sequence is 1.5 and the common ratio is 2.

    (a) Find the twelfth term of the sequence.

    (b) Find the sum of the first twelve terms.

3 The third term of a geometric sequence is 6 and the sixth term is 162.

    (a) Find the first term and the common ratio of the sequence.

    (b) Find the tenth term of the sequence.

    (c) Find the sum of the first ten terms of the sequence.

4 In a geometric sequence the fourth term is $\frac{1}{2}$ and the sixth term is $\frac{1}{8}$. All the terms are positive.

    (a) Show that the common ratio is $\frac{1}{2}$ and find the first term.

    (b) What is the sum of the first twenty terms?

5 The first term of a geometric sequence is 5 and the common ratio is -2. Find the sum of the first ten terms.

6 A geometric sequence has first term 3 and common ratio 2. The sequence has 8 terms.

    (a) Find the last term.

    (b) Find the sum of the terms in the sequence.

7 **(a)** Find how many terms there are in the geometric sequence:

$$8, 16, \ldots, 2048$$

    (b) Find the sum of the terms in this sequence.

8 **(a)** Find how many terms there are in the geometric sequence:

$$200, 50, \ldots, 0.195\,312\,5$$

    (b) Find the sum of the terms in this sequence.

9 The third term of a geometric sequence is 3 and the sixth term is -24. Find the sum of the eleventh to twentieth terms, inclusive.

10 Find how many terms of the geometric sequence:

$$1, \frac{1}{2}, \frac{1}{4}, \frac{1}{8}, \ldots$$

are needed for the sum to exceed 1.9995.

INFINITE GEOMETRIC SERIES

The series $1 + \frac{1}{2} + \frac{1}{4} + \frac{1}{8} + \frac{1}{16}$, ... is geometric, with common ratio $\frac{1}{2}$.

Summing the terms one by one gives $1, 1\frac{1}{2}, 1\frac{3}{4}, 1\frac{7}{8}, 1\frac{15}{16}$, ...

Clearly the more terms you take, the nearer the sum gets to 2. In the limit, as the number of terms tends to infinity, the sum tends to 2.

As $n \to \infty$, $S_n \to 2$.

This is an example of a *convergent* series. The sum to infinity is a finite number.

You can see this by substituting $a = 1$ and $r = \frac{1}{2}$ in the formula for the sum of the series:

$$S_n = \frac{a(1 - r^n)}{1 - r}$$

$$\text{giving } S_n = \frac{1 \times (1 - (\frac{1}{2})^n)}{(1 - \frac{1}{2})}$$

$$= 2 \times \left(1 - (\tfrac{1}{2})^n\right)$$

The larger the number of terms, n, the smaller $(\frac{1}{2})^n$ becomes and so the nearer S_n is to the limiting value of 2 (figure 8.2). Notice that $(\frac{1}{2})^n$ can never be negative, however large n becomes; so S_n can never exceed 2.

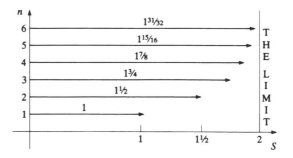

FIGURE 8.2

In the general geometric series $a + ar + ar^2 + ...$ the terms become progressively smaller in size if the common ratio r is between −1 and 1. This was the case above: r had the value $\frac{1}{2}$. In such cases, the geometric series is *convergent*.

If, on the other hand, the value of r is greater than 1 (or less than −1) the terms in the series become larger and larger in size and so it is described as *divergent*.

A series corresponding to a value of r of exactly +1 consists of the first term a repeated over and over again. A sequence corresponding to a value of r of exactly −1 oscillates between $+a$ and $-a$. Neither of these is convergent.

It only makes sense to talk about the sum of an infinite series if it is convergent. Otherwise the sum is undefined.

The condition for a geometric series to converge, $-1 < r < 1$, ensures that as $n \to \infty$, $r^n \to 0$, and so the formula for the sum of a geometric series:

$$S_n = \frac{a(1 - r^n)}{1 - r}$$

may be rewritten for an infinite series as:

$$S_\infty = \frac{a}{1 - r} \quad \text{provided } -1 < r < 1$$

EXAMPLE 8.6

Find the sum of the terms of the infinite sequence 0.2, 0.02, 0.002, ...

Solution This is a geometric sequence with $a = 0.2$ and $r = 0.1$.

Its sum is given by:

$$S_\infty = \frac{a}{1 - r}$$

$$= \frac{0.2}{1 - 0.1}$$

$$= \frac{0.2}{0.9}$$

$$= \frac{2}{9}$$

Note

You may have noticed that the sum of the series $0.2 + 0.02 + 0.002 + \ldots$ is $0.2\dot{2}$, and that this recurring decimal is indeed the same as $\frac{2}{9}$.

EXAMPLE 8.7

The first three terms of an infinite geometric sequence are 16, 12 and 9.

(a) Write down the common ratio.
(b) Find the sum of the terms of the sequence.
(c) After how many terms is the sum greater than 99.99% of that of the infinite series?

Solution (a) The common ratio is $\frac{3}{4}$.

(b) The sum of the terms of an infinite geometric series is given by:

$$S_\infty = \frac{a}{1 - r}$$

In this case $a = 16$ and $r = \frac{3}{4}$, so:

$$S_\infty = \frac{16}{1 - \frac{3}{4}} = 64$$

(c) The sum of a geometric series to n terms, S_n, is given by $\dfrac{a(1 - r^n)}{(1 - r)}$.

In this case:

$$S_n = \frac{16\left(1 - (\frac{3}{4})^n\right)}{\left(1 - (\frac{3}{4})\right)} = 64\left(1 - (\frac{3}{4})^n\right)$$

The value of n which gives a sum equal to 99.99% of 64 is given by:

$$64\left(1 - (\tfrac{3}{4})^n\right) = \frac{99.99}{100} \times 64$$

$$1 - (\tfrac{3}{4})^n = 0.9999$$

$$(\tfrac{3}{4})^n = 0.0001$$

By trial and improvement:

$(\tfrac{3}{4})^{32} = 0.000\,100\,45$
$(\tfrac{3}{4})^{33} = 0.000\,075\,34$

So after 33 terms the sum is greater than 99.99% of the sum to infinity of the sequence.

---

1 Which of the following sequences converge?

   (a) $2, 1, \frac{1}{2}, \frac{1}{4}, \ldots$ (b) $3, -1, \frac{1}{3}, -\frac{1}{9}, \ldots$ (c) $1, 1.1, 1.2, 1.3, \ldots$

   (d) $1, 0.9, 0.81, 0.729, \ldots$ (e) $1, \frac{1}{5}, \frac{1}{25}, \frac{1}{625}, \ldots$ (f) $32, -8, 2, -\frac{1}{2}, \ldots$

   (g) $\frac{1}{2}, \frac{1}{3}, \frac{1}{4}, \frac{1}{5}, \ldots$

2 Find the sum to infinity of each of the following geometric sequences.

   (a) $2, 1, \frac{1}{2}, \frac{1}{4}, \ldots$ (b) $3, -1, \frac{1}{3}, -\frac{1}{9}, \ldots$ (c) $5, 4, 3.2, 2.56, \ldots$

   (d) $\frac{2}{3}, \frac{5}{9}, \frac{25}{54}, \frac{125}{324}, \ldots$ (e) $0.2, 0.02, 0.002, 0.0002, \ldots$

   (f) $2, -1, \frac{1}{2}, -\frac{1}{4}, \ldots$ (g) $1 + r, r^2(1 + r), r^4(1 + r), r^6(1 + r), \ldots \; |r| < 1$

3 Evaluate $\displaystyle\sum_{r=1}^{n} \frac{1}{2^{r-1}}$ for $n = 1, 2, 3, 4$ and 5. To what value will the sum converge as n becomes very large?

4 The first three terms of an infinite geometric sequence are 4, 2 and 1.

   (a) State the common ratio of this sequence.

   (b) Calculate the sum to infinity of its terms.

5 The first three terms of an infinite geometric sequence are 0.7, 0.07, 0.007.

    (a) Write down the common ratio for this sequence.

    (b) Find, as a fraction, the sum to infinity of the terms of this sequence.

    (c) Find the sum to infinity of the geometric series $0.7 - 0.07 + 0.007, \ldots,$ and hence show that $\frac{7}{11} = 0.\overset{..}{6}\overset{..}{3}$.

6 The first three terms of a geometric sequence are 100, 90 and 81.

    (a) Write down the common ratio of the sequence.

    (b) Which is the position of the first term in the sequence that has value less than 1?

    (c) Find the sum to infinity of the terms of this sequence.

    (d) After how many terms is the sum of the sequence greater than 99% of the sum to infinity?

7 A geometric series has first term 4 and its sum to infinity is 5.

    (a) Find the common ratio.

    (b) Find the sum to infinity if the first term is excluded from the series.

8 Given the sequence:

$$6, 3, 1\tfrac{1}{2}, \tfrac{3}{4}, \ldots$$

find:

    (a) the sum to infinity

    (b) the number of terms required before the difference between the sum to infinity and the sum to n terms is less than 0.01.

9 The third term of a geometric sequence is $\frac{16}{27}$ and the fifth term is $\frac{64}{243}$. All the terms are positive.

    (a) Find the common ratio and show that the first term is $\frac{4}{3}$.

    (b) Find the sum to infinity.

    (c) How many terms are needed before their sum is greater than 3.95?

10 For a particular geometric sequence, 7 times the sum to infinity is equal to 8 times the sum of the first three terms. Given that the sum to infinity is 5, find:

    (a) the common ratio and the first term

    (b) the sum of the first seven terms.

BINOMIAL EXPANSIONS

A special type of polynomial is produced when a binomial (i.e. two part) expression like $(1 + x)$ is raised to a power. The resulting polynomial is often called a *binomial expansion*.

The simplest binomial expansion is $(1 + x)$ itself. This, and other expansions, are given below:

$$(1 + x)^1 = 1 + x$$
$$(1 + x)^2 = (1 + x)(1 + x) = 1 + 2x + x^2$$
$$(1 + x)^3 = (1 + x)(1 + 2x + x^2) = 1 + 3x + 3x^2 + x^3$$
$$(1 + x)^4 = (1 + x)(1 + 3x + 3x^2 + x^3) = 1 + 4x + 6x^2 + 4x^3 + x^4$$
$$(1 + x)^5 = (1 + x)(1 + 4x + 6x^2 + 4x^3 + x^4) = 1 + 5x + 10x^2 + 10x^3 + 5x^4 + x^5.$$

You will notice that the numbers, which are called coefficients, form the following pattern:

```
              1
          1       1
      1       2       1
   1       3       3       1
 1      4       6       4       1
1     5     10     10     5       1
```

This is called *Pascal's triangle* (or the *Chinese triangle*). Each number is obtained by adding the two above it, for example:

```
4    +    6
    = 10
```

Following the pattern the next few rows would be:

```
     1      6      15     20     15     6      1
  1     7      21     35     35     21     7     1
1     8     28     56     70     56     28     8     1
```

Note

Pascal's triangle can give you hours of fun. For example, if you look at diagonal lines from the top you first have a series of 1's, then the counting numbers 1, 2, 3, 4, 5 …, the next diagonal has triangular numbers, the next has pyramid numbers and so on. If you add all the numbers on each row you will have 1, 2, 4, 8, 16 … that is the powers of 2. There are many other fascinating patterns to be found. Can you spot the powers of 11 or can you find a way to generate the Fibonacci series?

You will use Pascal's triangle to give the coefficients in your expansions. For example:

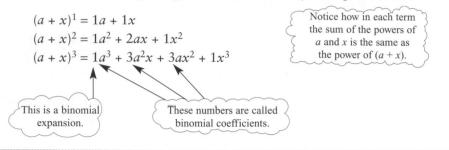

$(a + x)^1 = 1a + 1x$

$(a + x)^2 = 1a^2 + 2ax + 1x^2$

$(a + x)^3 = 1a^3 + 3a^2x + 3ax^2 + 1x^3$

Notice how in each term the sum of the powers of a and x is the same as the power of $(a + x)$.

This is a binomial expansion.

These numbers are called binomial coefficients.

EXAMPLE 8.8

Write out the binomial expansion of $(2 + x)^4$.

Solution   The binomial coefficients from Pascal's triangle are 1 4 6 4 1. In each term the sum of the powers of 2 and x must be equal to 4.

So the expansion is:

$(2 + x)^4 = 1 \times 2^4 + 4 \times 2^3 \times x + 6 \times 2^2 \times x^2 + 4 \times 2 \times x^3 + 1 \times x^4$

$\therefore (2 + x)^4 = 16 + 32x + 24x^2 + 8x^3 + x^4$

EXAMPLE 8.9

Write out the binomial expansion of $(2a - 3b)^5$.

Solution   The binomial coefficients for power 5 are   1   5   10   10   5   1

The expression $(2a - 3b)$ is treated as $(2a + (-3b))$.

So the expansion is:

$1 \times (2a)^5 + 5 \times (2a)^4 \times (-3b) + 10 \times (2a)^3 \times (-3b)^2 + 10 \times (2a)^2 \times (-3b)^3$
$$+ 5 \times (2a) \times (-3b)^4 + 1 \times (-3b)^5$$

Note that $(2a)^5 = 2^5 \times a^5 = 32a^5$ and $(-3b)^3 = (-3)^3 \times b^3 = -27b^3$, etc.

$\therefore (2a - 3b)^5 = 32a^5 - 240a^4b + 720a^3b^2 - 1080a^2b^3 + 810ab^4 - 243b^5$

EXAMPLE 8.10

Find the first four terms in the expansion of $(1 - 2x)^8$.

Solution   The binomial coefficients are 1 8 28 56 ... Hence:

$(1 - 2x)^8 = (1 + (-2x))^8$
$$= 1^8 + 8 \times 1^7 \times (-2x)^1 + 28 \times 1^6 \times (-2x)^2 + 56 \times 1^5 \times (-2x)^3 + \ldots$$
$$= 1 - 16x + 112x^2 - 448x^3 + \ldots$$

EXAMPLE 8.11

Find the term independent of x in the expansion of $\left(x + \dfrac{1}{x}\right)^6$.

Solution   Expanding gives:

$$\left(x + \frac{1}{x}\right)^6 = x^6 + 6x^5 \times \frac{1}{x} + 15x^4 \times \frac{1}{x^2} + 20x^3 \times \frac{1}{x^3} + \ldots$$

You do not need to look at any more terms because you see that in the last term written the x cancels out, i.e. the term independent of x is 20.

EXAMPLE 8.12

Expand $(1 + x + x^2)^2$.

Solution
$$\begin{aligned}
(1 + x + x^2)^2 &= (1 + (x + x^2))^2 \\
&= 1 + 2(x + x^2) + (x + x^2)^2 \\
&= 1 + 2(x + x^2) + x^2 + 2x^3 + x^4 \\
&= 1 + 2x + 3x^2 + 2x^3 + x^4
\end{aligned}$$

Historical note

Blaise Pascal has been described as the greatest might-have-been in the history of mathematics. Born in France in 1623, he was making discoveries in geometry by the age of 16 and had developed the first computing machine before he was 20.

Pascal suffered from poor health and religious anxiety, so that for periods of his life he gave up mathematics in favour of religious contemplation. The second of these periods was brought on when he was riding in his carriage: his runaway horses dashed over the parapet of a bridge, and he was only saved by the miraculous breaking of the traces. He took this to be a sign of God's disapproval of his mathematical work. A few years later a toothache subsided when he was thinking about geometry and this, he decided, was God's way of telling him to return to mathematics.

Pascal's triangle (and the binomial theorem) had actually been discovered by Chinese mathematicians several centuries earlier, and can be found in the works of Yang Hui (around 1270 A.D.) and Chu Shi-kie (in 1303 A.D.). Pascal is remembered for his application of the triangle to elementary probability, and for his study of the relationships between binomial coefficients.

Pascal died at the early age of 39.

EXERCISE 8D

1 Expand and simplify each expression.

(a) $(x + 3)^4$
(b) $(x - 2)^5$
(c) $(1 + 2x)^4$
(d) $(2 - 3x)^3$
(e) $(2 + x)^6$
(f) $(2x - 3)^4$

2 Find the first four terms in the expansions of each expression.

(a) $(1 + x)^7$
(b) $\left(x - \frac{1}{2}\right)^8$
(c) $\left(2 + \frac{x}{2}\right)^6$

3 Find the term independent of x in the following expansions.

(a) $\left(x + \frac{1}{x}\right)^4$
(b) $\left(x^2 - \frac{1}{x}\right)^3$
(c) $\left(\frac{1}{2x} + 2x\right)^8$

4 (a) Expand $(1 + x)^5$ and $(1 - x)^5$ as far as the term in x^3.
   (b) Expand $(1 - x^2)^5$ as far as the term in x^3.
   (c) Multiply your two expansions in part (a) together to give a series expansion as far as the term x^3. Explain why the answer is the same as that from part (b).

5 Expand $(1 + x + x^2)^3$.

THE FORMULA FOR A BINOMIAL COEFFICIENT

There will be times when you need to find binomial coefficients that are outside the range of your Pascal's triangle. What happens if you need to find the power of x^{17} in the expansion of $(x + 2)^{25}$? Clearly you need a formula that gives binomial coefficients.

The first thing you need is a notation for identifying binomial coefficients. It is usual to denote the power of the binomial expression by n, and the position in the row of binomial coefficients by r, where r can take any value from 0 to n. So for row 5 of Pascal's triangle:

$n = 5$:	1	5	10	10	5	1
	$r = 0$	$r = 1$	$r = 2$	$r = 3$	$r = 4$	$r = 5$

The general binomial coefficient corresponding to values of n and r is written as ^nC_r and said as 'N C R'.

An alternative notation is $\binom{n}{r}$. Thus $^5C_3 = \binom{5}{3} = 10$.

The next step is to find a formula for the general binomial coefficient ^nC_r or $\binom{n}{r}$. However, to do this you must be familiar with the term *factorial*.

The quantity '8 factorial', written 8!, is

$$8! = 8 \times 7 \times 6 \times 5 \times 4 \times 3 \times 2 \times 1 = 40\,320$$

Similarly, $12! = 12 \times 11 \times 10 \times 9 \times 8 \times 7 \times 6 \times 5 \times 4 \times 3 \times 2 \times 1 = 479\,001\,600$,

and $n! = n \times (n - 1) \times (n - 2) \times \ldots \times 1$, where n is a positive integer.

Note   0! is defined to be 1. You will see the need for this when you use the formula for $\binom{n}{r}$.

The table shows an alternative way of laying out Pascal's triangle.

		Column (r)								
		0	1	2	3	4	5	6	7	8
	1	1	1							
	2	1	2	1						
Row (n)	3	1	3	3	1					
	4	1	4	6	4	1				
	5	1	5	10	10	5	1			
	6	1	6	15	20	15	6	1		
	7	1	7	21	35	35	21	7	1	
	8	1	8	28	56	70	56	28	8	1
	...	...	...	...	...	...	...	...	...	...
	n	1	n	?	?	?	?	?	?	?

Note

The numbers in each row of Pascal's triangle are symmetrical about the middle number or middle pair of numbers.

From the table it is possible to see that:

when	the coefficient is:
$r = 0$	1
$r = 1$	n
$r = 2$	$\dfrac{n(n-1)}{2}$
$r = 3$	$\dfrac{n(n-1)(n-2)}{3 \times 2}$
$r = 4$	$\dfrac{n(n-1)(n-2)(n-3)}{4 \times 3 \times 2}$

and, in general, the rth term is:

$$\frac{n(n-1)(n-2)(n-3) \ldots (n-r+1)}{r!} = \frac{n!}{r!(n-r)!}$$

i.e. the rth term of the nth row of Pascal's triangle is:

$$^nC_r = \binom{n}{r} = \frac{n!}{r!(n-r)!}$$

THE FORMULA FOR THE BINOMIAL EXPANSION OF $(a + b)^n$

This allows us to write the binomial expansion when n is a positive integer as:

$$(a + b)^n = a^n + \binom{n}{1}a^{n-1}b + \binom{n}{2}a^{n-2}b^2 + \ldots + \binom{n}{r}a^{n-r}b^r + \ldots + b^n$$

....

Note   The values of the coefficients can be found from writing out Pascal's triangle, from using the formula for $\binom{n}{r}$ or by using the ^nC_r button on your calculator.

....

It follows that the expansion of $(a + bx)^n$ is:

$$(a + bx)^n = a^n + \binom{n}{1}a^{n-1}bx + \binom{n}{2}a^{n-2}b^2x^2 + \ldots + \binom{n}{r}a^{n-r}b^rx^r + \ldots + b^nx^n$$

EXAMPLE 8.13

Use the formula $\binom{n}{r} = \dfrac{n!}{r!(n-r)!}$ to calculate these coefficients.

(a) $\binom{5}{0}$ (b) $\binom{5}{1}$ (c) $\binom{5}{2}$

(d) $\binom{5}{3}$ (e) $\binom{5}{4}$ (f) $\binom{5}{5}$

Solution

(a) $\binom{5}{0} = \dfrac{5!}{0!(5-0)!} = \dfrac{120}{1 \times 120} = 1$

(b) $\binom{5}{1} = \dfrac{5!}{1!4!} = \dfrac{120}{1 \times 24} = 5$

(c) $\binom{5}{2} = \dfrac{5!}{2!3!} = \dfrac{120}{2 \times 6} = 10$

(d) $\binom{5}{3} = \dfrac{5!}{3!2!} = \dfrac{120}{6 \times 2} = 10$

(e) $\binom{5}{4} = \dfrac{5!}{4!1!} = \dfrac{120}{24 \times 1} = 5$

(f) $\binom{5}{5} = \dfrac{5!}{5!0!} = \dfrac{120}{120 \times 1} = 1$

....

Note   You can see that these numbers, 1, 5, 10, 10, 5, 1, are row 5 of Pascal's triangle.

EXAMPLE 8.14

Find the coefficient of x^{17} in the expansion of $(x + 2)^{25}$.

Solution $(x + 2)^{25} = \binom{25}{0}x^{25} + \binom{25}{1}x^{24}2^1 + \binom{25}{2}x^{23}2^2 + \dots + \binom{25}{8}x^{17}2^8 + \dots \binom{25}{25}2^{25}$

So the required term is $\binom{25}{8} \times 2^8 \times x^{17}$.

$$\binom{25}{8} = \frac{25!}{8!17!} = \frac{25 \times 24 \times 23 \times 22 \times 21 \times 20 \times 19 \times 18 \times \cancel{17!}}{8! \times \cancel{17!}}$$

$$= 1\,081\,575$$

So the coefficient of x^{17} is $1\,081\,575 \times 2^8 = 276\,883\,200$.

Note

Notice how 17! was cancelled in working out $^{25}C_8$. Factorials become large numbers very quickly and you should keep a look-out for such opportunities to simplify calculations.

EXAMPLE 8.15

Expand $(2 - x)^{10}$ as far as the term in x^3.

Solution $(2 - x)^{10} = 2^{10} + 10 \times 2^9 \times (-x) + \binom{10}{2} \times 2^8 \times (-x)^2 + \binom{10}{3} \times 2^7 \times (-x)^3 + \dots$

but $\binom{10}{2} = \dfrac{10 \times 9 \times 8!}{2! \times 8!} = 45$ and $\binom{10}{3} = \dfrac{10 \times 9 \times 8 \times 7!}{3! \times 7!} = 120$

so:

$(2 - x)^{10} = 1024 - 5120x + 11\,520x^2 - 15\,360x^3 + \dots$

EXAMPLE 8.16

Find the coefficient of x^6 in the expansion of $(1 + 2x)^{15}$.

Solution   You require the term when $r = 6$, that is:

$$\binom{15}{6} \times 1^9 \times (2x)^6 = \frac{15!}{6! \times 9!} \times 2^6 \times x^6 = 320\,320x^6$$

$\therefore$ the coefficient of x^6 is $320\,320$.

EXAMPLE 8.17

Given that the coefficient of x^4 in the expansion of $(3 - 2x)^n$ is 2160 find the value of n.

Solution   You require the term when $r = 4$ so:

$$\binom{n}{4} \times 3^{n-4} \times (-2)^4 = 2160$$

$$\text{so } \binom{n}{4} \times 3^{n-4} = 135$$

Try different values of n:

$$n = 4 \quad \binom{4}{4} \times 3^0 = 1 \qquad \text{too small}$$

$$n = 5 \quad \binom{5}{4} \times 3^1 = 15 \qquad \text{too small}$$

$$n = 6 \quad \binom{6}{4} \times 3^2 = 135 \quad \text{correct}$$

Hence the value of n is 6.

EXAMPLE 8.18

Find the binomial expansion of $\left(2 + \frac{x}{10}\right)^4$. Use your expansions with a suitable value of x to work out 2.1^4.

Solution $\left(2 + \frac{x}{10}\right)^4 = 2^4 + 4 \times 2^3 \times \frac{x}{10} + 6 \times 2^2 \times \left(\frac{x}{10}\right)^2 + 4 \times 2^1 \times \left(\frac{x}{10}\right)^3 + \left(\frac{x}{10}\right)^4$

$$= 16 + 3.2x + 0.24x^2 + 0.008x^3 + 0.0001x^4$$

To find 2.1^4 put $\frac{x}{10} = 0.1$ so $x = 1$ giving:

$$2.1^4 = 16 + 3.2 + 0.24 + 0.008 + 0.0001 = 19.4481$$

THE EXPANSION OF $(1 + x)^n$

When deriving the result for $\binom{n}{r}$ we found the binomial coefficients in the form:

$$1 \qquad n \qquad \frac{n(n-1)}{2!} \qquad \frac{n(n-1)(n-2)}{3!} \qquad \frac{n(n-1)(n-2)(n-3)}{4!} \cdots$$

This form is commonly used in the expansion of expressions of the type $(1 + x)^n$.

$$(1 + x)^n = 1 + nx + \frac{n(n-1)}{1 \times 2} x^2 + \frac{n(n-1)(n-2)}{1 \times 2 \times 3} x^3 + \frac{n(n-1)(n-2)(n-3)}{1 \times 2 \times 3 \times 4} x^4 + \ldots$$

$$+ \frac{n(n-1)}{1 \times 2} x^{n-2} + nx^{n-1} + 1x^n$$

or $(1 + x)^n = 1 + nx + \frac{1}{2!}n(n-1)x^2 + \frac{1}{3!}n(n-1)(u-2)x^3 + \ldots + x^n$

noting that n is a positive integer.

EXAMPLE 8.19

Use the binomial expansion to write down the first four terms of $(1 + x)^9$.

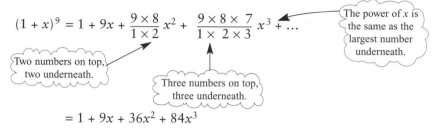

$$(1 + x)^9 = 1 + 9x + \frac{9 \times 8}{1 \times 2} x^2 + \frac{9 \times 8 \times 7}{1 \times 2 \times 3} x^3 + \dots$$

The power of x is the same as the largest number underneath.

Two numbers on top, two underneath.

Three numbers on top, three underneath.

$$= 1 + 9x + 36x^2 + 84x^3$$

EXAMPLE 8.20

Use the binomial expansion to write down the first four terms of $(1 - 3x)^7$.
Simplify the terms.

Solution   Think of $(1 - 3x)^7$ as $(1 + (-3x))^7$. Keep the brackets while you write out the terms.

$$(1 + (-3x))^7 = 1 + 7(-3x) + \frac{7 \times 6}{1 \times 2}(-3x)^2 + \frac{7 \times 6 \times 5}{1 \times 2 \times 3}(-3x)^3 + \dots$$

$$= 1 - 21x + 189x^2 - 945x^3 + \dots$$

Note how the signs alternate.

EXAMPLE 8.21

(a)   Write down the binomial expansion of $(1 - x)^4$.

(b)   Using $x = 0.03$ and the first three terms of the expansion find an approximate value for $(0.97)^4$.

(c)   Use your calculator to find the percentage error in your answer.

Solution   (a) $(1 - x)^4 = (1 + (-x))^4$

$$= 1 + 4(-x) + 6(-x)^2 + 4(-x)^3 + (-x)^4$$
$$= 1 - 4x + 6x^2 - 4x^3 + x^4$$

(b) $(0.97)^4 \approx 1 - 4(0.03) + 6(0.03)^2$
$$= 0.8854$$

(c) $(0.97)^4 = 0.885\,292\,81$

Error $= 0.8854 - 0.885\,292\,81$
$$= 0.000\,107\,19$$

Percentage error $= \dfrac{\text{error}}{\text{true value}} \times 100$

$$= \frac{0.000\,107\,19}{0.885\,292\,81} \times 100$$

$$= 0.0121\%$$

EXERCISE 8E

1 Write out the following binomial expansions.

(a) $(x + 1)^4$ (b) $(1 + x)^7$ (c) $(x + 2)^5$

(d) $(2x + 1)^6$ (e) $(2x - 3)^5$ (f) $(2x + 3y)^3$

2 Calculate the following binomial coefficients.

(a) $\binom{4}{2}$ (b) $\binom{6}{2}$ (c) $\binom{6}{3}$ (d) $\binom{6}{4}$ (e) $\binom{6}{0}$

(f) $\binom{12}{9}$ (g) $\binom{12}{3}$ (h) $\binom{15}{11}$ (i) $\binom{8}{8}$

3 Find:

(a) the coefficient of x^5 in the expansion of $(1 + x)^8$

(b) the coefficient of x^4 in the expansion of $(1 - x)^{10}$

(c) the coefficient of x^6 in the expansion of $(1 + 3x)^{12}$

(d) the coefficient of x^7 in the expansion of $(1 - 2x)^{15}$

(e) the value of the term in the expansion of $\left(x - \frac{1}{x}\right)^8$ which is independent of x.

4 (a) Write down the binomial expansion of $(1 + x)^4$.

(b) Use the first two terms of the expansion to find an approximate value for $(1.002)^4$, substituting $x = 0.002$.

(c) Find, using your calculator, the percentage error in making this approximation.

5 (a) Simplify $(1 + x)^3 - (1 - x)^3$.

(b) Show that $a^3 - b^3 = (a - b)(a^2 + ab + b^2)$.

(c) Substitute $a = 1 + x$ and $b = 1 - x$ in the result in part (b) and show that your answer is the same as that for part (a).

6 (a) Write down the binomial expansion of $(2 - x)^5$.

(b) By substituting $x = 0.01$ in the first three terms of your expansion, obtain an approximate value for 1.99^5.

(c) Use your calculator to find the percentage error in your answer.

7 Expand and simplify the following.

(a) $\left(x + \frac{1}{x}\right)^6$

(b) $\left(2x - \frac{1}{2x}\right)^4$

(c) $\left(1 + \frac{2}{x}\right)^5$

8 A sum of money, £P, is invested such that compound interest is earned at a rate of $r\%$ per year. The amount, £A, in the account n years later is given by

$$A = P\left(1 + \frac{r}{100}\right)^n.$$

   (a) Write down the first four terms in this expansion when $P = 1000$, $r = 10$, $n = 10$ and add them to get an approximate value for A.
   (b) Compare your result with what you get using your calculator for 1000×1.1^{10}.
   (c) Calculate the percentage error in using the sum of these four terms instead of the true value.

9 (a) Show that $(2 + x)^4 = 16 + 32x + 24x^2 + 8x^3 + x^4$ for all x.
   (b) Find the values of x for which $(2 + x)^4 = 16 + 16x + x^4$.

[MEI]

10 Given that $f(x) = (1 + 2x)^{12}$:
   (a) expand $f(x)$ as far as the term in x^3
   (b) use your expansion to evaluate 1.02^{12}
   (c) find the percentage error between your answer to part (b) and the answer from your calculator.

EXERCISE 8F   **Examination-style questions**

1 The fifth term of a geometric sequence is 48 and the 9th term is 768. All the terms are positive.

   (a) Find the common ratio.
   (b) Find the first term.
   (c) Find the sum of the first ten terms.

2 A pendulum is set swinging. Its first oscillation is through an angle of 30°, and each succeeding oscillation is through 95% of the angle of the one before it.

   (a) After how many swings is the angle through which it swings less than 1°?
   (b) What is the total angle it has swung through at the end of its 10th oscillation?

3 A ball is thrown vertically upwards from the ground. It rises to a height of 10 m and then falls and bounces. After each bounce it rises vertically to $\frac{2}{3}$ of the height from which it fell.

   (a) Find the height to which the ball bounces after the nth impact with the ground.
   (b) Find the total distance travelled by the ball from the first throw to the 10th impact with the ground.

4 A company offers a ten-year contract to an employee. This gives a starting salary of £15 000 a year with an annual increase of 8% of the previous year's salary.

(a) Show that the amounts of annual salary form a geometric sequence and write down its common ratio.

(b) How much does the employee expect to earn in the 10th year?

(c) Show that the total amount earned over the 10 years is nearly £217,500.

After considering the offer, the employee asks for a different scheme of payment. This has the same starting salary of £15 000 but with a fixed annual pay rise £d.

(d) Find d if the total amount paid out over 10 years is to be the same under the two schemes.

[MEI]

5 You are given that $u_n = 5\left(\frac{1}{2}\right)^n$.

(a) Find the values of u_1, u_2, u_3 and u_4.

(b) Evaluate $\sum_{n=1}^{16} 5\left(\frac{1}{2}\right)^n$.

(c) Show that $\sum_{n=1}^{\infty} 5\left(\frac{1}{2}\right)^n - \sum_{n=1}^{16} 5\left(\frac{1}{2}\right)^n = 0.000\,076$ correct to 2 significant figures.

[Edexcel]

6 A geometric series has fourth term 3 and seventh term $\frac{1}{9}$.

(a) Find the common ratio and show that the first term is 81.

(b) Find the sum to infinity.

(c) Find the number of terms required so that the difference between the sum to infinity and the sum of these terms is less than 0.0001.

7 A geometric series has non-equal terms. The sum of the first six terms is equal to nine times the sum of the first three terms. The fourth term is 48.

(a) Show that the value of the common ratio is 2.

(b) Find the first term.

(c) Find the least number of terms required for their sum to be greater than one million.

8 A competitor is running in a 25 km race. For the first 15 km, she runs at a steady rate of 12 km h^{-1}. After completing 15 km, she slows down and it is now observed that she takes 20% longer to complete each kilometre than she took to complete the previous kilometre.

(a) Find the time, in hours and minutes, the competitor takes to complete the first 16 km of the race.

The time taken to compelete the rth kilometre is u_r hours.

(b) Show that, for $16 \leqslant r \leqslant 25$, $u_r = \frac{1}{12}(1.2)^{r-15}$.

(c) Using the answer to (b) or otherwise, find the time, to the nearest minute, that she takes to complete the race.

[Edexcel]

9 The second term of a geometric series is 80 and the fifth term of the series is 5.12.

(a) Show that the common ratio of the series is 0.4.

Calculate:

(b) the first term of the series
(c) the sum to infinity of the series, giving your answer as an exact fraction
(d) the difference between the sum to infinity of the series and the sum of the first 14 terms of the series, giving your answer in the form $a \times 10^n$, where $1 \leqslant a < 10$ and n is an integer.

[Edexcel]

10 A geometric series is:

$$a + ar + ar^2 + \ldots$$

(a) Prove that the sum of the first n terms of this series is given by:

$$S_n = \frac{a(1 - r^n)}{1 - r}$$

The second and fourth terms of the series are 3 and 1.08 respectively.

Given that all the terms in the series are positive, find:

(b) the value of r and the value of a
(c) the sum to infinity of the series.

[Edexcel]

11 Expand and simplify $\left(3x - \dfrac{1}{3x}\right)^4$.

12 Given that $(2 + x)^{12} \equiv A + Bx + Cx^2 + Dx^3 + \ldots$ find the values of the integers A, B, C and D.

13 (a) Expand $(1 - 4x)^{11}$ as far as the term in x^3.
(b) Use your expansion to find an approximate value for 0.96^{11}.

14 (a) Expand $(3 + 2x)^4$ in ascending powers of x, giving each coefficient as an integer.
(b) Hence, or otherwise, write down the expansion of $(3 - 2x)^4$ in ascending powers of x.
(c) Hence, by choosing a suitable value for x show that $(3 + 2\sqrt{2})^4 + (3 - 2\sqrt{2})^4$ is an integer and state its value.

[Edexcel]

15 The coefficient of x^2 in the expansion of $\left(1 + \dfrac{x}{2}\right)^n$, where n is a positive integer, is 7.
(a) Find the value of n.
(b) Using the value of n found in (a), find the coefficient of x^4.

[Edexcel]

KEY POINTS   **1 Geometric Progressions**

A geometric progression (GP) is a series of terms in which each term can be found from the previous term by multiplying (or dividing) by a fixed number.

a = first term, r = common ratio

nth term is:

$$u_n = a\,r^{n-1}$$

Sum of the first n terms is:

$$S_n = \frac{a(1 - r^n)}{1 - r}$$

Sum to infinity is:

$$S_\infty = \frac{a}{1 - r} \qquad \text{provided} \quad -1 < r < 1$$

2 Binomial Expansions

$$(a + b)^n = a^n + \binom{n}{1}a^{n-1}b + \binom{n}{2}a^{n-2}b^2 + \ldots + \binom{n}{r}a^{n-r}b^r + \ldots + b^n$$

The coefficients are found using:

$$\binom{n}{r} = {}^nC_r = \frac{n!}{r!(n - r)!}$$

or using Pascal's triangle:

$$
\begin{array}{ccccccccc}
 & & & & 1 & & & & \\
 & & & 1 & & 1 & & & \\
 & & 1 & & 2 & & 1 & & \\
 & 1 & & 3 & & 3 & & 1 & \\
1 & & 4 & & 6 & & 4 & & 1 \\
\end{array}
$$

$$
\begin{array}{ccccccccccc}
1 & & 5 & & 10 & & 10 & & 5 & & 1 \\
\end{array}
$$

$$
1 \quad 6 \quad 15 \quad 20 \quad 15 \quad 6 \quad 1
$$

$$
1 \quad 7 \quad 21 \quad 35 \quad 35 \quad 21 \quad 7 \quad 1
$$

$$
1 \quad 8 \quad 28 \quad 56 \quad 70 \quad 56 \quad 28 \quad 8 \quad 1
$$

and for the case where n is a positive integer:

$$(1 + x)^n = 1 + nx + \frac{1}{2!}n(n - 1)x^2 + \frac{1}{3!}n(n - 1)(n - 2)x^3 + \ldots + x^n$$

Chapter nine

TRIGONOMETRY

Haste still pays haste, and leisure answers leisure;
Like doth quote like, and Measure still for Measure.

William Shakespeare

● ● ● ● ● ● ● ● ● ● ● ● ● ● ● ● ●

RIGHT-ANGLED TRIANGLES

You will recall from earlier studies of mathematics that for the right-angled triangle in figure 9.1:

$$\sin\theta = \frac{\text{opposite}}{\text{hypotenuse}}$$

$$\cos\theta = \frac{\text{adjacent}}{\text{hypotenuse}}$$

$$\tan\theta = \frac{\text{opposite}}{\text{adjacent}}$$

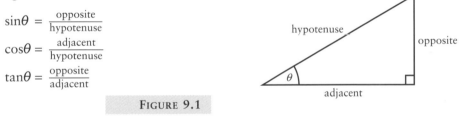

FIGURE 9.1

and using these trigonometric ratios you can calculate angles and sides.
If you cannot remember how to solve right-angled triangle problems you should revise them, referring to a GCSE text.

NON-RIGHT ANGLED TRIANGLES

You now need to be able to solve problems related to triangles which do not have a right angle. The rules above do not apply in the same way. You need to learn some new rules: the *sine rule*, the *cosine rule* and an equation for the *area of the triangle*.

THE SINE RULE

Look at the triangle ABC in figure 9.2.

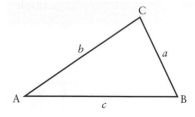

FIGURE 9.2

Notice that the sides opposite each angle are labelled with the respective lower case letter (i.e. the side a is the one opposite the angle A).
The sine rule, which will be proved later, is:

$$\frac{a}{\sin A} = \frac{b}{\sin B} = \frac{c}{\sin C}$$

This is effectively three different equations. You just choose the pair you need to use.

EXAMPLE 9.1

Find the length of AC in this triangle.

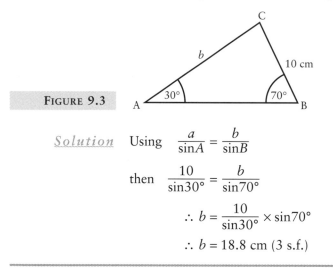

FIGURE 9.3

Solution Using $\dfrac{a}{\sin A} = \dfrac{b}{\sin B}$

then $\dfrac{10}{\sin 30°} = \dfrac{b}{\sin 70°}$

$\therefore b = \dfrac{10}{\sin 30°} \times \sin 70°$

$\therefore b = 18.8$ cm (3 s.f.)

EXAMPLE 9.2

Find the size of angle XYZ in figure 9.4.

Solution Using $\dfrac{y}{\sin Y} = \dfrac{z}{\sin Z}$

$\dfrac{7}{\sin Y} = \dfrac{10}{\sin 85°}$

$\sin Y = \dfrac{7 \sin 85°}{10}$

$\therefore Y = 44.2°$ (3 s.f.)

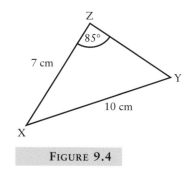

FIGURE 9.4

In example 9.2 the only possible value for angle XYZ was 44.2°. However, in some triangles it is possible to make two constructions satisfying the given data. Consider a triangle ABC where AB is 10 cm, BC is 8 cm and the angle BAC is 40°.

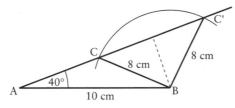

FIGURE 9.5

If an arc of a circle of radius 8 cm is drawn from B, then it cuts the line at 40° to AB in two places: C and C'. So there are two possible solutions to the problem. This is known as the *ambiguous case*.

....

Note   BC and BC' are symmetrical about the line through B perpendicular to the line AC.

....

EXAMPLE 9.3

Sketch two possible triangles XYZ in which XY is 12 cm, YZ is 8 cm and the angle YXZ is 35°. Find the size of angle XZY in each case.

Solution

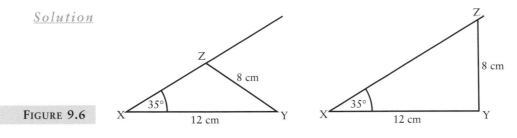

FIGURE 9.6

Using $\dfrac{x}{\sin X} = \dfrac{z}{\sin Z}$

$$\dfrac{8}{\sin 35°} = \dfrac{12}{\sin Z}$$

$$\sin Z = \dfrac{12\sin 35°}{8}$$

$$\therefore Z = 59.4°$$

This is the answer for the second diagram in figure 9.6.

To find the answer for the first diagram, use the fact that the two possibilities of YZ are symmetrical about the perpendicular to XZ through Y. This means that in the first diagram:

$$\angle XZY = 180° - 59.4°$$
$$= 120.6°$$

In general, for the *ambiguous case*, if one possible solution is θ then the other is $180° - \theta$.

You will learn later in this chapter that equations such as $\sin\theta = 0.5$ have, in general, more than one solution. This shows a different way of looking at the ambiguous case.

PROOF OF THE SINE RULE

Look at the triangle ABC in figure 9.7.

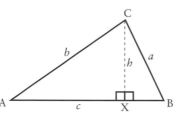

FIGURE 9.7

A line CX has been drawn perpendicular to AB. The length CX is labelled h (the height of the triangle).

From triangle AXC: $\sin A = \dfrac{h}{b}$ $\therefore h = b \sin A$

and from triangle BXC: $\sin B = \dfrac{h}{a}$ $\therefore h = a \sin B$

hence: $b \sin A = a \sin B$

which gives: $\dfrac{a}{\sin A} = \dfrac{b}{\sin B}$

The rule can then be extended, using a different height, to give the full sine rule.

$$\frac{a}{\sin A} = \frac{b}{\sin B} = \frac{c}{\sin C}$$

Note   Use the sine rule when you are working with two sides and two angles of a triangle.

Historical note   The word *sine* has a curious derivation. When the Arabic texts were being translated into Latin the term for what we now call *sine* was *jiba*. However, this was misread as *jaib* which was a word for breast and this was translated to the Latin *sinus* from which our current word *sine* has come.

EXERCISE 9A

1 Use the sine rule to find the lengths of the marked sides, giving answers correct to three significant figures.

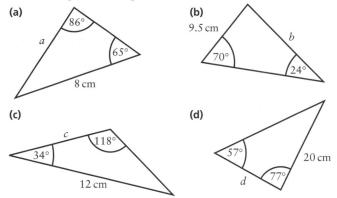

(a)

(b)

(c)

(d)

2 Use the sine rule to find the angles labelled θ, giving your answers correct to one decimal place.

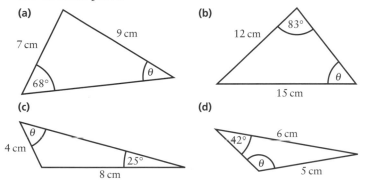

(a)

(b)

(c)

(d)

3 A yacht leaves a harbour H. It travels on a bearing of 070° for 7 km. It then travels on a bearing of 235° until its bearing to H is 320°.
   (a) At this point, how far is it from the harbour?
   (b) How long was the second leg of the journey?

4 A triangle PQR has side PQ of length x and side PR of length $x + 3$. The angle PQR is 62° and the angle PRQ is 47°.
   Find the value of x and the length of QR.

5 A piano lid is 1.2 metres wide. It is held open by a wooden stay of length 0.6 metre. The stay is hinged at a point 1.0 metres from the hinge on the lid, as shown in the diagram.

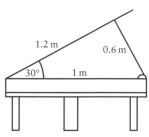

   The lid is to be held open by the stay fixing into slots on the lid. The lid is to be opened to an angle of 30°.
   (a) Find the two angles that the stay could make with the piano.
   (b) Find the distance along the lid of the two slots measured from the hinge on the lid.

COSINE RULE

Look again at the triangle ABC.
The cosine rule states that:

$$a^2 = b^2 + c^2 - 2bc\cos A$$

FIGURE 9.8

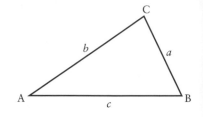

Note

Use the cosine rule when you are working with three sides and one angle of the triangle.

EXAMPLE 9.4

Find the length of BC in this triangle.

Solution   Let the length BC = a, then, using $a^2 = b^2 + c^2 - 2bc\cos A$:

$$a^2 = 10^2 + 7^2 - 2 \times 10 \times 7 \times \cos 35°$$

$$a^2 = 34.32$$

$$a = 5.86 \text{ cm (3 s.f.)}$$

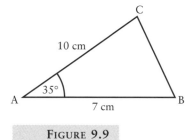

FIGURE 9.9

EXAMPLE 9.5

Find the size of angle XYZ.

FIGURE 9.10

Solution   Think of angle XYZ as angle A and therefore XZ as side a, hence:

$$a^2 = b^2 + c^2 - 2bc\cos A$$

becomes:

$$8^2 = 9^2 + 5^2 - 2 \times 9 \times 5 \times \cos A$$
$$64 = 81 + 25 - 90\cos A$$
$$64 = 106 - 90\cos A$$
$$90\cos A = 106 - 64$$
$$\cos A = \frac{42}{90}$$
$$\therefore A = 62.2° \text{ (3 s.f.)}$$

Take care!
You cannot subtract the 90 from 106.

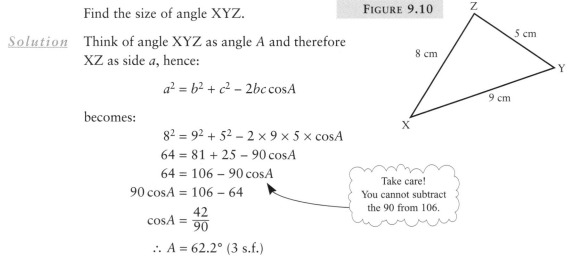

Note

You could have rearranged the cosine rule to give:

$$\cos A = \frac{b^2 + c^2 - a^2}{2bc}$$

You may wish to remember this and use it.

As it did not matter which corner was labelled A, B or C then you could have:

$$b^2 = a^2 + c^2 - 2ac\cos B$$
$$c^2 = a^2 + b^2 - 2ab\cos C$$

and:

$$\cos B = \frac{a^2 + c^2 - b^2}{2ac}$$

or:

$$\cos C = \frac{a^2 + b^2 - c^2}{2ab}$$

If you have a choice of using the sine rule or the cosine rule it is usually better to use the cosine rule as it avoids the ambiguous case.

EXERCISE 9B

1 Use the cosine rule to find the lengths of the sides labelled a.

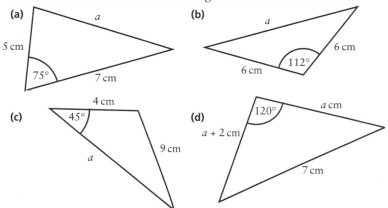

2 Use the cosine rule to find the angles labelled θ, correct to one decimal place.

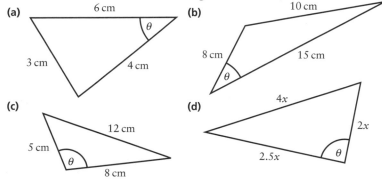

3 On an orienteering exercise Hugh starts at a point P. He runs 1500 metres in a straight line on a bearing of 230° to a point Q. He then turns to his left and runs 2000 metres to a point R which is 1800 metres from P.
   (a) What is the bearing of R from P?
   (b) Through what angle does Hugh turn at Q?

4 In triangle XYZ the length of XY is 8 cm. The length of YZ is 12 cm and the length of XZ is 16 cm. Find the sizes of the three angles of the triangle.

5 In triangle ABC the length of AB is 8 cm, the length of AC is 5 cm. The angle BAC is 60°.
   (a) Find the length of BC.
   (b) Show that $\cos C = \frac{1}{7}$.

AREA OF A NON-RIGHT ANGLED TRIANGLE

The area of triangle ABC in figure 9.11 is given by:

$$\Delta = \tfrac{1}{2}ab \sin C$$

As it does not matter how the triangle is labelled, you could have:

$$\Delta = \tfrac{1}{2}bc \sin A$$

$$\text{or } \Delta = \tfrac{1}{2}ca \sin B$$

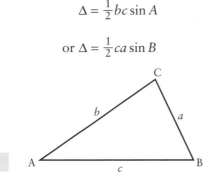

FIGURE 9.11

The best way to remember the equation is by saying:
area = half × product of any two sides × sine of the included angle.

EXAMPLE 9.6

Find the area of the triangle in figure 9.12.

Solution   Using $\Delta = \tfrac{1}{2}ab \sin C$

$$\Delta = \tfrac{1}{2} \times 10 \times 12 \times \sin 40°$$

$$= 38.6 \text{ cm}^2 \text{ (3 s.f.)}$$

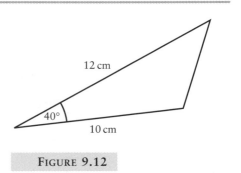

FIGURE 9.12

PROOF OF THE AREA OF A NON-RIGHT ANGLED TRIANGLE

Construct a perpendicular to BC through the corner A.
This gives a line which is the height h of the triangle.

$$\Delta = \frac{1}{2}ah$$

But the height h can be calculated using $h = b \sin C$.

Hence: $\boxed{\Delta = \frac{1}{2}ab \sin C}$

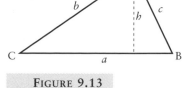

FIGURE 9.13

EXERCISE 9C

1 Find the area of each of these triangles.

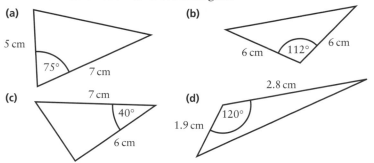

(a) 5 cm, 75°, 7 cm

(b) 6 cm, 112°, 6 cm

(c) 7 cm, 40°, 6 cm

(d) 2.8 cm, 120°, 1.9 cm

2 Find the area of each of of these triangles. Use the cosine rule or the sine rule if you need to.

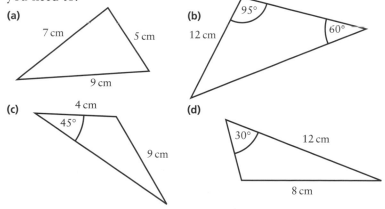

(a) 7 cm, 5 cm, 9 cm

(b) 12 cm, 95°, 60°

(c) 4 cm, 45°, 9 cm

(d) 30°, 12 cm, 8 cm

3 A triangle has vertices $(-2, -3)$, $(10, 2)$ and $(7, 6)$.
  (a) Find the length of each side of the triangle.
  (b) Find the angles of the triangle.
  (c) Find the area of the triangle.

4 Two sides of a triangle have lengths x cm and $(x + 2)$ cm with an included angle of 150°. Given that the area of the triangle is 12 cm² find:
  (a) the value of x
  (b) the lengths of the sides of the triangle
  (c) the other two angles of the triangle.

5 Find the area of a triangle with sides of lengths $\sqrt{3}$ cm, $\sqrt{5}$ cm and $\sqrt{7}$ cm.

CIRCULAR MEASURE

Have you ever wondered why angles are measured in degrees, and why there are 360° in one revolution?

There are various legends to support the choice of 360, most of them based in astronomy. One of these is that since the shepherd astronomers of Sumeria thought that the solar year was 360 days long, this number was then used by the ancient Babylonian mathematicians to divide one revolution into 360 equal parts.

Degrees are not the only way in which you can measure angles. Some calculators have modes called 'rad' and 'gra' (or 'grad'); if yours is one of these, you have probably noticed that these give different answers when you are using the sin, cos or tan keys. These answers are only wrong when the calculator mode is different from the units being used in the calculation.

The *grade* (mode 'gra') is a unit which was introduced to give a means of angle measurement which was compatible with the metric system. There are 100 grades in a right angle, so when you are in the grade mode, sin 100 = 1, just as when you are in the degree mode, sin 90 = 1. Grades are largely of historical interest and are only mentioned here to remove any mystery surrounding this calculator mode.

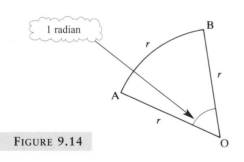

FIGURE 9.14

By contrast, radians are used extensively in mathematics because they simplify many calculations. The *radian* (mode 'rad') is sometimes referred to as the natural unit of angular measure.

If, as in figure 9.14, the arc AB of a circle centre O is drawn so that it is equal in length to the radius of the circle, then the angle AOB is 1 radian, about 57.3°.

You will sometimes see 1 radian written as 1ᶜ, just as 1 degree is written 1°.

Since the circumference of a circle is given by $2\pi r$, it follows that the angle of a complete turn is 2π radians.

$$360° \equiv 2\pi \text{ radians}$$

Consequently:

$$180° \equiv \pi \text{ radians}$$

$$\text{and: } 90° \equiv \frac{\pi}{2} \text{ radians}$$

$$60° \equiv \frac{\pi}{3} \text{ radians}$$

$$45° \equiv \frac{\pi}{4} \text{ radians}$$

$$30° \equiv \frac{\pi}{6} \text{ radians etc.}$$

To convert degrees into radians you multiply by $\frac{\pi}{180}$.

To convert radians into degrees you multiply by $\frac{180}{\pi}$.

Note

1   If an angle is a simple fraction or multiple of 180° and you wish to give its value in radians, it is usual to leave the answer as a fraction of π.

2   When an angle is given as a multiple of π it is assumed to be in radians.

EXAMPLE 9.7

1   Express these in radians.

(a) 30°   (b) 315°   (c) 29°

2   Express these angles given in radians in degrees.

(a) $\frac{\pi}{12}$ (b) $\frac{8\pi}{3}$ (c) 1.2

Solution   1   (a) $30° \equiv 30 \times \frac{\pi}{180} = \frac{\pi}{6}$

(b) $315° \equiv 315 \times \frac{\pi}{180} = \frac{7\pi}{4}$

(c) $29° \equiv 29 \times \frac{\pi}{180} = 0.506$ radians (to 3 s.f.)

2   (a) $\frac{\pi}{12} \equiv \frac{\pi}{12} \times \frac{180°}{\pi} = 15°$

(b) $\frac{8\pi}{3} \equiv \frac{8\pi}{3} \times \frac{180°}{\pi} = 480°$

(c)   1.2 radians $\equiv 1.2 \times \frac{180°}{\pi} = 68.8°$ (to 3 s.f.)

USING YOUR CALCULATOR IN RADIAN MODE

If you wish to find the value of, say, $\sin 1.4^c$ or $\cos \frac{\pi}{12}$, use the 'RAD' mode on your calculator. This will give the answers directly – in these examples 0.9854... and 0.9659...

You could alternatively convert the angles into degrees (by multiplying by $\frac{180}{\pi}$) but this would usually be a clumsy method. It is much better to get into the habit of working in radians.

EXERCISE 9D   **1** Express these angles in radians, leaving your answers in terms of π
where appropriate.

    (a) 45°   **(b)** 90°   **(c)** 120°   **(d)** 75°   **(e)** 300°

    (f) 23°   **(g)** 450°   **(h)** 209°   **(i)** 150°   **(j)** 7.2°

2 Express these angles in degrees, using a suitable approximation where necessary.

    (a) $\frac{\pi}{10}$ **(b)** $\frac{3\pi}{5}$ **(c)** 2 radians   **(d)** $\frac{4\pi}{9}$ **(e)** 3π

    (f) $\frac{5\pi}{3}$ **(g)** 0.4 radians   **(h)** $\frac{3\pi}{4}$ **(i)** $\frac{7\pi}{3}$ **(j)** $\frac{3\pi}{7}$

THE LENGTH (s) OF AN ARC OF A CIRCLE

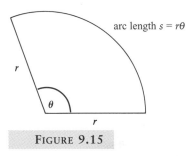

arc length $s = r\theta$

From the definition of a radian, an angle of 1 radian at the centre of a
circle corresponds to an arc of length r (the radius of the circle).
Similarly, an angle of 2 radians corresponds to an arc length of $2r$ and,
in general, an angle of θ radians corresponds to an arc length of θr,
which is usually written $r\theta$ (figure 9.15).

Defining the arc length as s then

FIGURE 9.15

$s = r\theta$, θ in radians.

THE AREA (A) OF A SECTOR OF A CIRCLE

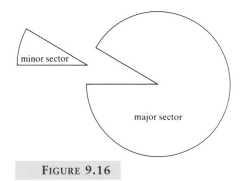

minor sector

major sector

A *sector* of a circle is the shape enclosed by an arc of the circle
and two radii. It is the shape of a piece of cake. If the sector is
smaller than a semicircle it is called a *minor sector*; if it is larger
than a semicircle it is a *major sector*, see figure 9.16.

The area of a sector is a fraction of the area of the whole circle.
The fraction is found by writing the angle θ as a fraction of one
revolution, i.e. 2π (figure 9.17).

FIGURE 9.16

Area

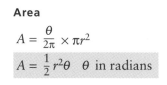

$$A = \frac{\theta}{2\pi} \times \pi r^2$$

$$A = \frac{1}{2} r^2 \theta \quad \theta \text{ in radians}$$

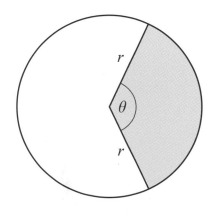

FIGURE 9.17

EXAMPLE 9.8

Calculate the arc length, perimeter, and area of a sector of angle $\frac{2\pi}{3}$ and radius 6 cm.

Solution

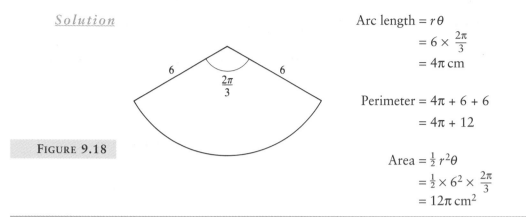

FIGURE 9.18

$$\text{Arc length} = r\theta$$
$$= 6 \times \frac{2\pi}{3}$$
$$= 4\pi \text{ cm}$$

$$\text{Perimeter} = 4\pi + 6 + 6$$
$$= 4\pi + 12$$

$$\text{Area} = \tfrac{1}{2}r^2\theta$$
$$= \tfrac{1}{2} \times 6^2 \times \frac{2\pi}{3}$$
$$= 12\pi \text{ cm}^2$$

EXAMPLE 4.9

Calculate the area of the shaded segment of angle $\frac{\pi}{3}$ and radius 4 cm.

Solution The area of the sector is:

$$\tfrac{1}{2}r^2\theta = \tfrac{1}{2} \times 4^2 \times \frac{\pi}{3} = \frac{8\pi}{3} = 8.3776$$

The area of the triangle OAB is:

$$\tfrac{1}{2}r^2\sin\theta = \tfrac{1}{2} \times 4^2 \times \sin\frac{\pi}{3} = 6.9282$$

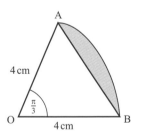

FIGURE 4.19

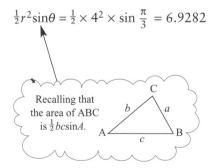

Recalling that the area of ABC is $\tfrac{1}{2}bc\sin A$.

The area of the shaded segment is:

$$8.3776 - 6.9282 = 1.45 \text{ cm}^2 \text{ (3 significant figures)}.$$

Note   It is useful to remember that the area of a triangle with two sides of length r and included angle θ is $\tfrac{1}{2}r^2\sin\theta$.

1 Each row of the table gives dimensions of a sector of a circle of radius r cm. The angle subtended at the centre of the circle is θ radians, the arc length of the sector is s cm and its area is A cm².

Copy and complete the table.

r (cm)	θ (rad)	s (cm)	A (cm²)
5	$\frac{\pi}{4}$		
8	1		
4		2	
	$\frac{\pi}{3}$	$\frac{\pi}{2}$	
5			10
	0.8	1.5	
	$\frac{2\pi}{3}$		4π

2 (a) (i) Find the area of the sector OAB in the diagram.

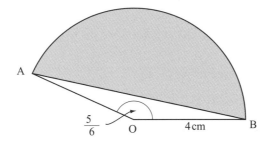

(ii) Find the area of triangle OAB.

(iii) Find the shaded area. (Note: this is called a *segment* of the circle.)

(b) The diagram below shows two circles, each of radius 4 cm, with each one passing through the centre of the other. Calculate the shaded area. (Hint: Add the common chord AB to the sketch.)

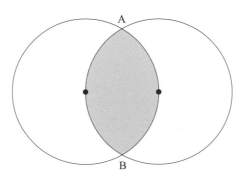

3 Calculate the area and the perimeter of the sector with radius 4 cm and angle 60° at the centre.

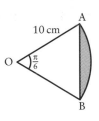

4 The diagram shows a sector of a circle AOB of radius 10 cm. The angle AOB is $\frac{\pi}{6}$ radians.

Calculate the exact value of the area of the shaded region of the sector, as shown in the diagram.

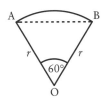

5 Find an exact value for the fraction of the sector represented by the triangle AOB in the sector AOB in the diagram.

6 The area of the sector AOB and the triangle AOB are in the ratio 3:2. The angle AOB is θ radians.

Show that $2\theta - 3\sin\theta = 0$.

By using trial and improvement, find, correct to 1 decimal place, the value of θ, other than zero, that satisfies this equation. Remember to set your calculator to radians.

7 The area of the sector of a circle of radius 12 cm is 48π cm². Find the perimeter of the sector.

8 Two circles of radius 6 cm overlap, with their centres 6 cm apart. Find the area of the overlap.

9 The diagram shows a circle of radius 5 cm overlapping a circle of radius 12 cm. The centres of the circles are 13 cm apart. Find the area of the shaded region.

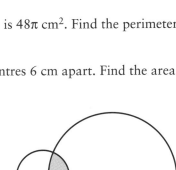

10 A straight tunnel is to be cut through a mountain. The length of the tunnel is 600 metres. The cross-section of the tunnel is part of a circle of radius 6.5 metres. The base chord has length 12 metres. The cross-section is shown in the diagram.

Calculate the volume of rock removed for the tunnel.

TRIGONOMETRICAL FUNCTIONS

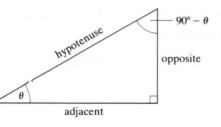

FIGURE 9.20

The simplest definitions of the trigonometrical functions are given in terms of the ratios of the sides of a right-angled triangle, for values of θ between 0 and 90°.

$$\sin\theta = \frac{\text{opposite}}{\text{hypotenuse}} \qquad \cos\theta = \frac{\text{adjacent}}{\text{hypotenuse}} \qquad \tan\theta = \frac{\text{opposite}}{\text{adjacent}}$$

You will see from the triangle in figure 9.20 that:

$$\sin\theta = \cos(90° - \theta) \text{ and } \cos\theta = \sin(90° - \theta)$$

SPECIAL CASES

Certain angles occur frequently in mathematics and you will find it helpful to know the value of their trigonometrical functions.

(a) The angles 30° and 60° ($\frac{\pi}{6}$ and $\frac{\pi}{3}$ radians)

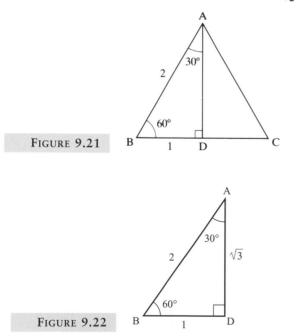

FIGURE 9.21

In figure 9.21, triangle ABC is an equilateral triangle with side 2 units, and AD is a line of symmetry.

Using Pythagoras' theorem:

$$AD^2 + 1^2 = 2^2 \Rightarrow AD = \sqrt{3}$$

From triangle ABD:

$$\sin 60° = \frac{\sqrt{3}}{2} \qquad \cos 60° = \frac{1}{2} \qquad \tan 60° = \sqrt{3}$$

$$\sin 30° = \frac{1}{2} \qquad \cos 30° = \frac{\sqrt{3}}{2} \qquad \tan 30° = \frac{1}{\sqrt{3}}$$

FIGURE 9.22

EXAMPLE 9.10

Without using a calculator, find the value of $\cos 60° \sin 30° + \cos^2 30°$.

(Note that $\cos^2 30°$ means $(\cos 30°)^2$.)

Solution

$$\cos 60° \sin 30° + \cos^2 30° = \frac{1}{2} \times \frac{1}{2} + \left(\frac{\sqrt{3}}{2}\right)^2$$

$$= \frac{1}{4} + \frac{3}{4}$$

$$= 1$$

You will recall that $180°$ is equivalent to π radians and that:

$$60° \equiv \frac{\pi}{3} \text{ radians}, \quad 30° \equiv \frac{\pi}{6} \text{ radians}$$

This gives the useful equations:

$$\sin \frac{\pi}{3} = \frac{\sqrt{3}}{2} \qquad \cos \frac{\pi}{3} = \frac{1}{2} \qquad \tan \frac{\pi}{3} = \sqrt{3}$$

$$\sin \frac{\pi}{6} = \frac{1}{2} \qquad \cos \frac{\pi}{6} = \frac{\sqrt{3}}{2} \qquad \tan \frac{\pi}{6} = \frac{1}{\sqrt{3}}$$

EXAMPLE 9.11

Give the exact value of $\tan \frac{\pi}{6} + \tan \frac{\pi}{3}$ expressing your answer as a single fraction.

Solution

$$\tan \frac{\pi}{6} + \tan \frac{\pi}{3} = \frac{1}{\sqrt{3}} + \sqrt{3}$$

$$= \frac{\sqrt{3}}{3} + \sqrt{3}$$

$$= \frac{\sqrt{3}}{3} + \frac{3\sqrt{3}}{3}$$

$$= \frac{4\sqrt{3}}{3}$$

(b) The angle 45° ($\frac{\pi}{4}$ radians)

In figure 9.23, triangle PQR is a right-angled isosceles triangle with equal sides of length 1 unit.

Using Pythagoras' theorem $PQ = \sqrt{2}$.

This gives:

$$\sin 45° = \frac{1}{\sqrt{2}} \qquad \cos 45° = \frac{1}{\sqrt{2}} \qquad \tan 45° = 1$$

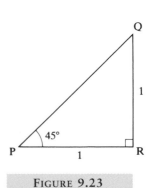

FIGURE 9.23

And in radians:

$$\sin\frac{\pi}{4} = \frac{1}{\sqrt{2}} \qquad \cos\frac{\pi}{4} = \frac{1}{\sqrt{2}} \qquad \tan\frac{\pi}{4} = 1$$

(c) The angles 0° and 90° (0 and $\frac{\pi}{2}$ radians)

Although you cannot have an angle of 0° in a triangle (because one side would be lying on top of another), you can still imagine what it might look like. In figure 9.24, the hypotenuse has length 1 unit and the angle at X is very small.

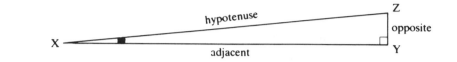

FIGURE 9.24

If you imagine the angle at X becoming smaller and smaller until it is 0, you can deduce that:

$$\sin 0° = \frac{0}{1} = 0 \qquad \cos 0° = \frac{1}{1} = 1 \qquad \tan 0° = \frac{0}{1} = 0$$

If the angle at X is 0°, then the angle at Z is 90°, and so you can also deduce that:

$$\sin 90° = \frac{1}{1} = 1 \qquad \cos 90° = \frac{0}{1} = 0$$

However, when you come to find $\tan 90°$, there is a problem. The triangle suggests this has value $\frac{1}{0}$, but you cannot divide by 0.

If you look at the triangle XYZ, you will see that what we actually did was to draw it with angle X not 0 but just very small, and to argue:

'We can see from this what will happen if the angle becomes smaller and smaller so that it is effectively 0.'

You saw this style of argument in Chapter 4 of this book where you meet differentiation. In this case we are looking at the *limits* of the values of $\sin\theta$, $\cos\theta$ and $\tan\theta$ as the angle θ approaches 0. The same approach can be used to look again at the problem of $\tan 90°$.

If the angle X is not quite 0, then the side ZY is also not quite 0, and $\tan Z$ is 1 (XY is almost 1) divided by a very small number and so is large. The smaller the angle X, the smaller the side ZY and so the larger the value of tan Z. We conclude that in the limit when angle X becomes 0 and angle Z becomes 90°, $\tan Z$ is infinitely large, and so we say:

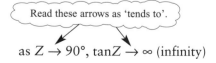

Read these arrows as 'tends to'.

as $Z \rightarrow 90°$, $\tan Z \rightarrow \infty$ (infinity)

You can see this happening in the table of values below.

Z	$\tan Z$
80°	5.67
89°	57.29
89.9°	572.96
89.99°	5729.6
89.999°	57296

When Z actually equals 90°, we say that $\tan Z$ is *undefined*.

EXERCISE 9F

1   Without using a calculator, find the exact value of the following, simplifying each answer.

(a) $\sin 30° + \sin 60°$

(b) $\cos\dfrac{\pi}{4} + \cos\dfrac{\pi}{6}$

(c) $\tan 45° + \tan 60°$

2   Without using a calculator, find the exact value of each of the following, rationalising your answers.

(a) $\dfrac{\tan 60° - \tan 45°}{1 + \tan 60° \tan 45°}$

(b) $\dfrac{\tan 45° - \tan 30°}{1 + \tan 45° \tan 30°}$

Comment on your answers.

3   Work out the value of this expression.

$$\sin\frac{\pi}{6}\,\cos\frac{\pi}{3} + \cos\frac{\pi}{6}\,\sin\frac{\pi}{3}$$

4   Show that $\sin^2 45° + \cos^2 45° = 1$.

5   Without using a calculator prove each of these relations.

(a) $2\sin\dfrac{\pi}{6}\,\cos\dfrac{\pi}{6} = \sin\dfrac{\pi}{3}$

(b) $\cos^2 45° - \sin^2 45° = \cos 90°$

(c) $\cos 45°\cos 30° + \sin 45°\sin 30° = \sin 45°\cos 30° + \cos 45°\sin 30°$

(d) $\sin 60°\cos 30° + \cos 60°\sin 30° = 1$

(e) $\sin^2 30° + \sin^2 45° = \sin^2 60°$

(f) $3\sin^2 30° = \cos^2 30°$

TRIGONOMETRICAL FUNCTIONS FOR ANGLES OF ANY SIZE

Is it possible to extend the use of the trigonometrical functions to angles greater than 90°, such as $\sin 120°$, $\cos 275°$ or $\tan 692°$? The answer is yes – provided we change the definition of sine, cosine and tangent to one that does not require the angle to be in a right-angled triangle. It is not difficult to extend the definitions, as follows.

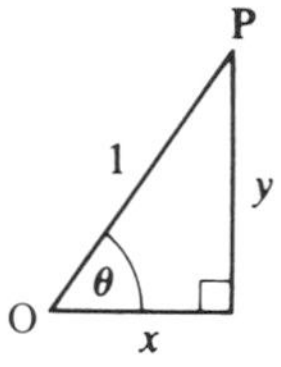

First look at the right-angled triangle in figure 9.25 which has hypotenuse of unit length.

This gives rise to the definitions:

FIGURE 9.25

$$\sin\theta = \frac{y}{1} = y \qquad \cos\theta = \frac{x}{1} = x \qquad \tan\theta = \frac{y}{x}$$

Now think of the angle θ being situated at the origin, as in figure 9.26, and allow θ to take any value. The vertex marked P has coordinates (x, y) and can now be anywhere on the unit circle.

You can now see that the definitions above can be applied to *any* angle θ, whether it is positive or negative, and whether it is less than or greater than 90°.

$$\sin\theta = y, \quad \cos\theta = x, \quad \tan\theta = \frac{y}{x}$$

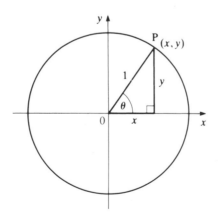

FIGURE 9.26

For some angles, x or y (or both) will take a negative value, so the sign of $\sin\theta$, $\cos\theta$ and $\tan\theta$ will vary accordingly.

THE SINE AND COSINE GRAPHS

In figure 9.27, angles have been drawn at intervals of 30° in the unit circle, and the resulting y-coordinates have been plotted relative to the axes on the right. They have been joined with a continuous curve to give the graph of $\sin \theta$ for $0° \leqslant \theta \leqslant 360°$.

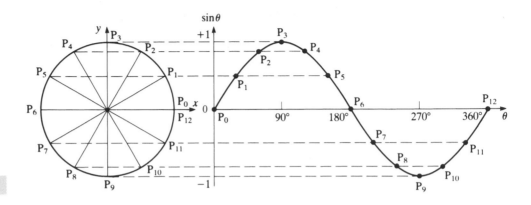

FIGURE 9.27

The angle 390° gives the same point P_1 on the circle as the angle 30°, the angle 420° gives point P_2 and so on. You can see that for angles from 360° to 720° the sine wave will simply repeat itself, as shown in figure 9.28. This is true also for angles from 720° to 1080° and so on.

Since the curve repeats itself every 360° the sine function is described as *periodic*, with *period* 360°.

The amplitude of the sine wave is 1.

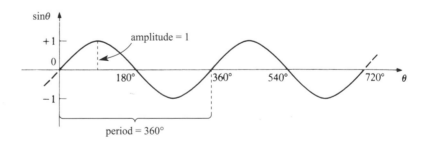

FIGURE 9.28

In a similar way you can transfer the *x*-coordinates onto a set of axes to obtain the graph of cosθ. This is most easily illustrated if you first rotate the circle through 90° anticlockwise.

Figure 9.29 shows the circle in this new orientation, together with the resulting graph.

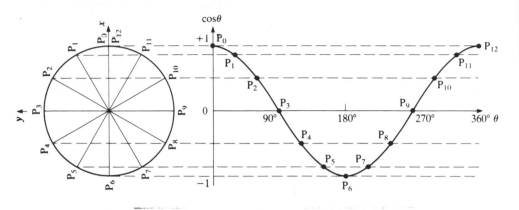

FIGURE 9.29

For angles in the interval $360° < \theta < 720°$, the cosine curve will repeat itself. You can see that the cosine function is also periodic with a period of 360°.

Notice that the graphs of sinθ and cosθ have exactly the same shape. The cosine graph can be obtained by translating the sine graph 90° to the left, as shown in figure 9.30.

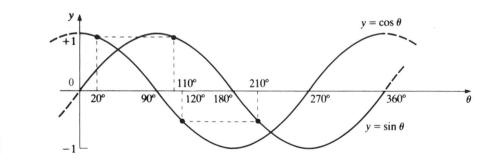

FIGURE 9.30

From the graphs it can be seen that, for example:

$$\cos 20° = \sin 110°, \quad \cos 90° = \sin 180°, \quad \cos 120° = \sin 210°$$

In general:

$$\cos\theta \equiv \sin(\theta + 90°)$$

THE TANGENT GRAPH

The value of tan θ can be worked out from the definition $\tan\theta = \frac{y}{x}$ or by using $\tan\theta = \frac{\sin\theta}{\cos\theta}$ (see page 241).

You have already seen that $\tan\theta$ is undefined for $\theta = 90°$. This is also the case for all other values of θ for which $\cos\theta = 0$, namely 270°, 450°, ..., and −90°, −270°, ...

The graph of $\tan\theta$ is shown in figure 9.31. The dotted lines $\theta = \pm90°$ and $\theta = 270°$ are *asymptotes*. They are not actually part of the curve. Its branches get closer and closer to them without ever quite reaching them.

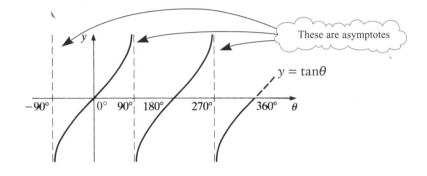

These are asymptotes

$y = \tan\theta$

FIGURE 9.31

Note

The graph of $\tan\theta$ is periodic, like those for $\sin\theta$ and $\cos\theta$, but in this case the period is 180°. Again, the curve for $0° \leqslant \theta < 90°$ can be used to generate the rest of the curve using rotations and translations.

ACTIVITY

Draw the graphs of $y = \sin\theta$, $y = \cos\theta$, and $y = \tan\theta$ for values of θ between −360° and 360°.

These graphs are very important. Keep them handy because they will be useful for solving trigonometrical equations.

Note

You may wish to revise transformations in Chapter 1 of *Pure Mathematics: Core 1* before moving on to the next section.

TRANSFORMATIONS OF TRIGONOMETRIC WAVES

y = a sinx

This is a stretch of $y = \sin x$ parallel to the y-axis, with the x-axis invariant and with a scale factor a.

The graph is as shown in figure 9.32. Its amplitude is a.

y = sin(nx)

This has n waves in 360°. The period of each wave is $\frac{360°}{n}$. For example, $y = \sin 2x$ has 2 waves in 360°, the period of each wave being 180° and having an amplitude of 1 (see figure 9.33).

y = sin(x − a)

This is a translation of $y = \sin x$ by the vector $\begin{pmatrix} a \\ 0 \end{pmatrix}$. In other words, the graph is moved a to the right. For example, the graph of $y = \sin(x − 30°)$ is shown in figure 9.34.

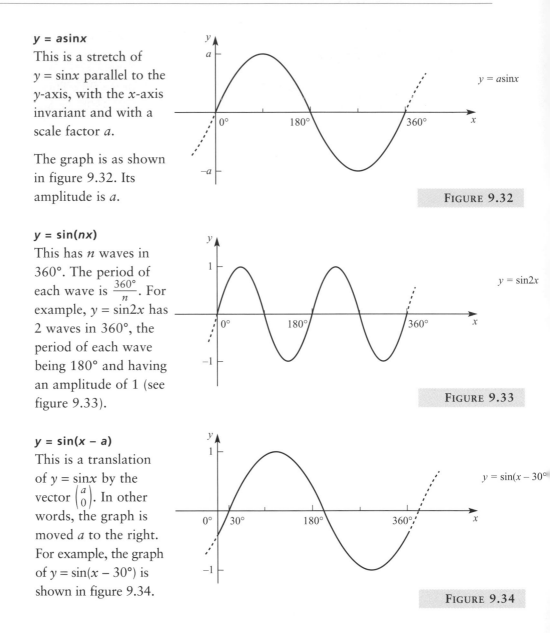

FIGURE 9.32

FIGURE 9.33

FIGURE 9.34

The above methods may be combined and may be applied to transformations of the cosine and tangent graphs.

EXAMPLE 9.12

Sketch, for $0 \leqslant x \leqslant 360°$, the graph of $y = 3\cos2x$.

Solution $y = 3\cos2x$

Amplitude = 3

Two waves in 360°

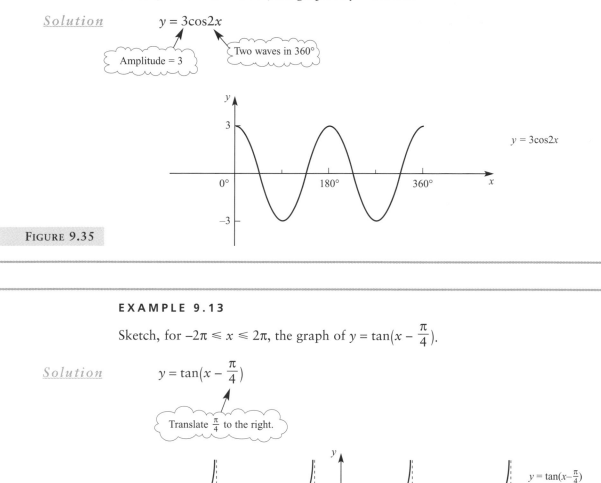

$y = 3\cos2x$

FIGURE 9.35

EXAMPLE 9.13

Sketch, for $-2\pi \leqslant x \leqslant 2\pi$, the graph of $y = \tan\left(x - \dfrac{\pi}{4}\right)$.

Solution $y = \tan\left(x - \dfrac{\pi}{4}\right)$

Translate $\dfrac{\pi}{4}$ to the right.

$y = \tan\left(x - \dfrac{\pi}{4}\right)$

FIGURE 9.36

EXERCISE 9G

1 On separate diagrams sketch the graphs of the following functions over the specified domains.

(a) $y = 3\sin x$ for $0° \leqslant x \leqslant 360°$

(b) $y = \cos\frac{1}{2}x$ for $-360° \leqslant x \leqslant 360°$

(c) $y = \tan 2x$ for $-180° \leqslant x \leqslant 180°$

(d) $y = 2\sin 3x$ for $-\pi \leqslant x \leqslant \pi$

(e) $y = 0.5\cos 2x$ for $-\frac{\pi}{2} \leqslant x \leqslant \frac{3\pi}{2}$

(f) $y = \tan(\frac{1}{2}x)$ for $-\pi \leqslant x \leqslant \pi$

(g) $y = 2\sin(x - 60°)$ for $0° \leqslant x \leqslant 720°$

(h) $y = 3\cos(x + 45°)$ for $0° \leqslant x \leqslant 360°$

(i) $y = \tan\left(x + \frac{\pi}{2}\right)$ for $0, \leqslant x \leqslant 2\pi$.

2 On the same axes, for $0° \leqslant x \leqslant 360°$, sketch the graphs of $y = \sin x$ and $y = \cos x$. Where do the graphs intersect?

3 (a) Draw a sketch of the graph $y = \sin x$ and use it to demonstrate why
$$\sin x = \sin(180° - x).$$

(b) By referring to the graphs of $y = \cos x$ and $y = \tan x$, state whether the following are true or false.

(i) $\cos x = \cos(180° - x)$ (ii) $\cos x = -\cos(180° - x)$

(iii) $\tan x = \tan(180° - x)$ (iv) $\tan x = -\tan(180° - x)$

4 Write down possible equations for the following graphs.

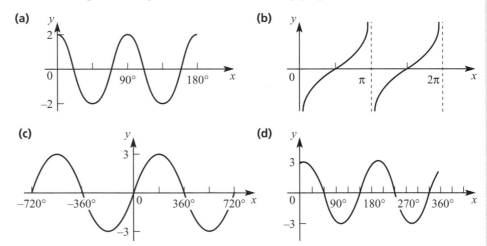

5 Sketch on separate diagrams, each over the domain $0 \leqslant x \leqslant 360°$, the graphs of these functions.

(a) $y = \sin x$ (b) $y = -\sin x$

(c) $y = \sin(-x)$ (d) $y = -\sin(-x)$

Comment on your answers.

SOLUTION OF EQUATIONS USING GRAPHS OF TRIGONOMETRICAL FUNCTIONS

Suppose that you want to solve the equation:

$$\cos\theta = 0.5$$

You press the calculator keys for $\cos^{-1} 0.5$ (or arccos 0.5 or invcos 0.5), and the answer comes up as 60°.

However, by looking at the graph of $y = \cos\theta$ (your own or figure 9.37) you can see that there are in fact infinitely many roots to this equation.

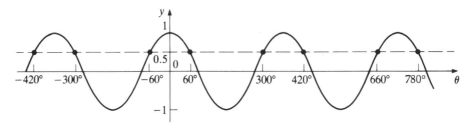

FIGURE 9.37

A calculator always gives the *principal value* of the solution, that is, the value in the range:

cosine	$0° \leqslant \theta \leqslant 180°$	or	$0 \leqslant \theta \leqslant \pi$ radians
sine	$-90° \leqslant \theta \leqslant 90°$	or	$-\frac{\pi}{2} \leqslant \theta \leqslant \frac{\pi}{2}$ radians
tangent	$-90° \leqslant \theta \leqslant 90°$	or	$-\frac{\pi}{2} < \theta < \frac{\pi}{2}$ radians

Other roots can be found by looking at the appropriate graph. Thus the roots for $\cos\theta = 0.5$ are seen (figure 9.37) to be:

$$\theta = ..., -420°, -300°, -60°, 60°, 300°, 420°, 660°, 780°, ...$$

EXAMPLE 9.14

Find values of θ in the interval $-360° \leqslant \theta \leqslant 360°$ for which $\sin\theta = 0.5$.

Solution $\sin\theta = 0.5 \Rightarrow \theta = 30°$ (principal value). Figure 9.38 shows the graph of $\sin\theta$.

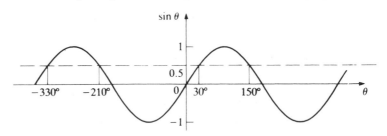

FIGURE 9.38

The values of θ for which $\sin\theta = 0.5$ are $-330°, -210°, 30°, 150°$.

EXAMPLE 9.15

Find values of θ in the range $-2\pi < \theta < 2\pi$ for which $\cos\theta = \dfrac{\sqrt{3}}{2}$, expressing your answers in terms of π.

Solution $\cos\theta = \dfrac{\sqrt{3}}{2} \Rightarrow \theta = \dfrac{\pi}{6}$ (principal value). The graph of $y = \cos\theta$ is shown in figure 9.39.

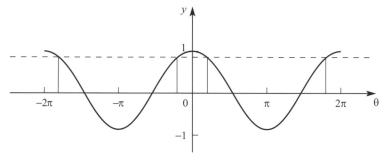

FIGURE 9.39

The values of θ for which $\cos\theta = \dfrac{\sqrt{3}}{2}$ are $\dfrac{\pi}{6}$ and $2\pi - \dfrac{\pi}{6} = \dfrac{11\pi}{6}$ and, by symmetry, $-\dfrac{\pi}{6}$ and $-\dfrac{11\pi}{6}$.

EXAMPLE 9.16

Solve $\tan 2x = 3$ for $0° \leq x \leq 180°$.

Solution $\tan 2x = 3$

$\Rightarrow 2x = 71.6°, 180° + 71.6°$ ← Note that if x lies between 0° and 180° then $2x$ lies between 0° and 360°.

$\Rightarrow 2x = 71.6°, 251.6°$

$\Rightarrow x = 35.8°, 125.8°$

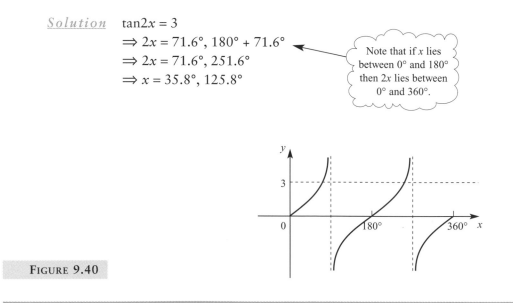

FIGURE 9.40

EXAMPLE 9.17

Solve $\cos\left(x + \frac{\pi}{6}\right) = \frac{1}{\sqrt{2}}$ for $0 \leqslant x \leqslant 2\pi$.

Solution

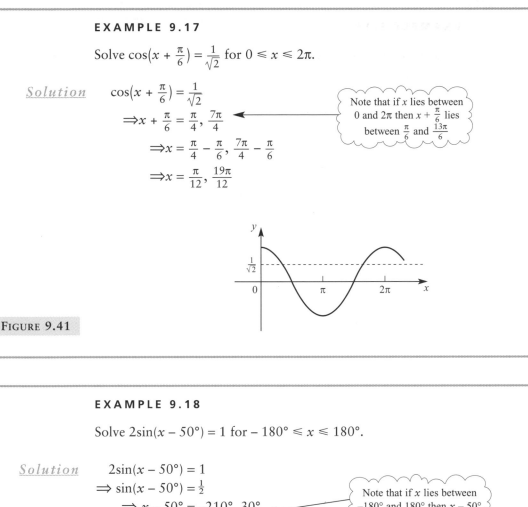

$$\cos\left(x + \frac{\pi}{6}\right) = \frac{1}{\sqrt{2}}$$

$$\Rightarrow x + \frac{\pi}{6} = \frac{\pi}{4}, \frac{7\pi}{4}$$

$$\Rightarrow x = \frac{\pi}{4} - \frac{\pi}{6}, \frac{7\pi}{4} - \frac{\pi}{6}$$

$$\Rightarrow x = \frac{\pi}{12}, \frac{19\pi}{12}$$

Note that if x lies between 0 and 2π then $x + \frac{\pi}{6}$ lies between $\frac{\pi}{6}$ and $\frac{13\pi}{6}$

FIGURE 9.41

EXAMPLE 9.18

Solve $2\sin(x - 50°) = 1$ for $-180° \leqslant x \leqslant 180°$.

Solution

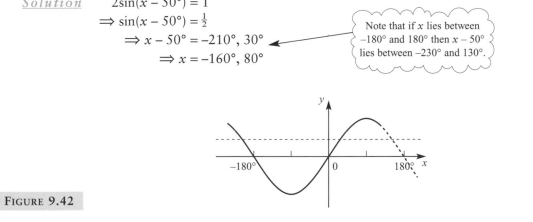

$$2\sin(x - 50°) = 1$$

$$\Rightarrow \sin(x - 50°) = \tfrac{1}{2}$$

$$\Rightarrow x - 50° = -210°, 30°$$

$$\Rightarrow x = -160°, 80°$$

Note that if x lies between $-180°$ and $180°$ then $x - 50°$ lies between $-230°$ and $130°$.

FIGURE 9.42

EXAMPLE 9.19

Solve $2\sin^2 x = 1$ for $-\pi \leqslant x \leqslant \pi$.

Solution

$2\sin^2 x = 1$

$\Rightarrow \sin x = \pm \dfrac{1}{\sqrt{2}}$

$\Rightarrow x = \dfrac{\pi}{4}, \pi - \dfrac{\pi}{4}, \pi + \dfrac{\pi}{4}, 2\pi - \dfrac{\pi}{4}$

$\Rightarrow x = \dfrac{\pi}{4}, \dfrac{3\pi}{4}, \dfrac{5\pi}{4}, \dfrac{7\pi}{4}$

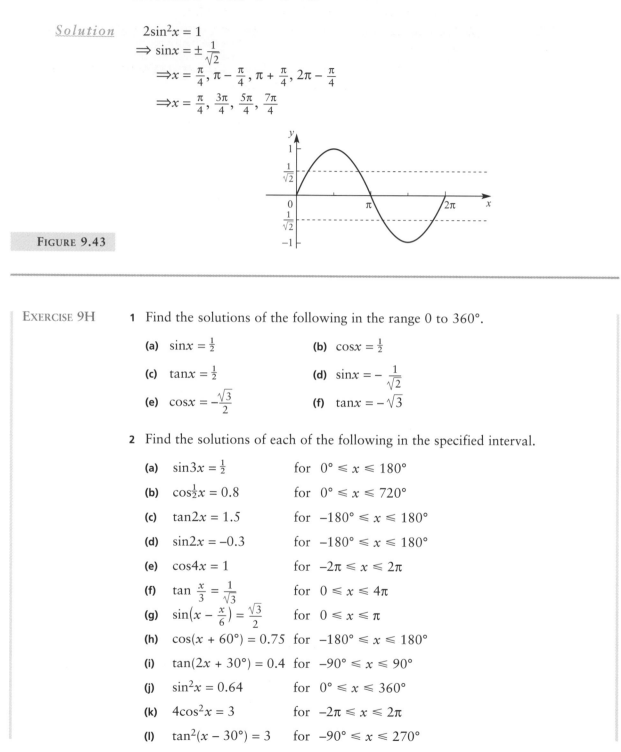

FIGURE 9.43

EXERCISE 9H

1 Find the solutions of the following in the range 0 to 360°.

(a) $\sin x = \frac{1}{2}$ **(b)** $\cos x = \frac{1}{2}$

(c) $\tan x = \frac{1}{2}$ **(d)** $\sin x = -\dfrac{1}{\sqrt{2}}$

(e) $\cos x = -\dfrac{\sqrt{3}}{2}$ **(f)** $\tan x = -\sqrt{3}$

2 Find the solutions of each of the following in the specified interval.

(a) $\sin 3x = \frac{1}{2}$ for $0° \leqslant x \leqslant 180°$

(b) $\cos \frac{1}{2}x = 0.8$ for $0° \leqslant x \leqslant 720°$

(c) $\tan 2x = 1.5$ for $-180° \leqslant x \leqslant 180°$

(d) $\sin 2x = -0.3$ for $-180° \leqslant x \leqslant 180°$

(e) $\cos 4x = 1$ for $-2\pi \leqslant x \leqslant 2\pi$

(f) $\tan \frac{x}{3} = \dfrac{1}{\sqrt{3}}$ for $0 \leqslant x \leqslant 4\pi$

(g) $\sin\left(x - \frac{x}{6}\right) = \dfrac{\sqrt{3}}{2}$ for $0 \leqslant x \leqslant \pi$

(h) $\cos(x + 60°) = 0.75$ for $-180° \leqslant x \leqslant 180°$

(i) $\tan(2x + 30°) = 0.4$ for $-90° \leqslant x \leqslant 90°$

(j) $\sin^2 x = 0.64$ for $0° \leqslant x \leqslant 360°$

(k) $4\cos^2 x = 3$ for $-2\pi \leqslant x \leqslant 2\pi$

(l) $\tan^2(x - 30°) = 3$ for $-90° \leqslant x \leqslant 270°$

3 (a) Sketch the curve $y = \cos x$ for $-90° \leqslant x \leqslant 450°$.

(b) Solve the equation $\cos x = 0.6$ for $-90° \leqslant x \leqslant 450°$, and illustrate all the roots on your sketch.

(c) Sketch the curve $y = \sin x$ for $-90° \leqslant x \leqslant 450°$.

(d) Solve the equation $\sin x = 0.8$ for $-90° \leqslant x \leqslant 450°$, and illustrate all the roots on your sketch.

(e) Explain why some of the roots of $\cos x = 0.6$ are the same as those for $\sin x = 0.8$, and why some are different.

4 In this question all the angles are in the interval $-180°$ to $180°$. Give all answers correct to 1 decimal place.

(a) Given that $\sin\alpha < 0$ and $\cos\alpha = 0.5$, find α.

(b) Given that $\tan\beta = 0.4463$ and $\cos\beta < 0$, find β.

(c) Given that $\sin\gamma = 0.8090$ and $\tan\gamma > 0$, find γ.

5 (a) For what values of α are $\sin\alpha$, $\cos\alpha$ and $\tan\alpha$ all positive?

(b) Are there any values of α for which $\sin\alpha$, $\cos\alpha$ and $\tan\alpha$ are all negative? Explain your answer.

(c) Are there any values of α for which $\sin\alpha$, $\cos\alpha$ and $\tan\alpha$ are all equal? Explain your answer.

TRIGONOMETRIC IDENTITIES

The unit circle was used earlier in the chapter. It looked like that shown in figure 9.44.

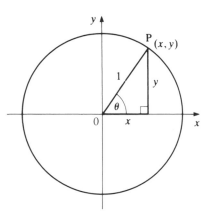

From the triangle with OP as hypotenuse you can see that $\cos\theta = x$, $\sin\theta = y$ and $\tan\theta = \frac{y}{x}$ so it follows that:

$$\tan\theta = \frac{\sin\theta}{\cos\theta}$$

It would be more accurate here to use the identity sign, $\equiv$, since the relationship is true for all values of θ.

$$\tan\theta \equiv \frac{\sin\theta}{\cos\theta}$$

FIGURE 9.44

An *identity* is different from an equation since an equation is only true for certain values of the variable, called the *solutions* of the equation. For example, $\tan\theta = 1$ is an equation: it is true when $\theta = 45°$ or $225°$, but not when it takes any other value in the range $0° \leqslant \theta \leqslant 360°$.

By contrast, an *identity* is true for all values of the variable, for example:

$$\tan 30° = \frac{\sin 30°}{\cos 30°} \qquad \tan 72° = \frac{\sin 72°}{\cos 72°} \qquad \tan(-339°) = \frac{\sin(-339°)}{\cos(-339°)},$$

and so on for all values of the angle.

In this book, as in mathematics generally, we often use an equals sign where it would be more correct to use an identity sign. The identity sign is kept for situations where we want to emphasise that the relationship is an identity and not an equation.

EXAMPLE 9.20

Solve $2\sin\theta = \cos\theta$ for $-180° \leqslant \theta \leqslant 180°$.

Solution $2\sin\theta = \cos\theta$

Divide both sides by $\cos\theta$ to give:

$$2\tan\theta = 1$$
$$\tan\theta = \tfrac{1}{2}$$
$$\theta = 26.6°, -153.4°$$

THE FUNDAMENTAL TRIGONOMETRIC IDENTITY

Another useful identity can be found by applying Pythagoras' theorem to any point $P(x, y)$ on the unit circle in figure 9.44:

$$y^2 + x^2 \equiv OP^2$$
$$(\sin\theta)^2 + (\cos\theta)^2 \equiv 1$$

This is written as the fundamental trigonometric identity:

$$\boxed{\sin^2\theta + \cos^2\theta \equiv 1}$$

from which:

$$\cos^2\theta \equiv 1 - \sin^2\theta \quad \text{and} \quad \sin^2\theta \equiv 1 - \cos^2\theta$$

EXAMPLE 9.21

Solve $2\cos^2\theta - \sin\theta = 1$ for $-360° \leqslant \theta \leqslant 360°$.

Solution

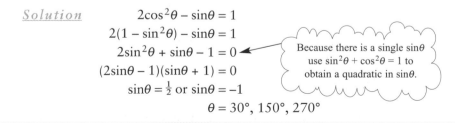

$$2\cos^2\theta - \sin\theta = 1$$
$$2(1 - \sin^2\theta) - \sin\theta = 1$$
$$2\sin^2\theta + \sin\theta - 1 = 0$$
$$(2\sin\theta - 1)(\sin\theta + 1) = 0$$
$$\sin\theta = \tfrac{1}{2} \text{ or } \sin\theta = -1$$
$$\theta = 30°, 150°, 270°$$

Because there is a single $\sin\theta$ use $\sin^2\theta + \cos^2\theta = 1$ to obtain a quadratic in $\sin\theta$.

EXAMPLE 9.22

Prove the identity $\dfrac{(1 + \cos\theta)(1 - \cos\theta)}{\cos^2\theta} \equiv \tan^2\theta$.

Solution   LHS $= \dfrac{(1 + \cos\theta)(1 - \cos\theta)}{\cos^2\theta}$ (LHS means left-hand side of the identity.)

$= \dfrac{1 - \cos^2\theta}{\cos^2\theta}$

Using $\sin^2\theta + \cos^2\theta = 1$

$= \dfrac{\sin^2\theta}{\cos^2\theta}$

$= \tan^2\theta = $ RHS   (RHS means right-hand side of the identity.)

EXERCISE 9I

1 Prove the following identities.

(a) $\sin^2x + \cos x + 1 \equiv 2 + \cos x - \cos^2x$

(b) $2\cos^2x + \sin x + 3 \equiv 5 + \sin x - 2\sin^2x$

(c) $3 + 4\sin x - 5\cos^2x \equiv 5\sin^2x + 4\sin x - 2$

(d) $1 - \tan^2x \cos^2x \equiv \cos^2x$

(e) $\dfrac{(1 + \sin x)(1 - \sin x)}{\sin^2x} \equiv \dfrac{1}{\tan^2x}$

2 Find the solutions of the following in the specified interval.

(a) $1 + \cos x - 2\sin^2x = 0$ for $0° \leqslant x \leqslant 360°$

(b) $2 + 2\sin x - 3\cos^2x = 0$ for $-180° \leqslant x \leqslant 180°$

(c) $\dfrac{3\sin x}{\cos x} = \dfrac{\cos x}{\sin x}$ for $-2\pi \leqslant x \leqslant 2\pi$

(d) $6\sin^2x + \cos x - 5 = 0$ for $0° \leqslant x \leqslant 360°$

(e) $4(1 + \cos x)(1 - \cos x) = 3$ for $-\pi \leqslant x \leqslant \pi$

(f) $\cos^2x + \sin x + 1 = 0$ for $-90° \leqslant x \leqslant 270°$

(g) $3\cos^2x = 1 + \sin x$ for $-180° \leqslant x \leqslant 180°$

(h) $1 - \sin^2x = \sin x \cos x$ for $0° \leqslant x \leqslant 2\pi$

(i) $3\cos^2x - \sin^2x = 2$ for $0° \leqslant x \leqslant 2\pi$

(j) $1 + 5\sin x = 3\cos^2x$ for $0° \leqslant x \leqslant 360°$.

3 In the triangle PQR, PQ = 17cm, QR = 15 cm and PR = 8 cm.

(a) Show that the triangle is right-angled.

(b) Write down the values of $\sin Q$, $\cos Q$ and $\tan Q$, leaving your answers as fractions.

(c) Use your answers to part (b) to show that:

(i) $\sin^2Q + \cos^2Q = 1$

(ii) $\tan Q = \dfrac{\sin Q}{\sin x}$

4 If angle A is obtuse and $\sin A = \frac{1}{5}$ obtain a value for $\cos A$ leaving it in surd form, simplifying your answer as much as possible.

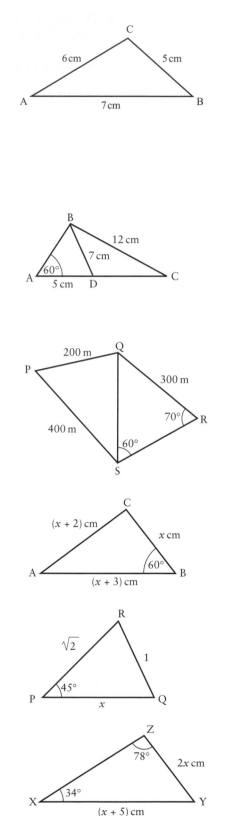

5 The triangle ABC has sides of length 5 cm, 6 cm and 7 cm, as shown. The cosine rule is $a^2 = b^2 + c^2 - 2bc \cos A$.

Use the cosine rule to obtain a value for $\cos A$.

Find a value for $\sin A$ and show that $\tan A = \dfrac{2\sqrt{6}}{5}$.

EXERCISE 9J   **Examination-style questions**

1 In the diagram the length of AD is 5 cm, the length of BD is 7 cm and the length of BC is 12 cm. The angle BAD is 60°. ADC is a straight line.

   (a) Find the length of AB.

   (b) Find the size of angle BCD.

2 In the quadrilateral PQRS the length of PG is 200 m, the length of QR is 300 m and the length of SP is 400 m. The angles QRS and QSR are 70° and 60° respectively.

   (a) Find the length of QS.

   (b) Find the size of angle PQR.

3 The triangle ABC has sides of length x cm, $(x + 2)$ cm and $(x + 3)$ cm, as shown. Angle ABC is 60°.

Find the value of x and hence show that the area of the triangle is $10\sqrt{3}$ cm².

4 In the triangle PQR, PQ = x cm, QR = 1 cm and PR = $\sqrt{2}$ cm. Angle QPR = 45°.

Find the value of x. Hence show that the triangle PQR is right-angled.

5 In the triangle XYZ, XY = $(x + 5)$ cm, YZ = $2x$ cm, angle YXZ = 34° and angle XZY = 78°.

Show that the area of the triangle is approximately 13 cm².

6 The diagram shows a minor sector OAB of a circle of radius 6 cm. The angle AOB is $\frac{\pi}{3}$ radians.

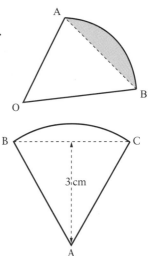

(a) Show that the perimeter of the sector is $2 + 2\pi$.

(b) Find the area of the sector and show that the area of the shaded segment is 3.3 cm², correct to 2 significant figures.

7 The shape of a badge is a sector ABC of a circle with a centre A and radius AB, as shown in the diagram. The triangle ABC is equilateral and has perpendicular height 3 cm.

(a) Find, in surd form, the length of AB.

(b) Find, in terms of π, the area of the badge.

(c) Prove that the perimeter of the badge is $\frac{2\sqrt{3}}{3}(\pi + 6)$ cm.

[Edexcel]

8 The diagram shows a gardener's design for the shape of a flowerbed, with a perimeter ABCD. AD is an arc of a circle with centre O and radius 5 metres, BC is an arc of a circle with centre O and radius 7 metres. OAB and ODC are straight lines and the size of angle AOD is θ radians.

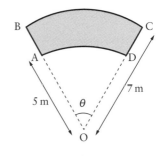

(a) Find, in terms of θ, an expression for the area of the flowerbed.

Given that the area of the flowerbed is 15 m²:

(b) show that $\theta = 1.25$

(c) calculate, in metres, the perimeter of the flowerbed.

The gardener now decides to replace arc AD with a straight line AD.

(d) Find, to the nearest centimetre, the reduction in the perimeter of the flowerbed.

[Edexcel]

9 The diagram shows a sector OAB of a circle, centre O, of radius 5 cm and a shaded segment of the circle. Given that $\angle AOB = 0.7$ radians, calculate:

(a) the area, in cm², of the sector OAB

(b) the area, in cm² to 2 significant figures, of the shaded segment.

[Edexcel]

10 (a) Find the coordinates of the point where the graph of $y = 2\sin\left(2x + \frac{5\pi}{6}\right)$ crosses the y-axis.

(b) Find the values of x, where $0 \leqslant x \leqslant 2\pi$, for which $y = \sqrt{2}$.

[Edexcel]

11 (a) Determine the solutions of the equation $\cos(2x - 30)° = 0$ for which $0° \leqslant x \leqslant 360°$.

(b) The diagram shows part of the curve $y = \cos(px - q)°$, where p and q are positive constants and $q < 180$. The curve cuts the x-axis at A, B and C, as shown. Given that the coordinates of A and B are $(100, 0)$ and $(220, 0)$ respectively:

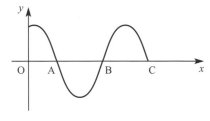

(i) write down the coordinates of C

(ii) find the value of p and the value of q.

[Edexcel]

12 Using the identity $\cos^2 x + \sin^2 x \equiv 1$ prove that:

$$5 + 2\cos x - 3\sin^2 x \equiv 3\cos^2 x + 2\cos x + 2$$

Hence find the value of x, in the range $0°$ to $360°$, for which:

$$5 + 2\cos x - 3\sin^2 x = 3$$

13 Solve the equations:

(a) $2\sin^2 x + 3\cos x = 0$

(b) $2\sin^2 x + 3\cos x + 1 = 0$,

giving your answers in degrees in the range $-90°$ to $270°$.

14 Find, to the nearest integer, the values of x in the interval $0° \leqslant x \leqslant 180°$ for which

$$3\sin^2 3x - 7\cos 3x - 5 = 0.$$

[Edexcel]

15 On the same diagram sketch the graphs of $y = \sqrt{3}\sin(x - 45°)$ and $y = \cos(x - 45°)$ over the set of x-values from $0°$ to $360°$.

Calculate the exact values of the coordinates of the points of intersection of these two curves.

KEY POINTS

1 **Sine rule** (two sides, two angles)

$$\frac{a}{\sin A} = \frac{b}{\sin B} = \frac{c}{\sin C}$$

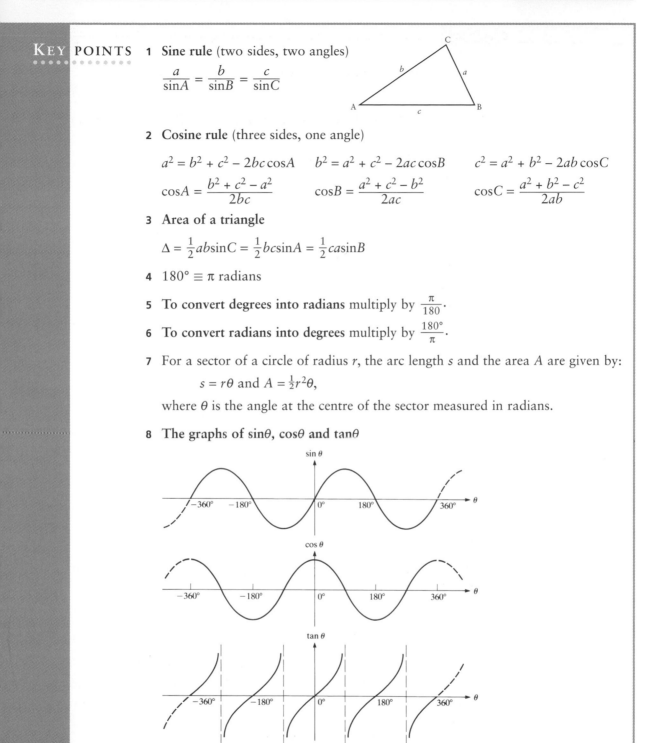

2 **Cosine rule** (three sides, one angle)

$$a^2 = b^2 + c^2 - 2bc\cos A \qquad b^2 = a^2 + c^2 - 2ac\cos B \qquad c^2 = a^2 + b^2 - 2ab\cos C$$

$$\cos A = \frac{b^2 + c^2 - a^2}{2bc} \qquad \cos B = \frac{a^2 + c^2 - b^2}{2ac} \qquad \cos C = \frac{a^2 + b^2 - c^2}{2ab}$$

3 **Area of a triangle**

$$\Delta = \tfrac{1}{2}ab\sin C = \tfrac{1}{2}bc\sin A = \tfrac{1}{2}ca\sin B$$

4 $180° \equiv \pi$ radians

5 **To convert degrees into radians** multiply by $\dfrac{\pi}{180}$.

6 **To convert radians into degrees** multiply by $\dfrac{180°}{\pi}$.

7 For a sector of a circle of radius r, the arc length s and the area A are given by:

$$s = r\theta \text{ and } A = \tfrac{1}{2}r^2\theta,$$

where θ is the angle at the centre of the sector measured in radians.

8 **The graphs of $\sin\theta$, $\cos\theta$ and $\tan\theta$**

9 $\tan\theta \equiv \dfrac{\sin\theta}{\cos\theta}$

10 $\sin^2\theta + \cos^2\theta \equiv 1$

Chapter ten

EXPONENTIALS AND LOGARITHMS

Prophesy upon these bones, and say unto them, O ye dry bones.

Bible

• • • • • • • • • • • • • • • • • •

THE FUNCTION a^x

Think of a pond with weed growing on its surface. Initially let $1\,\mathrm{m}^2$ of the surface be covered with weed and for the next few days, let the area double each day.
Then after one day there will be $2\,\mathrm{m}^2$ of weed, after two days there will be $4\,\mathrm{m}^2$ of weed and so on. The pond weed graph would look like the solid curve shown in figure 10.1.

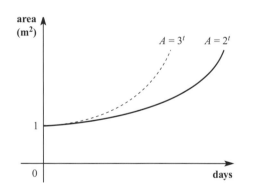

FIGURE 10.1

If A is the area and t the number of days then the equation connecting A and t is:

$$A = 2^t$$

where t is the exponent of 2 and the expression is an *exponential equation*.

If the weed had trebled in area each day then the equation would have been $A = 3^t$ and the curve would be like the dotted line shown in figure 10.1.

Both of these functions grow at an ever-increasing rate and this is describe as *exponential growth*.

The functions are both examples of the general one:

$$y = a^x \text{ with } a > 1$$

The graphs of these functions can be extended for negative values of x. Curves of this type all have the shape shown in figure 10.2.

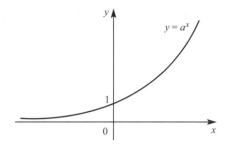

FIGURE 10.2

....

Note

There is a particularly interesting graph of this type. At every point on its curve the gradient of the graph is equal to the value of y. (You will recall that to find the gradient you draw a tangent to the curve.) From *Pure Mathematics: Core 1* the gradient is the derivative of the function.

....

Writing the equation of this particular graph in terms of x and y gives the exponential function $y = e^x$ and its graph looks like that in figure 10.3. The number e, like π, is an irrational number; it starts 2.7182818284590 ... and goes on for ever.

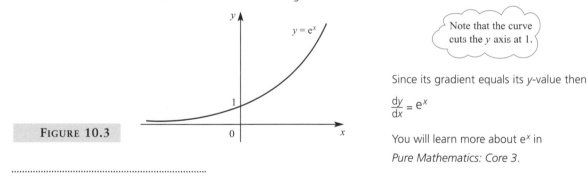

Note that the curve cuts the y axis at 1.

Since its gradient equals its y-value then

$$\frac{dy}{dx} = e^x$$

FIGURE 10.3

You will learn more about e^x in *Pure Mathematics: Core 3*.

....

If you reflect the graph of $y = a^x$ in the y-axis it looks like the curve in figure 10.4.

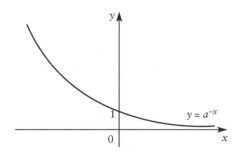

FIGURE 10.4

The equation for this graph is $y = a^{-x}$ and it describes *exponential decay*. You may have studied the half-life of radioactive substances which is an example of exponential decay.

Note

If $0 < a < 1$ the graph of $y = a^x$ will take the same shape as $y = a^{-x}$. You may wish to work out why. Also, think what the graph would look like if $a = 1$.

You will notice that the graphs of $y = a^x$ and $y = a^{-x}$ are both asymptotic to the x-axis (that is they get closer and closer to the axis but do not quite reach it). Consequently if you translate the graph in the y-direction you should draw a dotted line to indicate the position of the asymptote.

EXAMPLE 10.1

Sketch the graph of $y = a^x - 1$.

Solution   This is a translation of $y = a^x$ by $\begin{pmatrix} 0 \\ -1 \end{pmatrix}$. Since $y = a^x$ cuts the y-axis at 1 this graph will go through the origin when it is moved down one unit (see figure 10.5).

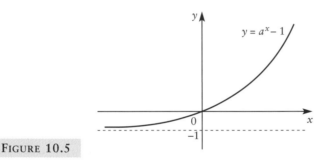

FIGURE 10.5

EXAMPLE 10.2

Sketch the graph of $y = 2 - a^{-x}$.

Solution   From *Pure Mathematics: Core 1* you will recognise that this is a reflection of $y = a^{-x}$ in the x-axis followed by a translation $\begin{pmatrix} 0 \\ 2 \end{pmatrix}$. See figure 10.6.

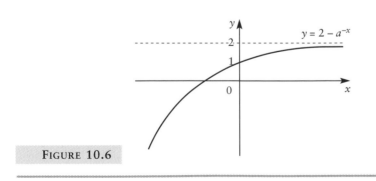

FIGURE 10.6

EXERCISE 10A

1 On separate diagrams sketch, for $a > 1$, the graphs of these functions.

(a) $y = a^x$ (b) $y = -a^x$ (c) $y = a^{-x}$ (d) $y = -a^{-x}$

2 On separate diagrams, sketch the following curves. Show asymptotes as dotted lines.

(a) $y = a^x + 1$ (b) $y = -a^x + 1$ (c) $y = a^{-x} - 2$

(d) $y = a^{x+1}$ (e) $y = a^{1-x}$

3 (a) On the same diagram sketch $y = 2^x$ and $y = 4^x$.

(b) Describe a transformation that maps $y = 2^x$ onto $y = 4^x$.

4 (a) On the same diagram, sketch the graphs of $y = 3^x$ and $y = 6x$.

(b) Show, by trial and improvement, that $y = 3^x$ and $y = 6x$ intersect where $x = 0.2$ (correct to one decimal place) and find, also correct to one decimal place, the x-value of the second point of intersection.

5 On the same diagram, sketch the graphs of:

$$y = 5^x - 1 \quad \text{and} \quad y = -5^{-x} + 1$$

Solve the simultaneous equations:

$$y = 5^x - 1$$
$$y = -5^{-x} + 1$$

Hence prove that the graphs touch at one point only. Find the coordinates of the point of contact.

THE FUNCTION $\log_a x$

You are now familiar with the function $y = a^x$ and its graph. Now consider what would happen if you swapped x and y around.

The equation becomes $x = a^y$.

Graphically, swapping x and y is equivalent to reflecting the function in the line $y = x$. The result is shown in figure 10.7.

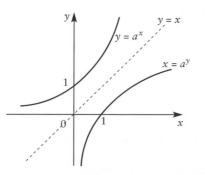

FIGURE 10.7

251

Note

$x = a^y$ is the *inverse function* of $y = a^x$. You will learn about inverse functions in more detail in *Pure Mathematics: Core 3*.

There is another way of writing $x = a^y$. Since y is the power to which a has to be raised to give the value x then y is referred to as the *logarithm of x to the base a*.

$$y = \log_a x$$

This is said as 'the log of x to the base a'. The most familiar and frequently used logarithm is the inverse of $y = 10^x$ which is:

$$y = \log_{10} x$$

This is abbreviated to $y = \log x$ and you will find a key for this, labelled 'log', on your calculator.

Whatever the value of a (the base of the logarithm), provided $a > 1$, the graph of $y = \log_a x$ has the same general shape (see figure 10.8).

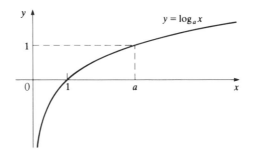

FIGURE 10.8

Note that:
- the curve crosses the x-axis at 1
- it passes through the point $(a, 1)$
- it only exists for $x > 0$
- the y-axis is an asymptote
- there is no limit to the height of the curve for large values of x though the gradient progressively decreases.

The function $y = \log_a x$ is also the inverse of $y = a^x$.

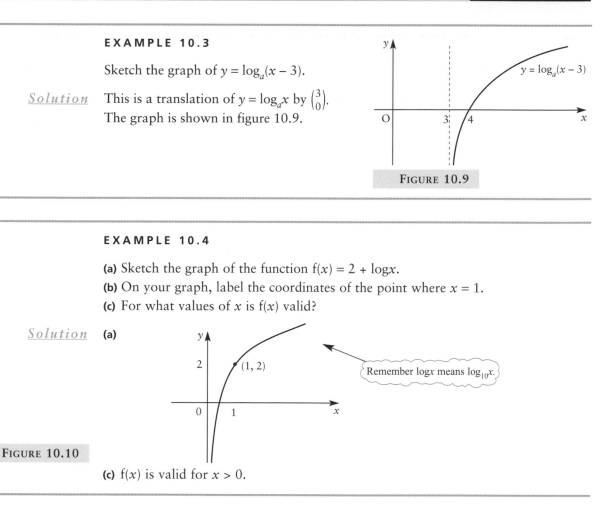

EXAMPLE 10.3

Sketch the graph of $y = \log_a(x - 3)$.

Solution This is a translation of $y = \log_a x$ by $\binom{3}{0}$.
The graph is shown in figure 10.9.

FIGURE 10.9

EXAMPLE 10.4

(a) Sketch the graph of the function $f(x) = 2 + \log x$.
(b) On your graph, label the coordinates of the point where $x = 1$.
(c) For what values of x is $f(x)$ valid?

Solution **(a)**

Remember $\log x$ means $\log_{10} x$.

FIGURE 10.10

(c) $f(x)$ is valid for $x > 0$.

Historical note

Logarithms were discovered independently by John Napier (1550–1617), who lived at Merchiston Castle in Edinburgh, and Jolst Bürgi (1552–1632) from Switzerland. It is generally believed that Napier had the idea first, and so he is credited with their discovery. Natural logarithms (based on the mathematical constant e) are also called Naperian logarithms but there is no basis for this since Napier's logarithms were definitely not the same as natural logarithms. Napier was deeply involved in the political and religious events of his day and mathematics and science were little more than hobbies for him. He was a man of remarkable ingenuity and imagination and also drew plans for war chariots that look very like modern tanks, and for submarines.

EXERCISE 10B On separate diagrams, sketch the following curves.

1 $y = \log x$ **2** $y = \log(-x)$

3 $y = -\log x$ **4** $y = -\log(-x)$

5 $y = \log(x - 2)$ **6** $y = \log(x + 3)$

7 $y = \log(1 - x)$ **8** $y = 3 + \log x$

9 $y = 2\log_3 x$ **10** $y = -4\log_a x$

LAWS OF LOGARITHMS

You have seen the definition:

$$x = a^y \iff y = \log_a x$$

It is important that you are able to change from the exponential form, $x = a^y$, to the logarithmic form $y = \log_a x$.

EXAMPLE 10.5

Express $64 = 4^3$ in logarithmic form.

Solution $64 = 4^3 \iff 3 = \log_4 64$

EXAMPLE 10.6

Express $\log_a p = q$ in exponential form.

Solution $\log_a p = q \iff p = a^q$

You are now in a position to develop the rules of logarithms.

Let $x = a^p$ so that $p = \log_a x$ and let $y = a^q$ so that $q = \log_a y$.

Then multiplying: $\qquad xy = a^p \times a^q = a^{p+q}$

$$\therefore \log_a xy = \log_a a^{p+q} = p + q = \log_a x + \log_a y$$

dividing: $\qquad \dfrac{x}{y} = \dfrac{a^p}{a^q} = a^{p-q}$

$$\therefore \log_a \frac{x}{y} = p - q = \log_a x - \log_a y$$

powers: $\qquad x^n = (a^p)^n = a^{pn}$

$$\therefore \log_a x^n = pn = n\log_a x$$

Further, since $a^0 = 1$ $(a \neq 0)$ and taking logarithms to the base a:

$$\log_a 1 = 0$$

which was a result seen on the graph of $y = \log_a x$.

Also $a^1 = a$ which, after taking logarithms, gives the result:

$$\log_a a = 1$$

The laws of logarithms are summarised in the following box.

$$n = a^x \iff x = \log_a n$$
$$\log_a xy \equiv \log_a x + \log_a y$$
$$\log_a \frac{x}{y} \equiv \log_a x - \log_a y$$
$$\log_a x^n \equiv n\log_a x$$
$$\log_a 1 = 0$$
$$\log_a a = 1$$

EXAMPLE 10.7

Write $\log_3(x + 1) - \log_3 x$ as a single logarithm. Hence solve:

$$\log_3(x + 1) - \log_3 x = 4$$

Solution   When subtracting logarithms you divide:

$$\log_3(x + 1) - \log_3 x = \log_3\left(\frac{x + 1}{x}\right)$$

then:

$$\log_3\left(\frac{x + 1}{x}\right) = 4$$

$$\therefore \frac{x + 1}{x} = 3^4$$

$$= 81$$

rearranging and solving gives:

$$x + 1 = 81x$$

$$\therefore x = \frac{1}{80}$$

EXAMPLE 10.8

Write down the values of:

(a) $\log_5 125$ **(b)** $\log_4\left(\frac{1}{16}\right)$

without using a calculator.

Solution   Express the numbers as powers of the base.

(a) $\log_5 125 = \log_5 5^3$

$$= 3\log_5 5 \quad \text{using } \log_a x^n \equiv n\log_a x$$

$$= 3 \quad\quad\quad \text{using } \log_a a \equiv 1$$

(b) $\log_4\left(\frac{1}{16}\right) = \log_4 4^{-2}$

$$= -2$$

EQUATIONS OF THE FORM $a^x = b$

Equations of this type are solved using the logarithm rule that:

$$\log_a x^n \equiv n\log_a x$$

and the method is illustrated in the following examples.

EXAMPLE 10.9

Solve the equation $2^n = 1000$.

Solution   Taking logarithms to the base 10 of both sides (since these can be found on a calculator):

$$\log(2^n) = \log 1000$$

$$n\log 2 = \log 1000$$

$$n = \frac{\log 1000}{\log 2}$$

> Remember $\log x$ means $\log_{10} x$.

$$= 9.97 \text{ to 3 significant figures.}$$

On your calculator you will find a key labelled ln. This calculates the logarithm to the base e, i.e. $\log_e x$. You will learn more about $\ln x$ in *Pure Mathematics: Core 3*.

EXAMPLE 10.10

Solve the equation $3^{2x} - 14 \times 3^x + 45 = 0$.

Solution   You should recognise this as a quadratic in 3^x, so let $y = 3^x$ then $y^2 = 3^{2x}$ and the equation becomes:

$$y^2 - 14y + 45 = 0$$

Factorising: $(y - 9)(y - 5) = 0$

Solving: $\qquad y = 3^x = 9 \quad \text{or} \quad y = 3^x = 5$

By inspection, the first solution is $x = 2$ but to find the second solution you have to take logarithms.

$$\log 3^x = \log 5$$

$$x\log 3 = \log 5$$

$$\therefore x = \frac{\log 5}{\log 3} = 1.465 \text{ to 3 decimal places.}$$

CHANGE OF BASE OF A LOGARITHM

When solving equations of the type $a^x = b$ it is sometimes convenient to be able to change the base of the logarithm.

Writing $a^x = b$ in logarithmic form gives:

$$x = \log_a b \qquad\qquad\qquad ①$$

Taking logarithms to base c of both sides of $a^x = b$ gives:

$$\log_c a^x = \log_c b$$

and so:

$$x\log_c a = \log_c b$$
$$\therefore x = \frac{\log_c b}{\log_c a}$$

and, using the result ①, this becomes the rule for changing the base of a logarithm.

$$\log_a b = \frac{\log_c b}{\log_c a}$$

This is particularly useful when $c = 10$, giving:

$$\log_a b = \frac{\log b}{\log a}$$

EXAMPLE 10.11

Work out the value of $\log_3 20$.

Solution　　$\log_3 20 = \dfrac{\log_{10} 20}{\log_{10} 3} = \dfrac{\log 20}{\log 3} = 2.73$ (3 s.f.).

EXAMPLE 10.12

Solve $2^x = 12$.

Solution　　Rewriting in logarithmic form gives:

$$x = \log_2 12 = \frac{\log_{10} 12}{\log_{10} 2} = \frac{\log 12}{\log 2} = 3.58 \text{ (3 s.f.)}$$

EXERCISE 10C

1 Write the following in logarithmic form.

(a) $81 = 3^4$ 　　　　　(b) $256 = 2^8$

(c) $1 = a^b$ 　　　　　(d) $x = y^z$

2 Write the following in exponential form.

(a) $\log_{10} 10\,000 = 4$ 　　　　(b) $\log_6 216 = 3$

(c) $q = \log_c b$ 　　　　(d) $x = \log_n y$

3 Write the following in terms of $\log p$, $\log q$ and $\log r$.

(a) $\log \dfrac{pq}{r}$

(b) $\log \dfrac{p}{qr}$

(c) $\log p^2 q^3$

(d) $\log \dfrac{p \times \sqrt[3]{r}}{\sqrt{q}}$

4 Write the following expressions in the form $\log x$ where x is a number.

(a) $\log 5 + \log 2$

(b) $\log 6 - \log 3$

(c) $2\log 6$

(d) $-\log 7$

(e) $\frac{1}{2}\log 9$

(f) $\frac{1}{4}\log 16 + \log 2$

(g) $\log 5 + 3\log 2 - \log 10$

(h) $\log 12 - 2\log 2 - \log 9$

(i) $\frac{1}{2}\log \sqrt{16} + 2\log(\frac{1}{2})$

(j) $2\log 4 + \log 9 - \frac{1}{2}\log 144$

5 Write down the values of the following without using a calculator.
Use your calculator to check your answers for those questions which use base 10.

(a) $\log_{10} 1000$

(b) $\log_{10}\left(\dfrac{1}{10\,000}\right)$

(c) $\log_{10} \sqrt{10}$

(d) $\log_{10} 1$

(e) $\log_3 81$

(f) $\log_3\left(\dfrac{1}{81}\right)$

(g) $\log_3 \sqrt{27}$

(h) $\log_3 \sqrt[4]{3}$

(i) $\log_4 2$

(j) $\log_5\left(\dfrac{1}{125}\right)$

6 Use logarithms to the base 10 to solve the following equations.

(a) $2^x = 1\,000\,000$

(b) $2^x = 0.001$

(c) $1.08^x = 2$

(d) $1.1^x = 100$

(e) $0.99^x = 0.000\,001$

(f) $2^{x+1} = 50$

(g) $5^{x-3} = 150$

(h) $3^{-0.2x} = 0.1$

(i) $5^x = 1\,000\,000$

(j) $5000 \times 4^{0.1x} = 2500$

7 Solve these equations.

(a) $4 \times 2^{2x} - 13 \times 2^x + 3 = 0$ (b) $5^{2x} - 23 \times 5^x - 50 = 0$

8 Express $\log_3(x + 1) - \log_3(2x)$ as a single logarithm.
Hence find the exact value of x if $\log_3(x + 1) - \log_3(2x) = 2$.

9 Simplify $\log_b a \times \log_c b \times \log_a c$.

10 At time t the population P is given by:
$$t = 500(\log P_0 - \log P).$$

(a) Find an expression for P in terms of t.

(b) Find the value of P_0 given that when $t = 0$ the population is 1000.

(c) Show that when the value of t is 278 the population is approximately 278.

EXERCISE 10D   **Examination-style questions**

1 **(a)** Draw the graph of $y = 2^x$.

   (b) On a separate diagram draw the graphs of $y = 2 \times 2^x$ and $y = 2^{-x} + 1$.

   (c) Prove that the point of intersection of the curves $y = 2 \times 2^x$ and $y = 2^{-x} + 1$ is $(0, 2)$.

2 Sketch the graphs of $y = a^x$ when:

   (a) $a > 1$. Find, correct to three significant figures, the value of a if, when $x = 4$, $a^x = 10$.

   (b) $0 < a < 1$. Find, correct to three significant figures, the value of a if, when $x = -4$, $a^x = 10$.

3 **(a)** Sketch the graph of $y = 1 - 3^x$.

   (b) Find, correct to three significant figures, the value of x where the graph of $y = 1 - 3^x$ intersects the graph of $y = -5$.

4 **(a)** Sketch the graph of $y = 5^x + 1$.

   (b) Find, correct to three significant figures, the value of x where the graph of $y = 5^x + 1$ intersects the graph of $y = 20$.

5 **(a)** Express $\log_2(x + 4) - \log_2(x - 1)$ as a single logarithm.

   (b) Solve $\log_2(x + 4) - \log_2(x - 1) = 4$.

   (c) Solve $\log_2(x + 4) - \log_2(x - 1) = 2\log_2 3$.

6 **(a)** Solve $5^x = 10$, giving your answer correct to 3 significant figures.

   (b) By making a suitable substitution, or otherwise, solve:

   $$5^{2x} - 25 \times 5^x + 150 = 0$$

   giving your answers correct to three significant figures.

7 Given that $y = 10^x$, show that:

   (a) $y^2 = 100^x$

   (b) $\dfrac{y}{10} = 10^{x-1}$

   (c) Using the results from parts (a) and (b) write the equation:

   $$100^x - 10\,001(10^{x-1}) + 100 = 0$$

   as an equation in y.

   (d) By first solving the equation in y, find the values of x which satisfy the given equation in x.

   [Edexcel]

8 You are given that $p = \log_b a$, $q = \log_c b$ and $r = \log_a c$.

    (a) Write $p = \log_b a$ in exponential form.

    (b) Show that $p = \dfrac{\log a}{\log b}$.

    (c) Using the results from parts (a) and (b) prove that $pqr = 1$.

    (d) Use the results of parts (a) and (b) to evaluate $\log_3 5$ to 3 significant figures.

9 Given that $p = \log_q 16$, express the following in terms of p.

    (a) $\log_q 2$ **(b)** $\log_q 8q$

10 **(a)** Given that $3 + 2\log_2 x = \log_2 y$, show that $y = 8x^2$.

    (b) Hence, or otherwise, find the roots α and β, where $\alpha < \beta$, of the equation
$3 + 2\log_2 x = \log_2(14x - 3)$.

    (c) Show that $\log_2 \alpha = -2$.

    (d) Calculate $\log_2 \beta$, giving your answer correct to three significant figures.

[Edexcel]

11 $f(x) = 3x^3 - 4x^2 - 5x + 2$

    (a) Show that $(x + 1)$ is a factor of $f(x)$.

    (b) Factorise $f(x)$ completely.

    (c) Solve, giving your answers correct to 2 decimal places, the equation
$3(\log 2x)^3 - 4(\log 2x)^2 - 5(\log 2x) + 2 = 0$, $x > 0$.

[Edexcel (adapted)]

12 Solve, giving your answers as exact fractions, these simultaneous equations.
$$8^y = 4^{2x + 3}$$
$$\log_2 y = 4 + \log_2 x$$

[Edexcel]

13 **(a)** Express $\log_a 36 + \dfrac{1}{2}\log_a 256 - 2\log_a 48$ as a single logarithm.

    (b) What is the exact value of the expression when $a = 2$?

14 **(a)** On the same diagram, sketch the graphs of $y = \log x$ and $y = \log(x - 3)$.
State the values of x where the graph cuts the x-axis.

    (b) Solve the equation $\log x + \log(x - 3) = 1$.

    (c) Solve the equation $\log_6 x + \log_6(x - 3) = 2$, giving your answer correct to three significant figures.

15 Solve the following equations, giving your answers correct to three significant figures.

    (a) $3^x = 15$ **(b)** $3^x = 5^{x-1}$ **(c)** $3^{2x+1} - 28 \times 3^x + 9 = 0$

    (d) $x = \log_3 5$ **(e)** $\log_3 x = 5$ **(f)** $\log_x 3 = 5$

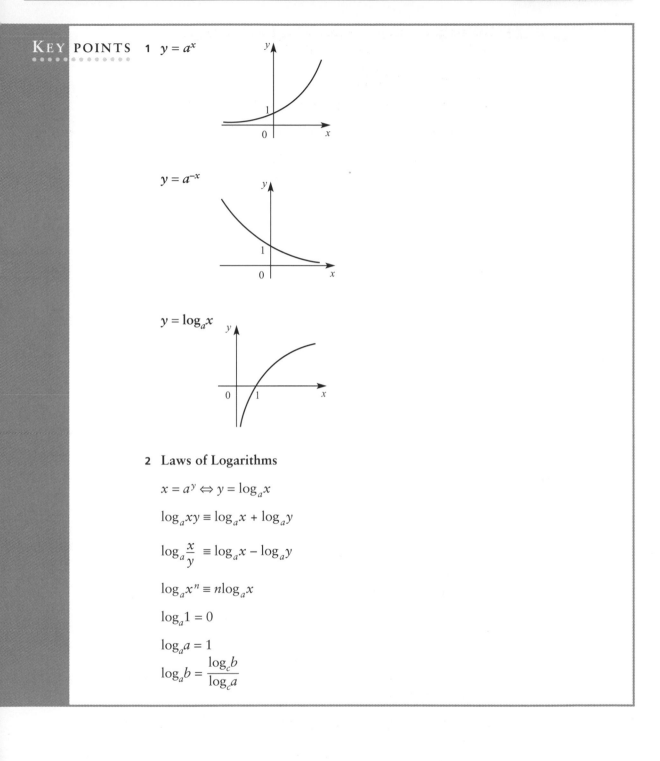

KEY POINTS

1   $y = a^x$

$y = a^{-x}$

$y = \log_a x$

2   **Laws of Logarithms**

$x = a^y \Leftrightarrow y = \log_a x$

$\log_a xy \equiv \log_a x + \log_a y$

$\log_a \dfrac{x}{y} \equiv \log_a x - \log_a y$

$\log_a x^n \equiv n\log_a x$

$\log_a 1 = 0$

$\log_a a = 1$

$\log_a b = \dfrac{\log_c b}{\log_c a}$

Chapter eleven

DIFFERENTIATION

The lady's not for turning.

Margaret Thatcher

• • • • • • • • • • • • • • •

In Chapter 4 of *Pure Mathematics: Core 1* you learned that the derivative $\dfrac{dy}{dx}$ of a function $y = f(x)$ gives the gradient of the tangent to the graph of $y = f(x)$.

You also saw that:

if $y = x^n$ then $\dfrac{dy}{dx} = nx^{n-1}$

and for a constant, k, you saw that:

if $y = kx$ then $\dfrac{dy}{dx} = k$

and if $y = k$ then $\dfrac{dy}{dx} = 0$.

You used the notation $f'(x)$ for the derivative of $f(x)$.

The notation for the second derivative is $\dfrac{d^2y}{dx^2}$ or $f''(x)$.

A *stationary point* is any point on a curve where the gradient is zero ($\dfrac{dy}{dx} = 0$ or $f'(x) = 0$).

There are three types of stationary point:

● a *maximum turning point*

● a *minimum turning point*

● a *point of inflection.*

Now you can look at how to find the positions and types of stationary points on a graph.

MAXIMUM AND MINIMUM POINTS

GRADIENT AT A TURNING POINT

Figure 11.1 shows the graph of $y = -x^2 + 16$. It has a *maximum point* at $(0, 16)$.

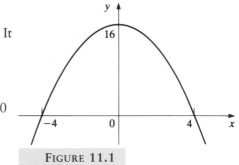

You will see that:

- at the maximum point the gradient $\dfrac{dy}{dx}$ is 0

- the gradient is positive to the left of the maximum and negative to the right of it.

FIGURE 11.1

This is true for any maximum point (figure 11.2).

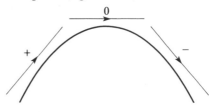

FIGURE 11.2

In the same way, for any minimum point (figure 11.3):

- the gradient is 0 at the minimum

- the gradient goes from negative to 0 to positive.

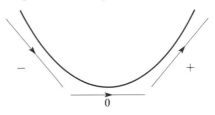

FIGURE 11.3

EXAMPLE 11.1

Find the turning points on the curve of $y = x^3 - 3x + 1$, and sketch the curve.

Solution　The gradient function for this curve is:

$$\frac{dy}{dx} = 3x^2 - 3$$

The x values for which $\dfrac{dy}{dx} = 0$ are given by:

$$3x^2 - 3 = 0$$
$$3(x^2 - 1) = 0$$
$$3(x + 1)(x - 1) = 0$$
$$\Rightarrow \qquad x = -1 \text{ or } 1$$

The signs of the gradient function just either side of these values tell you the nature of each turning point.

For $x = -1$, looking at points close to $x = -1$, but just to the left and right:

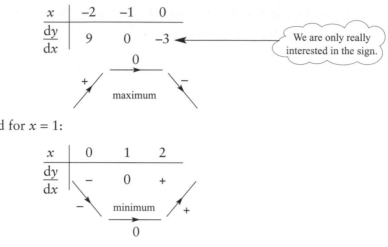

FIGURE 11.4

and for $x = 1$:

FIGURE 11.5

Thus the turning point at $x = -1$ is a maximum and the one at $x = 1$ is a minimum.

Substituting the x-values of the turning points into the original equation, $y = x^3 - 3x + 1$, gives:

    when $x = -1$, $y = (-1)^3 - 3(-1) + 1 = 3$
    when $x = 1$, $y = (1)^3 - 3(1) + 1 = -1$

There is a maximum at $(-1, 3)$ and a minimum at $(1, -1)$. Note, when $x = 0$, $y = 1$. The sketch can now be drawn (figure 11.6).

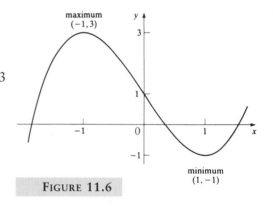

FIGURE 11.6

In Example 11.1 you knew the general shape of the cubic curve and the positions of all of the turning points, so it was easy to select values of x for which to test the sign of $\dfrac{dy}{dx}$. The curve of a more complicated function may have several maxima and minima close together, and even some points at which the gradient is undefined. To decide in such cases whether a particular turning point is a maximum or a minimum, you must look at points which are *just* either side of it.

GENERAL GUIDANCE FOR SKETCHING CURVES

It is helpful when sketching curves to do the following:

- see if you recognise the shape of the graph from the function (e.g. you should know the shape of quadratics, cubics and reciprocals)
- find where the curve cuts the x-axis by putting $y = 0$
- find where the curve cuts the y-axis by putting $x = 0$
- investigate the type and position of any turning points
- find any vertical asymptotes (where the denominator is 0)
- investigate the behaviour when x is large (both positive and negative).

EXAMPLE 11.2

The curve C has equation $y = -x^4 + 8x^2 - 4$.

(a) Find $\dfrac{dy}{dx}$.

(b) Find the coordinates of the stationary points.

(c) Determine the nature of each stationary point.

(d) Sketch the curve C.

Solution **(a)** $y = -x^4 + 8x^2 - 4 \implies \dfrac{dy}{dx} = -4x^3 + 16x$

(b) For a stationary point, $\dfrac{dy}{dx} = 0$.

$$\therefore -4x^3 + 16x = 0$$
$$-4x(x^2 - 4) = 0$$
$$\therefore x = 0 \quad \text{or} \quad x^2 = 4$$
$$x = \pm 2$$

When $x = 0$, $y = -4$.

When $x = \pm 2$, $y = -16 + 32 - 4 = 12$

The stationary points are $(-2, 12)$, $(0, -4)$ and $(2, 12)$.

(c) To find the nature of each stationary point, when:

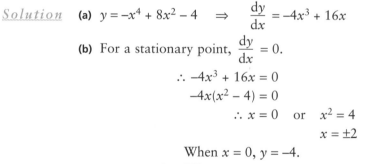

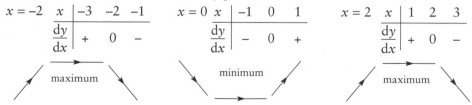

Hence $(-2, 12)$ and $(2, 12)$ are maximum turning points and $(0, -4)$ is a minimum turning point.

265

(d)

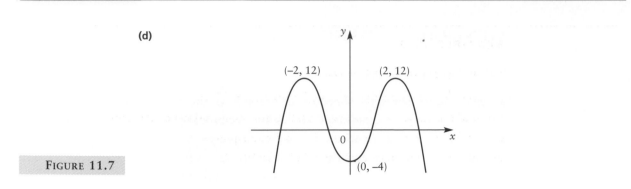

FIGURE 11.7

INCREASING AND DECREASING FUNCTIONS

In figure 11.7 you can see that from $x = -\infty$ to $x = -2$ the values of y increase as the values of x increase and that the gradient is positive. The function is said to be *increasing* for this set of x-values.

From $x = -2$ to $x = 0$ the values of y decrease as the values of x increase. The gradient is negative. The function is therefore *decreasing* for this set of x-values.

USING THE SECOND DERIVATIVE

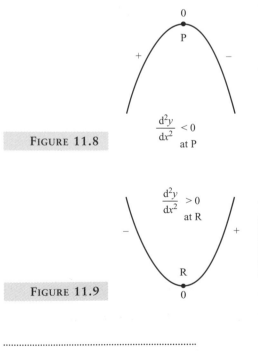

FIGURE 11.8

FIGURE 11.9

You can use the second derivative to identify the nature of a stationary point, instead of looking at the sign of $\frac{dy}{dx}$ just either side of it.

Notice that at P, $\frac{dy}{dx} = 0$ and $\frac{d^2y}{dx^2} < 0$. This tells you that the gradient, $\frac{dy}{dx}$, is 0 and decreasing. It must be going from positive to negative, so P is a maximum point (figure 11.8).

At R, $\frac{dy}{dx} = 0$ and $\frac{d^2y}{dx^2} > 0$. This tells you that the gradient, $\frac{dy}{dx}$ is 0 and increasing. It must be going from negative to positive, so R is a minimum point (figure 11.9).

Hence:

$$\frac{dy}{dx} = 0 \text{ and } \frac{d^2y}{dx^2} < 0 \Rightarrow \text{maximum point}$$

$$\frac{dy}{dx} = 0 \text{ and } \frac{d^2y}{dx^2} > 0 \Rightarrow \text{minimum point}$$

The next example illustrates the use of the second derivative to identify the nature of stationary points.

Note   The case when $\frac{d^2y}{dx^2} = 0$ is discussed on page 272 and 273.

EXAMPLE 11.3

Given that $y = 2x^3 + 3x^2 - 12x$:

(a) find $\frac{dy}{dx}$, and find the values of x for which $\frac{dy}{dx} = 0$

(b) find the value of $\frac{d^2y}{dx^2}$ at each stationary point and hence determine its nature

(c) find the y values of each of the stationary points

(d) sketch the curve given by $y = 2x^3 + 3x^2 - 12x$.

Solution (a) $\frac{dy}{dx} = 6x^2 + 6x - 12$

$\qquad = 6(x^2 + x - 2)$

$\qquad = 6(x + 2)(x - 1)$

$\frac{dy}{dx} = 0$ when $x = -2$ or 1

(b) $\frac{d^2y}{dx^2} = 12x + 6$

When $x = -2$: $\quad \frac{d^2y}{dx^2} = 12 \times (-2) + 6 = -18$

$\qquad\qquad\qquad \frac{d^2y}{dx^2} < 0 \Rightarrow$ a maximum

When $x = 1$: $\quad \frac{d^2y}{dx^2} = 12 \times 1 + 6 = 18$

$\qquad\qquad\qquad \frac{d^2y}{dx^2} > 0 \Rightarrow$ a minimum

(c) When $x - -2$: $\quad y - 2(-2)^3 + 3(-2)^2 - 12(-2)$

$\qquad\qquad\qquad\qquad = 20$

so $(-2, 20)$ is a maximum point.

When $x = 1$, $\; y = 2 + 3 - 12$

$\qquad\qquad\qquad = -7$

so $(1, -7)$ is a minimum point.

(d)

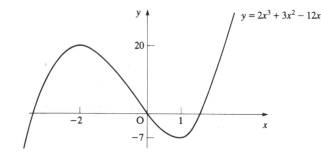

FIGURE 11.10

 Remember that you are looking for the value of $\frac{d^2y}{dx^2}$ at the turning point.

Note

On occasions when it is difficult or laborious to find $\frac{d^2y}{dx^2}$, remember that you can always classify the nature of a stationary point by looking at the sign of $\frac{dy}{dx}$ for points just either side of it.

EXERCISE 11A   Most questions in this exercise require you to find the turning points of a curve and use them as a guide for sketching it. Use a graphics calculator to check your final answers.

1   Given that $y = x^2 + 8x + 13$:

   (a) find $\frac{dy}{dx}$, and the value of x for which $\frac{dy}{dx} = 0$
   (b) showing your working clearly, decide whether the point corresponding to this x-value is a maximum or a minimum by considering the gradient either side of it
   (c) show that the corresponding y-value is -3
   (d) sketch the curve.

2   Given that $y = x^2 + 5x + 2$:

   (a) find $\frac{dy}{dx}$, and the value of x for which $\frac{dy}{dx} = 0$
   (b) classify the point that corresponds to this x-value as a maximum or a minimum
   (c) find the corresponding y-value
   (d) sketch the curve.

3   Given that $y = x^3 - 12x + 2$:

   (a) find $\frac{dy}{dx}$, and the values of x for which $\frac{dy}{dx} = 0$
   (b) classify the points that correspond to these x-values
   (c) find the corresponding y-values
   (d) sketch the curve.

4   (a) Find the coordinates of the turning points of the curve $y = x^3 - 6x^2$, and determine whether each one is a maximum or a minimum.
   (b) Use this information to sketch the graph of $y = x^3 - 6x^2$.
   (c) For what values of x is this function increasing?

5   Given that $y = x^3 + 3x^2 - 9x + 6$:

   (a) find $\frac{dy}{dx}$ and factorise the quadratic expression you obtain
   (b) write down the values of x for which $\frac{dy}{dx} = 0$
   (c) show that one of the points corresponding to these x values is a minimum and the other a maximum
   (d) show that the corresponding y values are 1 and 33 respectively
   (e) sketch the curve.

6   Given that $y = 9x + 3x^2 - x^3$:

   (a) find $\frac{dy}{dx}$ and factorise the quadratic expression you obtain
   (b) find the values of x for which the curve has turning points, and classify these turning points
   (c) find the corresponding y-values
   (d) sketch the curve.

7 **(a)** Find the coordinates and nature of each of the turning points of
$y = x^3 - 2x^2 - 4x + 3$.

(b) Sketch the curve.

8 **(a)** Find the coordinates and nature of each of the turning points of the
curve with equation $y = x^4 + 4x^3 - 36x^2 + 300$.

(b) Sketch the curve.

9 **(a)** Differentiate $y = x^3 + 3x$.

(b) What does this tell you about the number of turning points of the curve
with equation $y = x^3 + 3x$?

(c) Prove that the function is always increasing.

(d) Find the values of y corresponding to $x = -3, -2, -1, 0, 1, 2$ and 3.

(e) Hence sketch the curve and explain your answer to (b).

10 A curve has equation $y = x^4 - 8x^2 + 16$.

(a) Find $\dfrac{dy}{dx}$.

(b) Find the coordinates of any turning points and determine their nature.

(c) Sketch the curve.

(d) Factorise $z^2 - 8z + 16$, and hence factorise $x^4 - 8x^2 + 16$.

(e) Explain how knowing the factors of $x^4 - 8x^2 + 16$, and the value of y when
$x = 0$ could have enabled you to sketch the curve without using calculus.

11 You are given that $y = 2x^3 + 3x^2 - 72x + 130$.

(a) Find $\dfrac{dy}{dx}$.

P is the point on the curve where $x = 4$.

(b) Calculate the y-coordinate of P.

(c) Calculate the gradient at P.

(d) There are two stationary points on the curve. Find their coordinates.

[MEI]

12 A curve C has equation $y = x + \dfrac{4}{x}$.

(a) Show that C does not cross the x-axis.

(b) Explain why the y-axis is an asymptote.

(c) Find the coordinates and type of turning points of C.

(d) Given that $y = x$ is an asymptote of C sketch the curve.

13 Show that $y = \dfrac{x}{12} - \sqrt[3]{x}$ has two turning points. Find the coordinates of these
points and determine their nature.

14 It is given that $y = \sqrt{x} + \dfrac{4}{x}$.

(a) Find the values of x and y when $\dfrac{dy}{dx} = 0$.

(b) Show that the value of y you found in (a) is a minimum.

15 Find the position of the turning point on the graph of $y = 2x + \dfrac{27}{x^2}$ and
determine whether it is a minimum or maximum.

POINTS OF INFLECTION

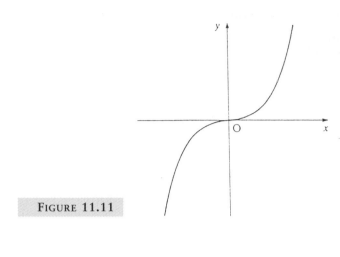

FIGURE 11.11

It is possible for the value of $\frac{dy}{dx}$ to be 0 at a point on a curve without it being a maximum or minimum. This is the case with the curve $y = x^3$, at the point (0, 0) (figure 11.11).

$$y = x^3 \quad \Rightarrow \quad \frac{dy}{dx} = 3x^2$$

and when $x = 0$, $\frac{dy}{dx} = 0$

This is an example of a *point of inflection*. In general, a point of inflection occurs where the tangent to a curve crosses the curve. This can happen also when $\frac{dy}{dx} \neq 0$, as shown in figure 11.12.

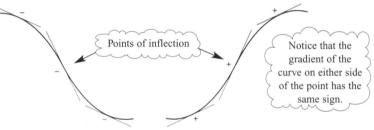

FIGURE 11.12

Points of inflection

Notice that the gradient of the curve on either side of the point has the same sign.

If you are a driver you may find it helpful to think of a point of inflection as the point at which you change from left lock to right lock, or vice versa. Another way of thinking about a point of inflection is to view the curve from one side and see it as the point where the curve changes from being concave to convex.

In this chapter you will be considering only *stationary* points of inflection, those at which $\frac{dy}{dx} = 0$. Either side of such points the gradient has the same sign, so that it goes through the sequence $+\quad 0\quad +$ or $-\quad 0\quad -$, as in figure 11.13.

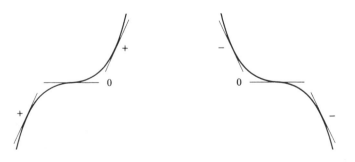

FIGURE 11.13

As stated earlier, a stationary point of inflection is *not* a turning point but, in common with a maximum and a minimum, it is called a *stationary point*.

EXAMPLE 11.4

Find and classify all the stationary points on the curve $y = 3x^5 - 5x^3$, then use the information to sketch the curve.

Solution
$$\frac{dy}{dx} = 15x^4 - 15x^2$$
$$= 15x^2(x^2 - 1)$$
$$= 15x^2(x - 1)(x + 1)$$

At a stationary point, $\frac{dy}{dx} = 0$.

Stationary points occur at $x = -1, 0$ and 1.

x		-2	-1	$-\frac{1}{2}$	0	$\frac{1}{2}$	1	2
Sign of $\frac{dy}{dx}$		$+$	0	$-$	0	$-$	0	$+$
Stationary point			max		inflection		min	

There is a maximum when $x = -1$, a stationary point of inflection when $x = 0$ and a minimum when $x = 1$.

When $x = -1$, $y = 3(-1)^5 - 5(-1)^3 = 2$: maximum at $(-1, 2)$.
When $x = 0$, $y = 0$: stationary point of inflection at $(0, 0)$.
When $x = 1$, $y = 3(1)^5 - 5(1)^3 = -2$: minimum at $(1, -2)$.

To draw a sketch of the graph you also need to know where the curve cuts the x-axis. This is at the point where $y = 0$, given by:

$$3x^5 - 5x^3 = x^3(3x^2 - 5) = 0$$

$$\therefore x = 0 \quad \text{or} \quad 3x^2 - 5 = 0 \quad \Rightarrow \quad x = \pm\sqrt{\tfrac{5}{3}} = \pm 1.29$$

The curve is shown in figure 11.14.

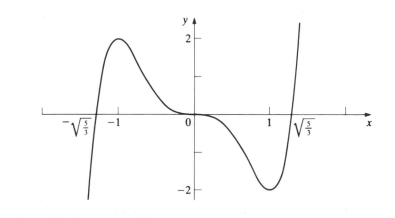

FIGURE 11.14

SECOND DERIVATIVES AND POINTS OF INFLECTION

When a curve is drawn on Cartesian axes, the points of inflection may be divided into two categories: non-stationary and stationary.

At a non-stationary point of inflection, $\frac{d^2y}{dx^2} = 0$ but $\frac{dy}{dx} \neq 0$ as shown in figure 11.15.

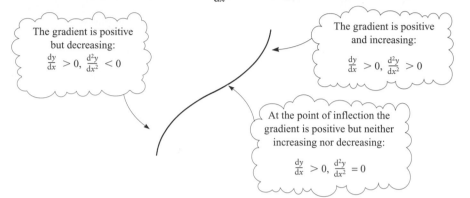

The gradient is positive but decreasing:

$$\frac{dy}{dx} > 0, \frac{d^2y}{dx^2} < 0$$

The gradient is positive and increasing:

$$\frac{dy}{dx} > 0, \frac{d^2y}{dx^2} > 0$$

At the point of inflection the gradient is positive but neither increasing nor decreasing:

$$\frac{dy}{dx} > 0, \frac{d^2y}{dx^2} = 0$$

FIGURE 11.15

At a stationary point of inflection, both $\frac{dy}{dx}$ and $\frac{d^2y}{dx^2}$ equal 0, as in the case of $y = x^3$ at the origin (figure 11.16).

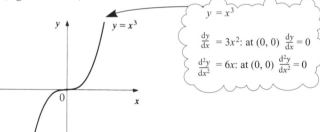

$y = x^3$

$\frac{dy}{dx} = 3x^2$: at $(0, 0)$ $\frac{dy}{dx} = 0$

$\frac{d^2y}{dx^2} = 6x$: at $(0, 0)$ $\frac{d^2y}{dx^2} = 0$

FIGURE 11.16

While it is true for all stationary points of inflection that $\frac{dy}{dx}$ and $\frac{d^2y}{dx^2} = 0$, it is also true for some turning points. For example, the curve $y = x^4$ has a minimum point at the origin (figure 11.17).

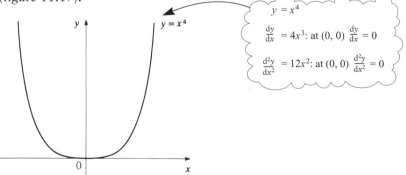

$y = x^4$

$\frac{dy}{dx} = 4x^3$: at $(0, 0)$ $\frac{dy}{dx} = 0$

$\frac{d^2y}{dx^2} = 12x^2$: at $(0, 0)$ $\frac{d^2y}{dx^2} = 0$

FIGURE 11.17

Consequently, if both $\frac{dy}{dx}$ and $\frac{d^2y}{dx^2}$ are 0 at a point, you still need to check the values of $\frac{dy}{dx}$ on either side of the point in order to determine its nature.

EXAMPLE 11.5

(a) Find and classify any stationary points on the curve $y = x^4 + x^3$.

(b) Sketch the curve $y = x^4 + x^3$.

Solution   (a) Finding the stationary points: $\dfrac{dy}{dx} = 4x^3 + 3x^2$

$$= x^2(4x + 3)$$

For a stationary point, $\dfrac{dy}{dx} = 0 \Rightarrow x = 0$ or $-\tfrac{3}{4}$

$$x = 0 \Rightarrow y = 0; \ x = -\tfrac{3}{4} \Rightarrow y = -\tfrac{27}{256}$$

The stationary points are at $(0, 0)$ and $\left(-\tfrac{3}{4}, -\tfrac{27}{256}\right)$.

Classifying the stationary points $\dfrac{d^2y}{dx^2} = 12x^2 + 6x$.

When $x = -\tfrac{3}{4}$, $\dfrac{d^2y}{dx^2} > 0$, so there is a minimum point at $\left(-\tfrac{3}{4}, -\tfrac{27}{256}\right)$.

When $x = 0$, $\dfrac{d^2y}{dx^2} = 0$, so you need to investigate further.

x	$-\tfrac{1}{4}$	0	$\tfrac{1}{4}$
$\dfrac{dy}{dx}$	$+$	0	$+$

Conclusion: there is a stationary point of inflection at $(0, 0)$.

(b)

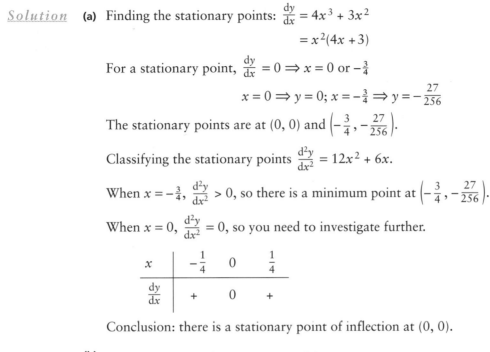

EXERCISE 11B   Most of the questions in this exercise require you to find the stationary points of
a curve and then to use them as a guide for sketching it. You will find it helpful
to use a graphics calculator or computer graph plotting package to check your
final answers.

1 Given that $y = 3x^4 - 4x^3$:

(a) find $\frac{dy}{dx}$ *and* factorise the quadratic expression you obtain

(b) write down the values of x for which $\frac{dy}{dx} = 0$

(c) classify the stationary points as maxima, minima, or stationary points of
inflection by considering the gradient of the curve either side of them

(d) find the corresponding y-values

(e) find the coordinates of A, B and C in the diagram.

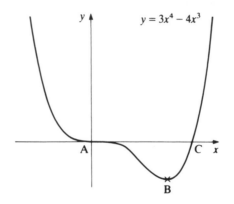

2 Given that $y = x^3 - x^2 - x + 4$:

(a) find $\frac{dy}{dx}$ and factorise the quadratic expression you obtain

(b) write down the values of x for which $\frac{dy}{dx} = 0$

(c) classify the stationary points as maxima, minima, or stationary points of
inflection by considering the gradient of the curve either side of them

(d) find the corresponding y-values

(e) sketch the curve.

3 Given that $y = 3x^4 - 8x^3 + 6x^2 + 2$:

(a) find $\frac{dy}{dx}$

(b) find the coordinates of any stationary points and determine their nature

(c) sketch the curve.

4 Given that $y = x^5 - 3$:

   (a) find $\dfrac{dy}{dx}$

   (b) find the coordinates of any stationary points and determine their nature

   (c) sketch the curve.

5 Given that $y = -x^4 + 4x^3 - 8$:

   (a) find the coordinates of any stationary points and determine their nature

   (b) sketch the curve.

6 **(a)** Find the position and nature of the stationary points of the curve
$$y = 3x^5 - 10x^3 + 15x.$$

   (b) Sketch the curve.

7 Given that $y = (x - 1)^2(x - 3)$:

   (a) multiply out the right-hand side and find $\dfrac{dy}{dx}$

   (b) find the position and nature of any stationary points

   (c) sketch the curve.

8 Given that $y = x^2(x - 2)^2$:

   (a) multiply out the right-hand side and find $\dfrac{dy}{dx}$

   (b) find the position and nature of any stationary points

   (c) sketch the curve.

9 The function $y = px^3 + qx^2$, where p and q are constants, has a stationary point at $(1, -1)$.

   (a) Using the fact that $(1, -1)$ lies on the curve, form an equation involving p and q.

   (b) Differentiate y and, using the fact that $(1, -1)$ is a stationary point, form another equation involving p and q.

   (c) Solve these two equations simultaneously to find the values of p and q.

10 Given the function $y = 3x^4 + 4x^3$:

   (a) find $\dfrac{dy}{dx}$

   (b) show that the graph of the function y has stationary points at $x = 0$ and $x = -1$ and find their coordinates

   (c) determine whether each of the stationary points is a maximum, minimum or point of inflection, giving reasons for your answers

   (d) sketch the graph of the function y, giving the coordinates of the stationary points and the points where the curve cuts the axes.   [MEI]

APPLICATIONS

There are many situations in which you need to find the maximum or minimum value of an expression. The examples which follow, and those in Exercise 11C, illustrate a few of them.

EXAMPLE 11.6

Judith's father has agreed to let her have part of his garden as a vegetable plot. He says that she can have a rectangular plot with one side against an old wall. He hands her a piece of rope 5 m long, and invites her to mark out the part she wants. Judith wants to enclose the largest area possible. What dimensions would you advise her to use?

Solution   Let the dimensions of the bed be x m $\times$ y m as shown in figure 11.19.

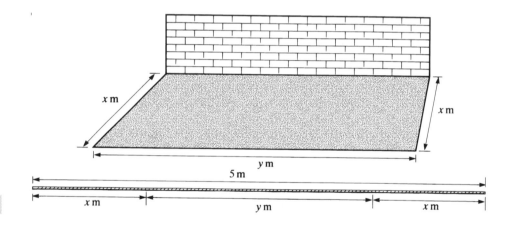

FIGURE 11.19

The area, A m^2, to be enclosed is given by:

$$A = xy$$

Since the rope is 5 m long:

$$2x + y = 5 \qquad \text{or} \qquad y = 5 - 2x$$

Writing A in terms of x only:

$$A = x(5 - 2x)$$
$$= 5x - 2x^2$$

To maximise A, which is now written as a function of x, differentiate A with respect to x:

$$\frac{\mathrm{d}A}{\mathrm{d}x} = 5 - 4x$$

At a turning point, $\frac{dA}{dx} = 0$, so:

$$5 - 4x = 0$$
$$x = \tfrac{5}{4} = 1.25$$

When x is just less than 1.25, $\frac{dA}{dx}$ is positive, and when x is just greater than 1.25, $\frac{dA}{dx}$ is negative. Therefore when $x = 1.25$ m the area is a maximum (or $\frac{d^2y}{dx^2} = -4 < 0$, therefore a maximum).

The corresponding value of y is $5 - 2(1.25) = 2.5$ m. Judith should mark out a rectangle 1.25 m wide and 2.5 m long.

EXAMPLE 11.7

A stone is projected vertically upwards with a speed of $30\,\mathrm{m\,s^{-1}}$. Its height, h m, above the ground after t seconds ($t < 6$) is given by $h = 30t - 5t^2$.

(a) Find $\frac{dh}{dt}$.

(b) Find the maximum height reached.

(c) Sketch the graph of h against t.

Solution **(a)** $\frac{dh}{dt} = 30 - 10t = 10(3 - t)$

(b) For a turning point: $\frac{dh}{dt} = 0$

$$\Rightarrow 10(3 - t) = 0$$
$$\Rightarrow t = 3$$

Either $\frac{d^2h}{dt^2} = -10$ which is negative, so the height is a maximum at $t = 3$.

or:

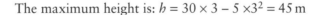

t	2	3	4
$\frac{dh}{dt}$	+	0	−

i.e. the height is a maximum at $t = 3$.

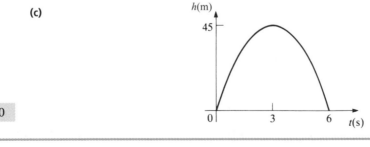

maximum

The maximum height is: $h = 30 \times 3 - 5 \times 3^2 = 45$ m

(c)

FIGURE 11.20

Note   The gradient, $\frac{dh}{dt}$ of a position–time graph, such as this one, is the velocity.

Exercise 11C

1 A farmer wants to construct a temporary rectangular enclosure of length x m and width y m for his prize bull while he works in the field. He has 120 m of fencing and wants to give the bull as much room to graze as possible.

(a) Write down an expression for y in terms of x.

(b) Write down an expression in terms of x for the area, A, to be enclosed.

(c) Find $\frac{dA}{dx}$, and so find the dimensions of the enclosure that give the bull the maximum area in which to graze. State this maximum area.

2 A square sheet of card of side 12 cm has four equal squares of side x cm cut from the corners. The sides are then turned up to make an open rectangular box to hold drawing pins as shown in the diagram.

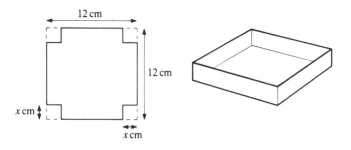

(a) Form an expression for the volume, V, of the box in terms of x.

(b) Find $\frac{dV}{dx}$, and show that the volume is a maximum when the depth is 2 cm.

3 The sum of two numbers, x and y, is 8.

(a) Write down an expression for y in terms of x.

(b) Write down an expression for S, the sum of the squares of these two numbers, in terms of x.

(c) By considering $\frac{dS}{dx}$, find the least value of the sum of their squares.

4 A new children's slide is to be built with a cross-section as shown in the diagram.

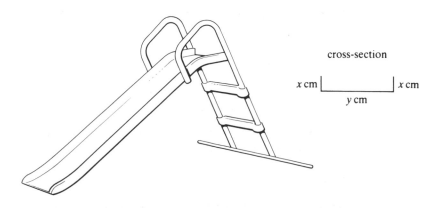

A long strip of metal 80 cm wide is available for the shute and will be bent to form the base and two sides.

The designer thinks that for maximum safety the area of the cross-section should be as large as possible.

(a) Write down an equation linking x and y.
(b) Using your answer to (a) form an expression for the cross-sectional area, A, in terms of x.
(c) By considering $\frac{dA}{dx}$, find the dimensions which make the slide as safe as possible.

5 A carpenter wants to make a box to hold toys. The box is to be made so that its volume is as large as possible. A rectangular sheet of thin plywood measuring 1.5 m by 1 m is available to cut into pieces as shown.

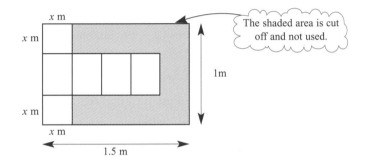

(a) Write down the dimensions of one of the four rectangular faces in terms of x.
(b) Form an expression for the volume, V, of the made-up box, in terms of x.
(c) Find $\frac{dV}{dx}$.
(d) Hence find the dimensions of a box with maximum volume, and the corresponding volume.

6 A piece of wire 16 cm long is cut into two pieces. One piece is $8x$ cm long and is bent to form a rectangle measuring $3x$ cm by x cm. The other piece is bent to form a square.

Find, in terms of x:

(a) the length of a side of the square
(b) the area of the square.

(c) Show that the combined area of the rectangle and the square is A cm^2 where $A = 7x^2 - 16x + 16$.

Find:

(d) the value of x for which A has its minimum value
(e) the minimum value of A.

7 Open display boxes are to be 12 cm high and have a base area of 1000 cm^2. Sketches of one of the boxes and its net are shown below.

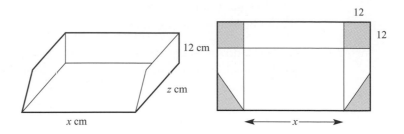

(a) The box is x cm wide. The depth from front to back is z cm as shown. Express z in terms of x.

The net of the box is stamped onto a rectangular piece of card, the shaded corners are removed and the card is folded and stuck to form the box.

(b) Find, in terms of x, the dimensions of the rectangle of card.
(c) Show that the total area, A cm^2, of the rectangle of card is given by
$$A = 1288 + 12x + \frac{24\,000}{x}.$$

(d) Find $\frac{dA}{dx}$ and use it to find the minimum area A showing clearly why the value is a minimum.

[MEI]

8 A cuboid has base of width x cm, length $2x$ cm and height h cm. Its volume is 72 cm^2.

(a) Show that its surface area is given by:
$$SA = 4x^2 + \frac{216}{x}$$

(b) Find the value of x for which the surface area is a minimum.
(c) Prove that the answer to part (b) gives a minimum surface area.

9 A cylinder of radius r cm has a volume of 100 cm^2.

(a) Find an expression for the height of the cylinder in terms of π and r.
(b) Show that the surface area of the cylinder is given by:
$$SA = \pi r^2 + \frac{200}{r}$$

(c) Find the minimum surface area of the cylinder. (You do not have to prove that it is a minimum.)

10 A box has sides of length $3l$ cm, l cm and h cm. The total length of all the edges is 84 cm.

(a) Show that the volume, V, of the box is given by:
$$V = 63l^2 - 12l^3 \text{ cm}^3$$

(b) Find the dimensions of the box with maximum volume.

EXERCISE 11D   **Examination-style questions**

1 Given that $f(x) = 4x^3 - 12x - \frac{3}{x}$ $(x \neq 0)$:

    (a) find $f'(x)$

    (b) prove that $f(x)$ is an increasing function.

2 A cylindrical biscuit tin has a close-fitting lid which overlaps the tin by 1 cm, as shown in the diagram. The radii of the tin and the lid are both x cm. The tin and the lid are made from a thin metal of area 80π cm^2 and there is no wastage. The volume of the tin is V cm^3.

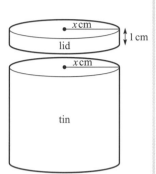

    (a) Show that $V = \pi(40x - x^2 - x^3)$.

    Given that x can vary:

    (b) use differentiation to find the positive value of x for which V is stationary

    (c) prove that this value of x gives a maximum value of V

    (d) find this maximum value of V

    (e) determine the percentage of the sheet metal used in the lid when V is a maximum.

                                   [Edexcel]

3 The volume of a closed cylindrical tank is 1000 cm^3. The radius of the base is r cm.

    (a) Find, in terms of r, an expression for the surface area S of the tank.

    (b) Determine the value of r for which S is a minimum.

    (c) Prove that this value of r gives a minimum value of S.

    (d) Calculate the surface area for this value of r.

4 The diagram shows part of the graph of the function $f(x)$ where $f(x) = x + \frac{4}{x}$.

    (a) Show that the minimum and maximum turning points are at $(2, 4)$ and $(-2, -4)$ respectively.

    (b) For what values of x is the function increasing?

5 It is given that $y = x^{\frac{3}{2}} + \frac{48}{x}$, $x > 0$.

    (a) Find the value of x and the value of y when $\frac{dy}{dx} = 0$.

    (b) Show that the value of y found in (a) is a minimum.

                                   [Edexcel]

6 Given that $f(x) = 15 - 7x - 2x^2$:

  (a) find the coordinates of all the points at which the graph of $y = f(x)$ crosses the coordinate axes

  (b) sketch the graph of $y = f(x)$

  (c) calculate the coordinates of the stationary points of $f(x)$.

  [Edexcel]

7 The function, f, defined for $x \in \Re$, $x > 0$, is such that:
$$f'(x) = x^2 - 2 + \frac{1}{x^2}$$

  (a) Find the value of $f''(x)$ at $x = 4$.

  (b) Given that $f(3) = 0$, find $f(x)$.

  (c) Prove that f is an increasing function.

  [Edexcel]

8 For the curve C with equation $y = x^4 - 8x^2 + 3$:

  (a) find $\frac{dy}{dx}$

  (b) find the coordinates of each of the stationary points

  (c) determine the nature of each stationary point

  (d) sketch the graph of C.

  [Edexcel, adapted]

9 A rectangular sheet of metal measures 50 cm by 40 cm. Squares of side x cm are cut from each corner of the sheet and the remainder is folded along the dotted lines to make an open tray, as shown in the diagram.

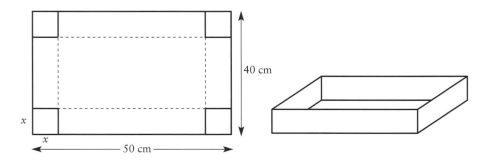

  (a) Show that the volume, V cm^3, of the tray is given by:
$$V = 4x(x^2 - 45x + 500)$$

  (b) State the range of possible values of x.

  (c) Find the value of x for which V is a maximum.

  (d) Hence find the maximum value for V.

  (e) Justify that the value of V you have found in part (d) is a maximum.

  [Edexcel]

10 The diagram shows an open tank for storing water, ABCDEF. The sides ABFE and CDEF are rectangles. The triangular ends, ADE and BCF are isosceles and $\angle AED = \angle BFC = 90°$. The ends ADE and BCF are vertical and EF is horizontal.

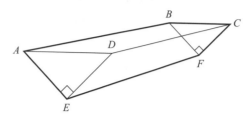

Given that AD = x metres:

(a) show that the area of $\triangle ADE$ is $\frac{1}{4}x^2$ m^2.

Given also that the capacity of the container is 4000 m^3 and that the total area of the two triangular and two rectangular sides of the container is S m^2:

(b) show that $S = \frac{x^2}{2} + \frac{16\,000\,\sqrt{2}}{x}$.

Given that x can vary:

(c) use calculus to find the minimum value of S

(d) justify that the value of S you have found is a minimum.

[Edexcel]

KEY POINTS

1 Differentiation

$$\left. \begin{array}{l} \text{If } y = kx^n \;\; \Rightarrow \;\; \dfrac{\mathrm{d}y}{\mathrm{d}x} = nkx^{n-1} \\[2mm] \text{If } y = kx \;\;\; \Rightarrow \;\; \dfrac{\mathrm{d}y}{\mathrm{d}x} = k \\[2mm] \text{If } y = k \;\;\;\; \Rightarrow \;\; \dfrac{\mathrm{d}y}{\mathrm{d}x} = 0 \end{array} \right\} \quad \begin{array}{l} \text{where } n \text{ is a rational number} \\ \text{and } k \text{ is a constant.} \end{array}$$

2 Stationary Points

- At a stationary point $\dfrac{\mathrm{d}y}{\mathrm{d}x} = 0$.
- The nature of a stationary point can be determined by looking at the sign of the gradient at points just on either side of it.

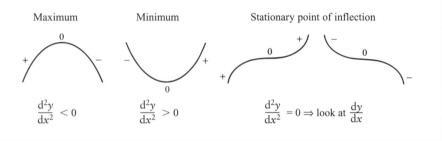

Maximum	Minimum	Stationary point of inflection
$\dfrac{\mathrm{d}^2 y}{\mathrm{d}x^2} < 0$	$\dfrac{\mathrm{d}^2 y}{\mathrm{d}x^2} > 0$	$\dfrac{\mathrm{d}^2 y}{\mathrm{d}x^2} = 0 \Rightarrow$ look at $\dfrac{\mathrm{d}y}{\mathrm{d}x}$

Chapter twelve

INTEGRATION

Think you mid all this mighty sum of things for ever speaking

That nothing of itself will ever come, but we must still be seeking.

William Wordsworth

● ● ● ● ● ● ● ● ● ● ● ● ● ●

You will recall from *Pure Mathematics: Core 1*, Chapter 5, that integration was the reverse of differentiation. You used the integration rule:

$$\int x^n \mathrm{d}x = \frac{1}{n+1}x^{n+1} + c \qquad n \neq -1$$

EXAMPLE 12.1

Find $\int (x^2 + \frac{1}{\sqrt{x}})\mathrm{d}x$

Solution
$$\int (x^2 + \frac{1}{\sqrt{x}})\,\mathrm{d}x = \int (x^2 + x^{-\frac{1}{2}})\,\mathrm{d}x$$
$$= \frac{1}{3}x^3 + 2x^{\frac{1}{2}} + c$$

Now you will extend your integrating skills, learning how to use limits and finding areas under curves.

DEFINITE INTEGRALS

In example 12.1 you reviewed how to find an indefinite integral in which the answer includes an arbitrary constant, c.

A *definite integral* has *limits* and it can be calculated by taking its value at the upper limit and subtracting its value at the lower limit.

For example:

$$\int_1^2 (2x + 3)\mathrm{d}x = [x^2 + 3x]_1^2 \qquad \text{2 is the } \textit{upper limit}$$
$$\text{1 is the } \textit{lower limit.}$$

$$= \text{value at 2} - \text{value at 1}$$
$$= (2^2 + 3 \times 2) - (1^2 + 3 \times 1)$$
$$= 10 - 4$$
$$= 6$$

There is no need for the constant c because if it were included it would simply cancel out.

In general:

$$\text{if} \quad \int f(x)dx = F(x) + c$$

$$\text{then} \int_a^b f(x)dx = F(b) - F(a)$$

It follows that if you reverse the limits you will reverse the sign of the answer, that is:

$$\int_b^a f(x)dx = -\int_a^b f(x)dx$$

EXAMPLE 12.2

Evaluate this integral.
$$\int_1^4 (2x + \sqrt{x})dx$$

Solution $\int_1^4 (2x + \sqrt{x})dx = [x^2 + \frac{2}{3}x^{\frac{3}{2}}]_1^4$

$$= (16 + \frac{16}{3}) - (1 + \frac{2}{3})$$

$$= 19\frac{2}{3}$$

EXERCISE 12A

1 Evaluate the following definite integrals.

(a) $\int_1^2 2x\,dx$ (b) $\int_0^3 2x\,dx$ (c) $\int_0^3 3x^2\,dx$

(d) $\int_2^5 x\,dx$ (e) $\int_5^6 (2x+1)\,dx$ (f) $\int_{-1}^2 (2x+4)\,dx$

(g) $\int_3^5 (3x^2+2x)\,dx$ (h) $\int_0^1 x^5\,dx$ (i) $\int_{-2}^{-1}(x^4+x^3)\,dx$

(j) $\int_{-1}^1 x^3\,dx$ (k) $\int_{-5}^4 (x^3+3x)\,dx$ (l) $\int_{-3}^{-2} 5dx$

2 Evaluate the following definite integrals.

(a) $\int_1^2 (4x^{-5} - 3x^{-4})dx$ (b) $\int_2^5 (3 + x^{-2})dx$

(c) $\int_8^{27} x^{\frac{2}{3}}\,dx$ (d) $\int_1^4 \frac{x+1}{2\sqrt{x}}dx$

(e) $\int_0^5 (x+1)(x-1)dx$ (f) $\int_0^3 (2x-3)^2dx$

(g) $\int_1^8 (\frac{\sqrt[3]{x}}{3} - \frac{3}{\sqrt[3]{x}})dx$ (h) $\int_1^{16}(1 + \frac{1}{2}x^2 + 3x^{-\frac{1}{4}})dx$

(i) $\int_1^2 (\frac{1}{x^2} - \frac{1}{x^3})dx$ (j) $\int_0^2 x(1+x)(2-x)dx$

(k) $\int_1^8 (2x^{\frac{1}{2}} + 3x^{-\frac{1}{3}})dx$ (l) $\int_2^3 (1 - 2x - 3x^2 - 4x^3)dx$

3 Evaluate the following definite integrals. Give answers as fractions or to 3 significant figures as appropriate.

(a) $\int_1^4 3x^{-2}dx$

(b) $\int_2^4 8x^{-3}dx$

(c) $\int_2^4 12x^{\frac{1}{2}}dx$

(d) $\int_{-3}^{-1} \frac{6}{x^3}dx$

(e) $\int_1^8 \left(\frac{x^2+3x+4}{x^4}\right)dx$

(f) $\int_4^9 \left(\sqrt{x} - \frac{1}{\sqrt{x}}\right)dx$

FINDING THE AREA UNDER A CURVE

Figure 12.1 shows a curve $y = f(x)$ with the area under it (the required area) shaded.

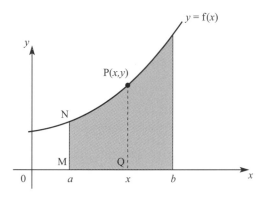

FIGURE 12.1

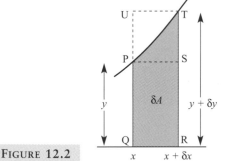

FIGURE 12.2

P is a point on the curve and its x-coordinate is between a and b. Let A denote the area bounded by MNPQ. As P moves, the values of A and x change, so you can see that the area A depends on the value of x. Figure 12.2 enlarges part of figure 12.1 and introduces T to the right of P.

If T is close to P it is appropriate to use the notation δx (a small change in x) for the difference in their x-coordinates and δy for the difference in their y-coordinates. The area shaded in figure 12.2 is then referred to as δA (a small change in A).

This area δA must lie between the areas of the rectangles PQRS and UQRT.

$$y\delta x < \delta A < (y + \delta y)\delta x$$

Dividing by δx:

$$y < \frac{\delta A}{\delta x} < y + \delta y$$

In the limit as $\delta x \to 0$, δy also approaches 0 so δA is sandwiched between y and something which tends to y.

But $\displaystyle\lim_{\delta x \to 0} \frac{\delta A}{\delta x} = \frac{dA}{dx}$

This gives $\frac{dA}{dx} = y$.

Note

This important result is known as the fundamental theorem of calculus: the rate of change of the area under a curve is equal to the length of the moving boundary

It follows that:

$$\text{area } A = \int y\,dx$$

STANDARDISING THE PROCEDURE

Suppose that you want to find the area between the curve $y = f(x)$, the x-axis, and the lines $x = a$ and $x = b$. This is shown shaded in figure 12.3.

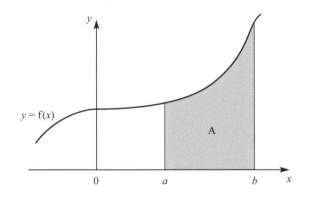

FIGURE 12.3

The shaded area is given by:

$$A = \int y\,dx = \int f(x)\,dx = F(x) + c$$

When the value of x is a the shaded area is 0, that is $F(a) + c = 0$ so $c = -F(a)$. When the value of x is b then the shaded area A is:

$$F(b) + c = F(b) - F(a) = \int_a^b f(x)\,dx$$

So it follows that:

> the area between the curve $y = f(x)$ and the x-axis from $x = a$ to $x = b$ is
> $$\int_a^b f(x)\,dx$$

Note

Over the interval $x = a$ to $x = b$ the function $y = f(x)$ must be continuous, that is have no breaks, and must be above the x-axis. You will look at cases where the function is below or crosses the x-axis later in the chapter.

EXAMPLE 12.3

Find the area under the curve $y = x^2$ from $x = 1$ to $x = 3$.

Solution

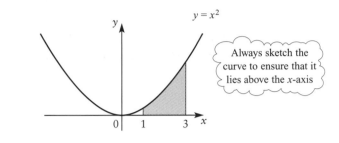

FIGURE 12.4

Always sketch the curve to ensure that it lies above the x-axis

$$\text{Area} = \int_1^3 x^2 \mathrm{d}x = \left[\tfrac{1}{3}\, x^3\right]_1^3 = \tfrac{1}{3} \times 3^3 - \tfrac{1}{3} \times 1^3 = 8\tfrac{2}{3} \text{ square units}$$

Note

The term 'square units' is used since area is a square measure and the units are unknown.

EXAMPLE 12.4

Find the area between the curve $y = 20 - 3x^2$, the x-axis and the lines $x = 1$ and $x = 2$.

Solution

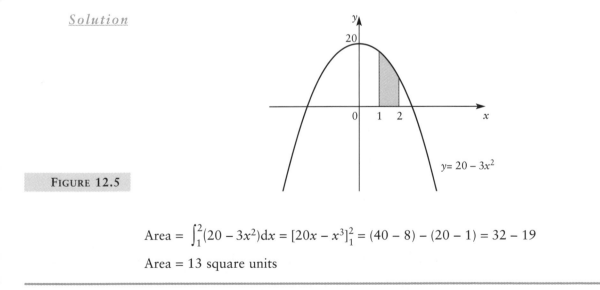

FIGURE 12.5

$$\text{Area} = \int_1^2 (20 - 3x^2)\mathrm{d}x = [20x - x^3]_1^2 = (40 - 8) - (20 - 1) = 32 - 19$$

Area = 13 square units

EXERCISE 12B   1   The graph of $y = 2x$ is shown opposite.

The shaded region is bounded by $y = 2x$,
the x-axis and the lines $x = 2$ and $x = 3$.

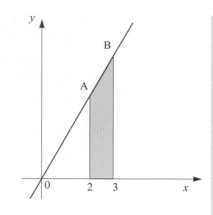

(a) Find the coordinates of the points A
and B in the diagram.

(b) Use the formula for the area of a
trapezium to find the area of the
shaded region.

(c) Find the area of the shaded region
as $\int_2^3 2x \, dx$, and confirm that your
answer is the same as that for (b).

(d) The method of (b) cannot be used to find the area under the curve $y = x^2$
bounded by the lines $x = 2$ and $x = 3$. Why?

2   (a) Sketch the curve $y = x^2$ for $-1 \leqslant x \leqslant 3$ and shade in the area bounded by
the curve, the lines $x = 1$ and $x = 2$ and the x-axis.

(b) Find, by integration, the area of the region you have shaded.

3   (a) Sketch the curve $y = 4 - x^2$ for $-3 \leqslant x \leqslant 3$.

(b) For what values of x is the curve above the x-axis?

(c) Find the area between the curve and the x-axis when the curve is above
the x-axis.

4   (a) Sketch the graph of $y = \sqrt{x}$ for $x = 0$ to $x = 9$.

(b) Calculate the area enclosed between the curve, the x-axis and the lines
$x = 1$ and $x = 4$.

5   (a) Sketch the graph of $y = \frac{1}{x^2}$ for values of x between -4 and 4.

(b) Calculate the area enclosed between the curve, the x-axis and the lines
$x = 1$ and $x = 3$.

(c) With reference to your sketch, state the area between the curve, the x-axis
and the lines $x = -1$ and $x = -3$.

(d) Explain why you cannot find the area enclosed between the curve, the
x-axis and the lines $x = -1$ and $x = 1$.

6   (a) Sketch the graph of $y = (x - 2)^2$ for values of x between $x = -1$ and $x = +5$.
Shade in the area under the curve, between $x = 0$ and $x = 2$.

(b) Calculate the area you have shaded.

[MEI]

7   (a) Sketch the graph of $y = (x + 1)^2$ for values of x between $x = -1$ and $x = 4$.

(b) Shade in the area under the curve between $x = 1$, $x = 3$ and the x-axis.
Calculate this area.

[MEI]

8 **(a)** Sketch the curves $y = x^2$ and $y = x^3$ for $0 \leqslant x \leqslant 2$.

(b) Which is the higher curve within the region $0 < x < 1$?

(c) Find the area under each curve for $0 \leqslant x \leqslant 1$.

(d) Which would you expect to be greater, $\int_1^2 x^2 \, dx$ or $\int_1^2 x^3 \, dx$?

Explain your answer in terms of your sketches, and confirm it by calculation.

9 **(a)** Sketch the curve $y = x^2 - 1$ for $-3 \leqslant x \leqslant 3$.

(b) Find the area of the region bounded by $y = x^2 - 1$, the line $x = 2$ and the x-axis.

(c) Sketch the curve $y = x^2 - 2x$ for $-2 \leqslant x \leqslant 4$.

(d) Find the area of the region bounded by $y = x^2 - 2x$, the line $x = 3$ and the x-axis.

(e) Comment on your answers to (b) and (d).

10 **(a)** Shade, on a suitable sketch, the region for which the area is given by

$$\int_{-1}^2 (9 - x^2) \, dx$$

(b) Find the area of the shaded region.

11 **(a)** Sketch the curve with equation $y = x^2 + 1$ for $-3 \leqslant x \leqslant 3$.

(b) Find the area of the region bounded by the curve, the lines $x = 2$ and $x = 3$, and the x-axis.

(c) Predict, with reasons, the value of $\int_{-3}^{-2} (x^2 + 1) \, dx$.

(d) Evaluate $\int_{-2}^{-3} (x^2 + 1) \, dx$.

12 **(a)** Sketch the curve with equation $y = x^2 - 2x + 1$ for $-1 \leqslant x \leqslant 4$.

(b) State, with reasons, which area you would expect from your sketch to be larger:

$$\int_{-1}^3 (x^2 - 2x + 1) \, dx \qquad \text{or} \qquad \int_0^4 (x^2 - 2x + 1) \, dx.$$

(c) Calculate the values of the two integrals. Was your answer to (b) correct?

13 **(a)** Sketch the curve with equation $y = x^3 - 6x^2 + 11x - 6$ for $0 \leqslant x \leqslant 4$.

(b) Shade the regions with areas given by:

(i) $\int_1^2 (x^3 - 6x^2 + 11x - 6) \, dx$

(ii) $\int_3^4 (x^3 - 6x^2 + 11x - 6) \, dx$

(c) Find the values of these two areas.

(d) Find the value of $\int_1^{1.5} (x^3 - 6x^2 + 11x - 6) \, dx$.

What does this, taken together with one of your answers to (c), indicate to you about the position of the maximum point between $x = 1$ and $x = 2$?

14 (a) Differentiate $y = -x^2 + 2x + 3$.

(b) Find the maximum value of y.

A gardener is considering a new design for his garden. He has a rectangular lawn measuring 5 m by 3 m, and wants to dig up part of it to include a flowerbed. He draws a plan of the lawn and flowerbed on graph paper, taking the bottom and left-hand edges as the axes, and chooses the scale so that 1 unit along each axis represents 1 metre on the ground.

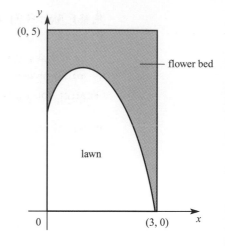

(c) The equation of the curved edge of the flower bed is:

$$y = -x^2 + 2x + 3$$

Calculate the area of the flowerbed.

[MEI]

15 The area under a velocity–time graph is the distance travelled. The graph below shows the relationship $v = t^2$ for the first 3 seconds of motion for a body moving from rest.

(a) Find the shaded area, and hence the distance travelled (in metres), during the first 3 seconds of motion.

(b) Did the body travel further in the third second than it did in the first 2 seconds together?

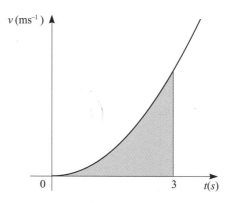

16 (a) Show that $x^3 - 6x^2 + 11x - 6$ has factor $x - 3$. Hence factorise the expression completely.

(b) Sketch the graph of $y = x^3 - 6x^2 + 11x - 6$.

(c) Find the area of the enclosed regions between the curve and the x-axis.

AREAS BELOW THE *x*-AXIS

When a graph goes below the *x*-axis, the corresponding *y*-value is negative and so the value of $y\delta x$ is negative (figure 12.6). So when an integral turns out to be negative you know that the area is below the *x*-axis.

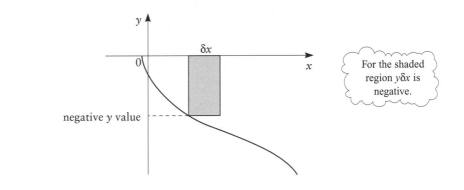

For the shaded region $y\delta x$ is negative.

FIGURE 12.6

EXAMPLE 12.5

Find the area of the region bounded by the curve with equation $y = x^2 - 4x$, the lines $x = 1$ and $x = 2$, and the *x*-axis.

Solution   The region in question is shaded in figure 12.7.

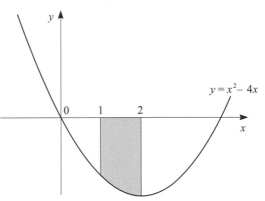

FIGURE 12.7

The shaded area is:

$$\text{Area} = \int_1^2 (x^2 - 4x)\,dx$$

$$= \left[\frac{x^3}{3} - 2x^2\right]_1^2$$

$$= \left(\frac{8}{3} - 8\right) - \left(\frac{1}{3} - 2\right)$$

$$= -\frac{11}{3}$$

Therefore the shaded area is $\frac{11}{3}$ square units, and it is below the *x*-axis.

EXAMPLE 12.6

Find the area between the curve and the x-axis for the function
$y = x^2 + 3x$ between $x = -1$ and $x = 2$.

Solution   The first step is to draw a sketch of the function to see whether the curve goes below
the x-axis (see figure 12.8).

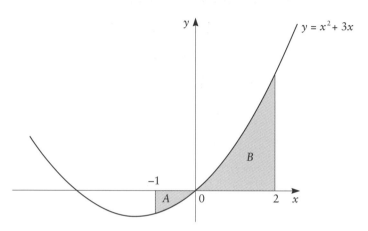

FIGURE 12.8

This shows that the y-values are positive for $0 < x < 2$ and negative for
$-1 < x < 0$. You therefore need to calculate the area in two parts.

$$\text{Area } A = \int_{-1}^{0} (x^2 + 3x)\,dx$$

$$= \left[\frac{x^3}{3} + \frac{3x^2}{2} \right]_{-1}^{0}$$

$$= 0 - \left(-\frac{1}{3} + \frac{3}{2} \right)$$

$$= -\frac{7}{6}$$

$$\text{Area } B = \int_{0}^{2} (x^2 + 3x)\,dx$$

$$= \left[\frac{x^3}{3} + \frac{3x^2}{2} \right]_{0}^{2}$$

$$= \left(\frac{8}{3} + 6 \right) - 0$$

$$= \frac{26}{3}$$

$$\text{Total area} = \frac{26}{3} + \frac{7}{6}$$

$$= \frac{59}{6} \text{ square units}$$

EXERCISE 12C   In each of questions 1 to 10 you are given the equation of a curve. Sketch the curve and find the area between the curve and the x-axis between the given bounds.

1 $y = x^3$ between $x = -3$ and $x = 0$.

2 $y = x^2 - 4$ between $x = -1$ and $x = 2$.

3 $y = x^5 - 2$ between $x = -1$ and $x = 0$.

4 $y = 3x^2 - 4x$ between $x = 0$ and $x = 1$.

5 $y = x^4 - x^2$ between $x = -1$ and $x = 1$.

6 $y = 4x^3 - 3x^2$ between $x = -1$ and $x = 0.5$.

7 $y = x^5 - x^3$ between $x = -1$ and $x = 1$.

8 $y = x^2 - x - 2$ between $x = -2$ and $x = 3$.

9 $y = x^3 + x^2 - 2x$ between $x = -3$ and $x = 2$.

10 $y = x^3 + x^2$ between $x = -2$ and $x = 2$.

11

The diagram shows a sketch of part of the curve for which the equation is:

$$y = 5x^4 - x^5$$

(a) Find $\frac{dy}{dx}$.

    Calculate the coordinates of the stationary points.

(b) Calculate the area of the shaded region enclosed by the curve and the x-axis.

(c) Evaluate $\int_0^6 x^4(5 - x)\,dx$ and comment on your result.

[MEI]

THE AREA BETWEEN TWO CURVES

EXAMPLE 12.7

Find the area enclosed by the line $y = x + 1$ and the curve $y = x^2 - 2x + 1$.

Solution First draw a sketch showing where these graphs intersect (see figure 12.9).

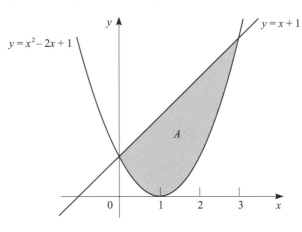

FIGURE 12.9

When they intersect:

$$x^2 - 2x + 1 = x + 1$$
$$\Rightarrow \quad x^2 - 3x = 0$$
$$\Rightarrow \quad x(x - 3) = 0$$
$$\Rightarrow \quad x = 0 \text{ or } 3$$

The shaded area can now be found in one of two ways.

Method 1
Area A can be treated as the difference between the two areas, B and C, shown in figure 12.10.

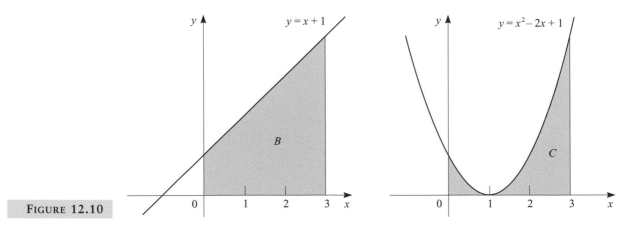

FIGURE 12.10

$$A = B - C$$
$$= \int_0^3 (x + 1)\,dx - \int_0^3 (x^2 - 2x + 1)\,dx$$
$$= \left[\frac{x^2}{2} + x\right]_0^3 - \left[\frac{x^3}{3} - x^2 + x\right]_0^3$$
$$= \left[\left(\frac{9}{2} + 3\right) - 0\right] - \left[\left(\frac{27}{3} - 9 + 3\right) - 0\right]$$
$$= \frac{9}{2} \text{ square units}$$

Method 2

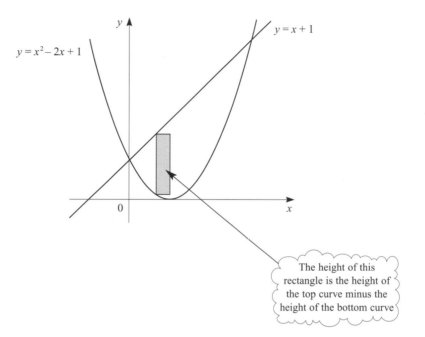

The height of this rectangle is the height of the top curve minus the height of the bottom curve

FIGURE 12.11

$$A = \int_0^3 \{\text{top curve} - \text{bottom curve}\}\,dx$$
$$= \int_0^3 ((x + 1) - (x^2 - 2x + 1))\,dx$$
$$= \int_0^3 (3x - x^2)\,dx$$
$$= \left[\frac{3x^2}{2} - \frac{x^3}{3}\right]_0^3$$
$$= \left(\frac{27}{2} - 9\right) - (0)$$
$$= \frac{9}{2} \text{ square units}$$

EXERCISE 12D

1 The diagram below shows the curve $y = x^2$ and the line $y = 9$. The enclosed region has been shaded.

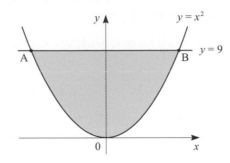

(a) Find the two points of intersection (labelled A and B).

(b) Using integration, show that the area of the shaded region is 36 square units.

2 (a) Sketch the curve $y = x^3$ and the line $y = 4x$ on the same axes.

(b) Find the coordinates of the points of intersection of the curve $y = x^3$ and the line $y = 4x$.

(c) Find the total area of the region bounded by $y = x^3$ and $y = 4x$.

3 (a) Sketch the curves $y = x^2$ and $y = 8 - x^2$ and the line $y = 4$ on the same axes.

(b) Find the area of the region enclosed by the line $y = 4$ and the curve $y = x^2$.

(c) Find the area of the region enclosed by the line $y = 4$ and the curve $y = 8 - x^2$.

(d) Find the area enclosed by the curves $y = x^2$ and $y = 8 - x^2$.

4 (a) Sketch the curve $y = x^2 - 6x$ and the line $y = -5$.

(b) Find the coordinates of the points of intersection of the line and the curve.

(c) Find the area of the region enclosed by the line and the curve.

5 (a) Sketch the curve $y = x(4 - x)$ and the line $y = 2x - 3$.

(b) Find the coordinates of the points of intersection of the line and the curve.

(c) Find the area of the region enclosed by the line and the curve.

6 (a) Sketch the curve with equation $y = x^3 + 1$ and the line $y = 4x + 1$.

(b) Find the areas of the two regions enclosed by the line and the curve.

7 (a) On the same graph, sketch the line with equation $y = x + 1$ and the curve with equation $y = 5x - x^2 + 6$. Values of x should be taken from $x = -2$ to $x = +8$. Shade in the region between the line and the curve.

(b) Calculate the points of intersection of the line $y = x + 1$ and the curve $y = 5x - x^2 + 6$.

(c) Use integration to calculate the area of the region which you have shaded in (a).

[MEI]

8 A shelf support is made of wood, with its cross-section in the shape shown in the diagram. The top and bottom edges of the support are horizontal.

The shape of the curved edge of the shelf support is modelled by the equation

$$8y = (2x - 3)^3 \quad \text{for} \quad -1 \leqslant y \leqslant 1.$$

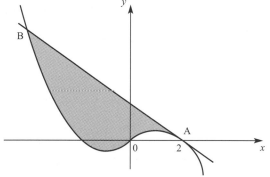

(a) Find the coordinates of the point P.

(b) Using Pascal's triangle, or otherwise, write down the expansion of $(2x - 3)^3$.

(c) Write down the integral which should be calculated in order to find the area AQR between the curve and the x-axis.

(d) Carry out the integration to find this area.

(e) Hence or otherwise calculate the area of cross-section of the shelf support.

[MEI]

9 The diagram shows the graph of $y = 4x - x^3$. The point A has coordinates $(2, 0)$.

(a) Find $\frac{dy}{dx}$.
Then find the equation of the tangent to the curve at A.

(b) The tangent at A meets the curve again at the point B. Show that the x-coordinate of B satisfies the equation $x^3 - 12x + 16 = 0$.
Find the coordinates of B.

(c) Calculate the area of the shaded region between the straight line AB and the curve.

[MEI]

10 (a) On the same diagram, sketch $y = (x - 2)^3$ and the straight line $y = 4x - 8$.
Shade the region enclosed between the curve and the line.

(b) Show that $(x - 2)^3 = x^3 - 6x^2 + 12x - 8$.

(c) Find the size of the areas you shaded in part (a).

NUMERICAL INTEGRATION

There are times when you need to find the area under a graph but cannot do this by the integration methods you have met already.

- The function may be one that cannot be integrated algebraically. (There are many such functions.)
- The function may be one that can be integrated algebraically but which requires a technique with which you are unfamiliar.
- It may be that you do not know the function in algebraic form, but just have a set of points (perhaps derived from an experiment).

In these circumstances you can always find an approximate answer using a numerical method, but you must:

(a) have a clear picture in your mind of the graph of the function, and how your method estimates the area beneath it

(b) understand that a numerical answer without any estimate of its accuracy, or error bounds, is valueless.

THE TRAPEZIUM RULE

In this chapter just one numerical method of integration is introduced, namely the *trapezium rule*. As an illustration of the rule, we shall use it to find the area under the curve $y = \sqrt{5x - x^2}$ for values of x between 0 and 4.

It is in fact possible to integrate this function algebraically, but not using the techniques that you have met so far.

Note

You should not use a numerical method when an algebraic (sometimes called analytic) technique is available to you. Numerical methods should be used only when other methods fail.

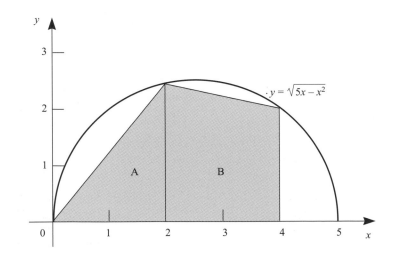

FIGURE 12.12

300

Figure 12.12 shows the area approximated by two trapezia of equal width.

Remember the formula for the area of a trapezium:

$$\text{Area} = \tfrac{1}{2}h(a + b)$$

where a and b are the lengths of the parallel sides and h the distance between them.

In the cases of the trapezia A and B, the parallel sides are vertical. The left-hand side of trapezium A has 0 height, and so the trapezium is also a triangle.

When $x = 0$ $\Rightarrow$ $y = \sqrt{0} = 0$

when $x = 2$ $\Rightarrow$ $y = \sqrt{6} = 2.4495$ (to 4 d.p.)

when $x = 4$ $\Rightarrow$ $y = \sqrt{4} = 2$

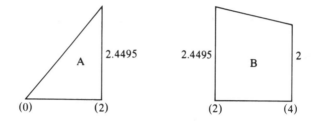

FIGURE 12.13

The area of trapezium A $= \tfrac{1}{2} \times 2 \times (0 + 2.4495) = \;\underline{2.4495}$

The area of trapezium B $= \tfrac{1}{2} \times 2 \times (2.4495 + 2) = \;\underline{4.4495}$

$$\textit{Total} \quad \underline{6.8990}$$

For greater accuracy you can use four trapezia, P, Q, R and S, each of width 1 unit as in figure 12.14. The area is estimated in just the same way.

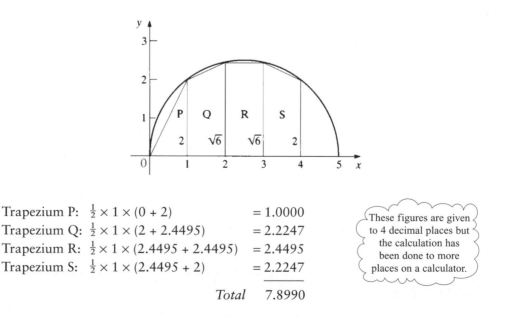

FIGURE 12.14

Trapezium P: $\tfrac{1}{2} \times 1 \times (0 + 2)$ $= 1.0000$

Trapezium Q: $\tfrac{1}{2} \times 1 \times (2 + 2.4495)$ $= 2.2247$

Trapezium R: $\tfrac{1}{2} \times 1 \times (2.4495 + 2.4495)$ $= 2.4495$

Trapezium S: $\tfrac{1}{2} \times 1 \times (2.4495 + 2)$ $= 2.2247$

$$\textit{Total} \quad \underline{7.8990}$$

These figures are given to 4 decimal places but the calculation has been done to more places on a calculator.

Accuracy

In this example, the first two estimates are 6.8989... and 7.8989... . You can see from figure 7.14 that the trapezia all lie underneath the curve, and so in this case the trapezium rule estimate of 7.8989... must be too small. You cannot, however, say by how much. To find that out you will need to take progressively more strips and see how the estimate homes in. Using 8 strips gives an estimate of 8.2407..., and 16 strips gives 8.3578... . The first figure, 8, looks reasonably certain but it is still not clear whether the second is 3, 4 or even 5. You need to take even more strips to be able to decide. In this example, the convergence is unusually slow because of the high curvature of the curve.

The rule

In general, the area that the integral $\int_a^b f(x)\,dx$ represents is divided into trapezia. The width of each trapezium is h. This is illustrated in figure 12.15.

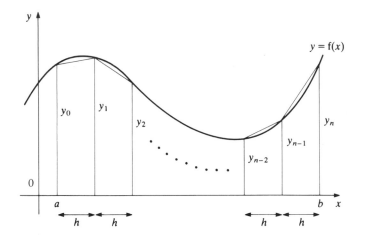

FIGURE 12.15

If y_0 is the value of the function at the left-hand limit $x = a$, y_1 is the value at $x = a + h$, y_2 at $a + 2h$ up to y_n at $x = b$ then $f(x)dx$ is approximated by the sum of the areas A of all the trapezia.

So:

$$A \approx \tfrac{1}{2}(y_0 + y_1)h + \tfrac{1}{2}(y_1 + y_2)h + \tfrac{1}{2}(y_2 + y_3)h + \ldots + \tfrac{1}{2}(y_{n-1} + y_n)h$$

and factorising gives:

$$A \approx \tfrac{1}{2}h(y_0 + y_1 + y_1 + y_2 + y_2 + y_3 + \ldots y_{n+1} + y_n)$$

and this simplifies to the trapezium rule:

$$A \approx \tfrac{1}{2}h\{y_0 + y_n + 2(y_1 + y_2 + y_3 + \ldots + y_{n-1})\}$$

This is often said in words as:

$$A \approx \tfrac{1}{2} \times \text{strip width} \times \{\text{first} + \text{last} + 2 \times \text{rest}\}$$

You should learn the proof of the derivation of the trapezium rule.

You have to decide from the shape of the graph of $y = f(x)$ as to whether the trapezium rule gives an overestimate, an underestimate or whether you cannot tell.

The following diagrams illustrate different situations.

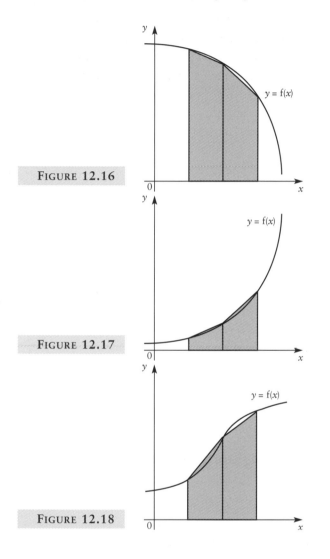

FIGURE 12.16

The trapezium rule will give an underestimate.

FIGURE 12.17

The trapezium rule will give an overestimate.

FIGURE 12.18

You cannot tell whether the trapezium rule gives an underestimate or an overestimate.

EXAMPLE 12.8

Use the trapezium rule with 5 ordinates to evaluate $\int_1^2 \frac{1}{x^2}\, dx$, correct to 4 significant figures.

By drawing a sketch state whether the trapezium rule gives an overestimate or an underestimate to the exact value.

Calculate the percentage error between the trapezium rule value and the exact value.

Solution   The word 'ordinate' means y-value, so 5 ordinates means 4 strips. As you wish to find the integral from $x = 1$ to $x = 2$ the width of each strip, h, is 0.25.
Tabulating the values gives:

x	$\dfrac{1}{x^2}$	
1	1	
1.25	0.64	
1.5	0.444 44	} 1.410 975
1.75	0.032 65	
2	0.25	

and these are used in the trapezium rule to give:

$$\int_1^2 \frac{1}{x^2}\, dx \approx \tfrac{1}{2} \times 0.25 \times \{1 + 0.25 + 2 \times 1.410\,975\} = 0.508\,993...$$

$$\approx 0.5090 \text{ (4 significant figures)}$$

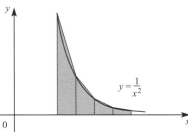

FIGURE 12.19

From figure 12.19 you can see that the trapezium rule gives an overestimate of the integral.

The exact value is:

$$\int_1^2 \frac{1}{x^2}\, dx = \left[-\frac{1}{x}\right]_1^2 = -\frac{1}{2} - (-1) = 0.5$$

So the percentage error is:

$$\frac{0.5090 - 0.5}{0.5} \times 100 = 1.8\%$$

Note   For any question using trigonometric functions, calculations must be carried out in radians.

EXERCISE 12E   In questions 1–3, use the trapezium rule to estimate the areas shown.

Start with 2 strips, then 4 and then 8 in each case.

State, with reasons, whether your final estimate is an overestimate or an underestimate.

1 $y = \dfrac{3}{(x + 1)}$, $x > -1$ 2 $y = x + \dfrac{1}{x}$

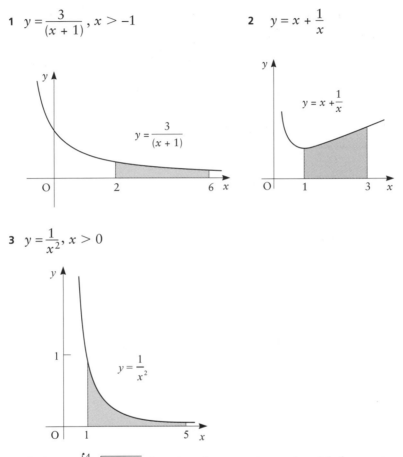

3 $y = \dfrac{1}{x^2}$, $x > 0$

4 Estimate $\displaystyle\int_0^4 \sqrt{2x + 1}\; dx$ using the trapezium rule with four strips.

5 (a) Draw the graph of $y = \frac{1}{x}$ for values of x between 1 and 10. Shade the region A bounded by the curve, the lines $x = 1$ and $x = 2$ and the x-axis, and the region B bounded by the curve, the lines $x = 4$ and $x = 8$ and the x-axis.
   (b) For each of the regions A and B, find estimates of the area using the trapezium rule with (i) 2   (ii) 4   (iii) 8 strips.
   (c) Which region has the greater area?

6 The speed v in m s^{-1} of a train is given at time t seconds in the following table.

t	0	10	20	30	40	50	60
v	0	5.0	6.7	8.2	9.5	10.6	11.6

The distance the train has travelled is given by the area under the graph of the speed (vertical axis) against time (horizontal axis).

(a) Estimate the distance the train travels in this 1-minute period.

(b) Give two reasons why your method cannot give a very accurate answer.

7 The definite integral $\int_0^1 \frac{1}{1 + x^2} \, dx$ is known to equal $\frac{\pi}{4}$.

(a) Using the trapezium rule for 5 strips, find an approximation for π.

(b) Repeat your calculation with 10 and 20 strips to obtain closer estimates.

(c) If you did not know the value of π, what value would you give it with confidence on the basis of your estimates in parts (a) and (b)?

8 The table below gives the values of a function $f(x)$ for different values of x.

x	0	0.5	1.0	1.5	2.0	2.5	3.0
$f(x)$	1.000	1.225	1.732	2.345	3.000	3.674	4.359

(a) Apply the trapezium rule to the values in this table to obtain an approximation for $\int_0^3 f(x) \, dx$.

(b) By considering the shape of the curve $y = f(x)$, explain whether the approximation calculated in part (a) is likely to be an overestimate or an underestimate of the true area under the curve $y = f(x)$ between $x = 0$ and $x = 3$.

[MEI]

9

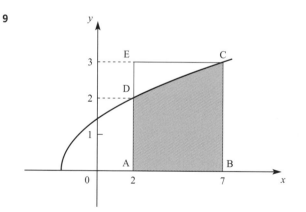

The graph of the function $y = \sqrt{2 + x}$ is given in the diagram. The area of the shaded region ABCD is to be found.

(a) Make a table of values for y, for integer values of x from $x = 2$ to $x = 7$, giving each value of y correct to 4 decimal places.

(b) Use the trapezium rule with 5 strips, each 1 unit wide, to calculate an estimate for the area ABCD. State, giving a reason, whether your estimate is too large or too small.

Another method is to consider the area ABCD as the area of the rectangle ABCE minus the area of the region CDE.

(c) Show that the area CDE is given by $\int_2^3 (y^2 - 4)\, dy$. Calculate the exact value of this integral.

(d) Find the exact value of the area ABCD. Hence find, as a percentage, the relative error in using the trapezium rule.

[MEI]

10 (a) The table gives the values of the function $(1 + x^2)^5$ for $x = 0$, 0.2 and 0.4.

x	0	0.1	0.2	0.3	0.4
$(1 + x^2)^5$	1.00000		1.21665		2.10034

Complete the table for $x = 0.1$ and 0.3.

Use the trapezium rule with 4 strips to estimate the value of the integral $\int_0^{0.4} (1 + x^2)^5\, dx$.

(b) Use the binomial theorem to expand $(1 + x^2)^5$ as a polynomial in ascending powers of x.

The integral $\int_0^{0.4} (1 + 5x^2 + 10x^4)\, dx$ is to be used to estimate the integral in part (a).

Evaluate $\int_0^{0.4} (1 + 5x^2 + 10x^4)\, dx$.

(c) The diagram below shows a sketch of the graph of $y = (1 + x^2)^5$. State whether the estimate in part (a) is too high or too low and use the graph to explain your answer.

State also whether the estimate in part (b) is too high or too low and explain your answer.

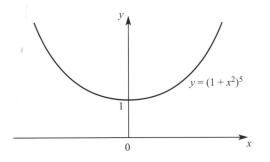

(d) How could the methods for estimating $\int_0^{0.4} (1 + x^2)^5\, dx$ used in parts (a) and (b) be improved to give greater accuracy?

[MEI]

EXERCISE 12F   **Examination-style questions**

1   Evaluate these integrals.

(a) $\int_{4}^{9}(\sqrt{x} + 2 - \frac{1}{\sqrt{x}})dx$ (b) $\int_{3}^{\infty}\frac{1}{x^3}dx$

2   (a)   Evaluate $\int_{3}^{5}(x^2 - 3x - 4)dx$.
   (b)   Find the position of the turning point of the graph of $y = x^2 - 3x - 4$.
   (c)   Sketch the graph of $y = x^2 - 3x - 4$.
   (d)   Explain why your answer to part (a) is not equal to the area enclosed between the curve and the x-axis from $x = 3$ to $x = 5$.

3   The curve, C, has equation $y = (x - 3)(x - 1)(x + 1)$.
   (a)   Sketch the graph of C.
   (b)   Find the magnitude of the total area enclosed between C and the x-axis.

4   $f(x) = \frac{(2 + 3\sqrt{x})^2}{\sqrt{x}}$, $x > 0$
   (a)   Express $f(x)$ in the form $Ax^{\frac{1}{2}} + B + Cx^{-\frac{1}{2}}$, stating the values of A, B and C.
   (b)   Find $\int f(x)dx$.
   (c)   Show that the area enclosed between $y = f(x)$ and the lines $y = 0$, $x = 4$ and $x = 9$ is 182 square units.

5   You are given that $f(x) = (2x^{\frac{3}{2}} - 3x^{-\frac{3}{2}})^2 + 5$, $x > 0$.
   (a)   Find, correct to 3 significant figures, the value of x for which $f(x) = 5$.
   (b)   Show that $f(x)$ may be written in the form $Ax^3 + \frac{B}{x^3} + C$, where A, B and C are constants to be found.
   (c)   Hence evaluate $\int_{1}^{2}f(x)dx$.

[Edexcel]

6   The diagram shows part of the curve with equation $y = x^3 - 6x^2 + 9x$.

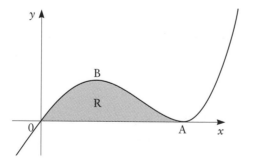

The curve touches the x-axis at A and has a maximum turning point at B.

(a)   Show that the equation of the curve may be written as $y = x(x - 3)^2$ and hence write down the coordinates of A.
(b)   Find the coordinates of B.
The shaded region R is bounded by the curve and the x-axis.
(c)   Find the area of R.

[Edexcel]

7 The diagram shows the line with equation $y = x + 1$ and the curve with equation $y = 6x - x^2 - 3$.
The line and the curve intersect at the points A and B, and O is the origin.

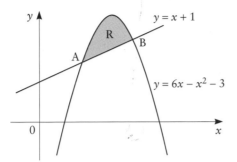

(a) Calculate the coordinates of A and the coordinates of B.
The shaded region R is bounded by the line and the curve.
(b) Calculate the area of R.

[Edexcel]

8 The diagram shows the line with equation $y = 9 - x$ and the curve with equation $y = x^2 - 2x + 3$. The line and the curve intersect at the points A and B, and O is the origin.

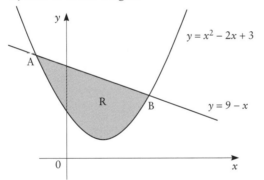

(a) Calculate the coordinates of A and the coordinates of B.
The shaded region R is bounded by the line and the curve.
(b) Calculate the area of R.

[Edexcel]

9 Evaluate $\int_0^1 \sqrt{4x + 1}\ dx$, using the trapezium rule with four strips, giving your answer correct to 4 significant figures.

10 It is given that $\int_0^1 (\frac{1}{1 + x^2})dx = \frac{\pi}{4}$.
Using the trapezium rule with five ordinates, find an approximate value for π, correct to 4 significant figures.
How would you find a better approximation?

11 (a) Evaluate $\int_1^3 (x^2 + 3)\,dx$, using the trapezium rule with five strips.

(b) Sketch the graph of $y = x^2 + 3$ and use your sketch to explain whether the answer to part (a) is an overestimate or an underestimate of the exact area.

(c) Find the exact value of the area.

12 Using the trapezium rule with five values of x, find an approximate value of $\int_2^3 \dfrac{1}{2\sqrt{(x-1)}}\,dx$.

13 A measure of the effective voltage, M volts, in an electrical circuit is given by

$$M^2 = \int_0^1 V^2 dt$$

where V volts is the voltage at time t seconds. Pairs of values of V and t are given in the table.

t	0	0.25	0.5	0.75	1
V	-48	207	37	-161	-29
V^2					

Use the trapezium rule, with five values of V^2, to estimate the value of M.

[Edexcel]

14 The table shows values for $y = \sqrt{1 + \sin x}$, where x is in radians.

x	0	0.5	1	1.5	2
y	1	1.216	p	1.413	q

(a) Find the value of p and the value of q.

(b) Use the trapezium rule and all the values of y in the completed table to obtain an estimate of I, where:

$$I = \int_0^2 \sqrt{1 + \sin x}\,dx$$

[Edexcel]

15 The table shows values for $y = \tan x$, where x is in radians.

x	0	$\dfrac{\pi}{24}$	$\dfrac{\pi}{12}$	$\dfrac{\pi}{8}$	$\dfrac{\pi}{6}$	$\dfrac{5\pi}{24}$	$\dfrac{\pi}{4}$
y	0	0.131	a	0.414	b	0.767	c

(a) Find the values of a, b and c, correct to 3 significant figures.

(b) Estimate the value of $\int_0^{\frac{\pi}{4}} \tan x\,dx$ by using the trapezium rule with all the values from your completed table.

KEY POINTS

1 $\int x^n dx = \frac{1}{n+1} x^{n+1} + c \quad n \neq -1$

2 If $\int f(x)dx = F(x) + c$ then $\int_a^b f(x)dx = F(b) - F(a)$

$\int_b^a f(x)dx = -\int_a^b f(x)dx$

3 **Areas**

Area $A = \int_a^b y\,dx = \int_a^b f(x)dx$

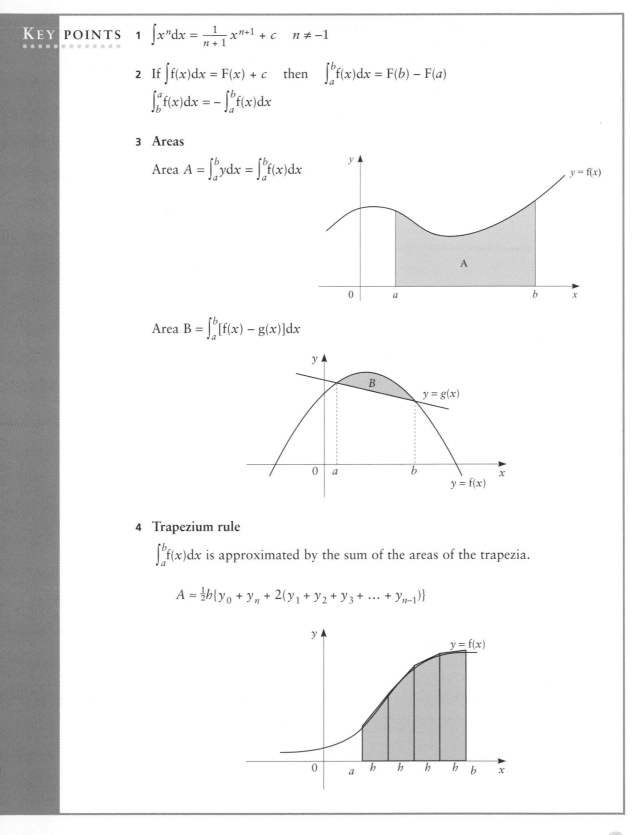

Area $B = \int_a^b [f(x) - g(x)]dx$

4 **Trapezium rule**

$\int_a^b f(x)dx$ is approximated by the sum of the areas of the trapezia.

$A \approx \tfrac{1}{2}h\{y_0 + y_n + 2(y_1 + y_2 + y_3 + \ldots + y_{n-1})\}$

Answers

C1 C2

ANSWERS (CORE 1)

CHAPTER 1

Exercise 1A (Page 9)

1 (a) 3^4
(b) 3^0
(c) 3^3
(d) 3^{-3}
(e) $3^{\frac{1}{2}}$
(f) 3^1

2 (a) 4^2
(b) $4^{\frac{1}{2}}$
(c) 4^{-1}
(d) $4^{-\frac{1}{2}}$
(e) $4^{\frac{3}{2}}$
(f) $4^{-\frac{3}{2}}$

3 (a) 10^3
(b) 10^{-4}
(c) 10^{-3}
(d) $10^{\frac{3}{2}}$
(e) $10^{-\frac{1}{2}}$
(f) 10^{-6}

4 (a) $\frac{1}{32}$
(b) 3
(c) $\frac{1}{3}$
(d) 5
(e) $\frac{1}{5}$
(f) 1

5 (a) 9
(b) 10
(c) $\frac{1}{5}$
(d) 1
(e) $5^5 = 3125$
(f) 27

6 (a) 49
(b) 3
(c) 3
(d) $3^6 = 729$

(e) 8
(f) 4

7 (a) 4
(b) 0
(c) 0
(d) $1\frac{15}{16}$
(e) 0
(f) $2\frac{2}{3}$

8 (a) $3\sqrt{2}$
(b) $-3\sqrt[3]{4}$
(c) $4\sqrt[3]{7}$
(d) $(2 + x)(1 + x)^{\frac{1}{2}}$
(e) $(10 + 9x)(2 + 3x)^{\frac{1}{2}}$
(f) $[6 - 5(x + y)](x + y)^{\frac{1}{2}}$
(g) $\left(1 + \frac{1}{x}\right)\sqrt{x}$
(h) $(x^2 - 2x + 3)^{\frac{1}{2}}(-2 + 2x - x^2)$

9 (a) $2^4 = 16$
(b) $3^2 = 9$
(c) $5^3 = 125$
(d) $2^{\frac{5}{12}}$

10 (a) 18
(b) $4a^4b^8$
(c) $\dfrac{x^2 y^3}{z^6}$
(d) $a + 2\sqrt{ab} + b$

Exercise 1B (Page 13)

1 (a) $2\sqrt{2}$
(b) $3\sqrt{3}$
(c) $10\sqrt{2}$
(d) $12\sqrt{2}$
(e) $3\sqrt{7}$
(f) $7\sqrt{2}$
(g) $8\sqrt{5}$
(h) $\frac{3}{7}$ (i) $\frac{2}{3}$ (j) $\frac{8}{5}$

2 (a) $\sqrt{7}$
(b) $\sqrt{3}$
(c) $\sqrt{2} + 1$
(d) 1
(e) $2(\sqrt{3} - \sqrt{2})$

3 (a) $8 + 3\sqrt{2}$
(b) $5\sqrt{2}$
(c) $19 + 16\sqrt{5}$
(d) $8\sqrt{2} + 2\sqrt{5}$
(e) $6\sqrt{2}$

4 (a) $10 + \sqrt{2}$
(b) $7 - 4\sqrt{3}$
(c) $10 - 2\sqrt{5}$
(d) -4
(e) $1 + \frac{1}{2}\sqrt{3}$
(f) $3\sqrt{7} - 9$
(g) 14
(h) $\sqrt{30}$
(i) 2

5 (a) $1 + \frac{1}{2}\sqrt{2}$
(b) $\frac{1}{2}(3\sqrt{3} - 5)$
(c) $1 - \frac{2}{5}\sqrt{5}$
(d) $-\frac{1}{4}(1 + \sqrt{5})$
(e) $-3 - \sqrt{3}$
(f) $\frac{1}{3}(\sqrt{7} + 1)$
(g) $\frac{1}{7}(9 + 4\sqrt{2})$
(h) $11 + 2\sqrt{30}$

Exercise 1C (Page 17)

1 $(x + 2)(x + 4)$
2 $(x - 3)(x - 5)$
3 $(x - 2)(x - 8)$
4 $(x + 3)(x - 4)$
5 $(x + 3)(x + 8)$

6 $(x - 12)(x + 2)$

7 $(x - 9)(x + 4)$

8 $(x + 4)(x + 8)$

9 $(x + 9)(x - 9)$

10 $(x - 5)^2$

11 $(x + 6)^2$

12 $(x + 13)(x - 13)$

13 $(2x + 1)(x - 3)$

14 $(2x - 3)(x - 2)$

15 $(2x + 5)(x + 4)$

16 $(2x + 3)(x - 4)$

17 $(3x - 2)(x - 4)$

18 $(3x + 5)(x + 5)$

19 $(3x - 2)(2x + 3)$

20 $(6x - 5)(x + 4)$

21 $(x - 1)(x - 5)$

22 $(x + 4)(x + 6)$

23 $(3x - 2)(x - 3)$

24 $(4x + 1)(x - 5)$

25 $(4x + 1)(x + 3)$

Exercise 1D (Page 18)

1 $1, 2$

2 $2, -5$

3 $4, 6$

4 $-3, -5$

5 $-2, 7$

6 $-8, 6$

7 $-4, -12$

8 7

9 $-7, 7$

10 $-4, 8$

11 $\frac{1}{2}, 4$

12 $2\frac{1}{2}, -3$

13 $-3, -\frac{2}{3}$

14 $\frac{5}{6}, 2$

15 $\frac{3}{2}, \frac{5}{2}$

16 $\frac{3}{4}, -2$

17 $-\frac{2}{5}, -\frac{1}{2}$

18 $\frac{3}{5}, 2$

19 $\frac{2}{3}, \frac{3}{2}$

20 $3, 5$

21 $-5, 4$

22 $4, 9$

23 $\frac{3}{2}, \frac{4}{3}$

24 $-8, 5$

25 $-5, 10$

Exercise 1E (Page 20)

1 $0.4, 4.6$

2 $-4.8, 0.8$

3 $0.6, 5.4$

4 $2.3, 5.7$

5 $-0.4, 5.4$

6 $-0.6, 4.6$

7 $0.6, 2.4$

8 $-1.5, 1.1$

9 $-0.9, 2.4$

10 $-0.2, 1.8$

Exercise 1F (Page 21)

1 (a) $(x + 3)^2 - 2$

　　(b) $(x + 5)^2 + 5$

　　(c) $(x - 6)^2 - 12$

　　(d) $(x - 4)^2 - 19$

　　(e) $(x + 8)^2 + 6$

　　(f) $(x + \frac{5}{2})^2 - \frac{1}{4}$

　　(g) $(x - \frac{9}{2})^2 + \frac{3}{4}$

　　(h) $(x + \frac{1}{2})^2 + \frac{3}{4}$

　　(i) $(x + \frac{1}{4})^2 - \frac{5}{16}$

　　(j) $(x + \frac{p}{2})^2 + q - \frac{p^2}{4}$

2 (a) $2(x + 2)^2 - 1$

　　(b) $2(x + \frac{5}{2})^2 + \frac{13}{2}$

　　(c) $3(x - \frac{3}{2})^2 + \frac{13}{4}$

　　(d) $5(x + 2)^2 - 21$

　　(e) $4(x + \frac{5}{4})^2 - \frac{5}{4}$

　　(f) $-(x - 1)^2 + 6$

　　(g) $-(x + \frac{3}{2})^2 + \frac{33}{4}$

　　(h) $-2(x - \frac{5}{4})^2 + \frac{1}{8}$

　　(i) $-3(x - \frac{2}{3})^2 + \frac{52}{3}$

　　(j) $5(x - 4)^2 - 12$

3 Min $f(x) = -\frac{23}{4}$, $x = \frac{3}{2}$

4 $x = 2$

5 $a(x + \frac{b}{2a})^2 + c - \frac{b^2}{4a}$

Exercise 1G (Page 22)

1 $2 \pm \sqrt{2}$

2 $-2 \pm 2\sqrt{2}$

3 $3 \pm \sqrt{6}$

4 $4 \pm \sqrt{3}$

5 $\frac{5}{2} \pm \frac{1}{2}\sqrt{33}$

6 $2 \pm \sqrt{7}$

7 $2 \pm \frac{1}{2}\sqrt{10}$

8 $\frac{3}{2} \pm \frac{1}{2}\sqrt{3}$

9 $3 \pm \frac{1}{2}\sqrt{46}$

10 $\frac{3}{4} \pm \frac{1}{4}\sqrt{41}$

Exercise 1H (Page 23)

1 $-12, -8$

2 $7, 9$

3 $-8, 10$

4 12

5 No real roots

6 -5

7 $2 \pm \sqrt{2}$

8 $-2 \pm 2\sqrt{2}$

9 $3 \pm \sqrt{6}$

10 $4 \pm \sqrt{3}$

11 $\frac{5}{2} \pm \frac{1}{2}\sqrt{33}$

12 $2 \pm \sqrt{7}$

13 $2 \pm \frac{1}{2}\sqrt{10}$

14 $\frac{3}{2} \pm \frac{1}{2}\sqrt{3}$

15 $3 \pm \frac{1}{2}\sqrt{46}$

16 $\frac{3}{4} \pm \frac{1}{4}\sqrt{41}$

17 No real roots.

18 $-\frac{2}{3} \pm \frac{1}{3}\sqrt{19}$

19 $\frac{3}{2} \pm \sqrt{2}$

20 No real roots.

Exercise 1I (Page 24)

1 (a) 8 or 3

　　(b) -8 or -3

　　(c) 2 or 9

　　(d) 3 (repeated)

　　(e) -8 or 8

2 (a) $\frac{2}{3}$ or 1

　　(b) $-\frac{2}{3}$ or -1

　　(c) $-\frac{1}{3}$ or 2

　　(d) $-\frac{4}{5}$ or $\frac{4}{5}$

　　(e) $\frac{2}{3}$ (repeated)

3 (a) $-4 \pm \sqrt{11}$

　　(b) No real solutions.

　　(c) $\frac{1}{2}(5 \pm \sqrt{101})$

　　(d) $2 \pm \sqrt{5}$

　　(e) $\frac{1}{4}(-2 \pm \sqrt{3})$

4 (a) $w(w + 30)$

　　(b) $80\,\text{m}, 380\,\text{m}$

5 (a) $t = 1$ and 2

　　(b) $t = 3.065$

　　(c) $12.25\,\text{m}$

6 (a) $A = 2\pi rh + 2\pi r^2$
  (b) 3 cm   **(c)** 5 cm

7 (b) 14   **(c)** 45

8 $x^2 + (x + 1)^2 = 29^2$; 20 cm,
   21 cm, 29 cm

9 $\frac{600}{x}$, $\frac{600}{x} - 5$; £24

10 $\frac{160}{n}$; $170 = (n + 2)\left(\frac{160}{n} - 3\right)$; 8

Exercise 1J (Page 27)

1 (a) 2
  (b) 0
  (c) 1
  (d) 2
  (e) 2
  (f) 2
  (g) 2
  (h) 0
  (i) 1
  (j) 2
2 $k < -24$, $k > 24$
3 $p = 9$
4 $-30 < b < 30$
5 $-1 \leqslant k \leqslant 4$

Exercise 1K (Page 30)

1 (i) (a) $x = -2$, $(-2, -3)$
  (b) $y = x^2 + 4x + 1$
  (c)

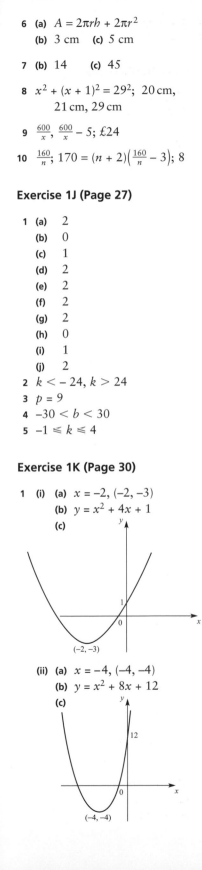

  (ii) (a) $x = -4$, $(-4, -4)$
  (b) $y = x^2 + 8x + 12$
  (c)

(iii) (a) $x = 1$, $(1, 2)$
  (b) $y = x^2 - 2x + 3$
  (c)

(iv) (a) $x = 10$, $(10, 12)$
  (b) $y = x^2 - 20x + 112$
  (c)

(v) (a) $x = \frac{1}{2}$, $\left(\frac{1}{2}, \frac{3}{4}\right)$
  (b) $y = x^2 - x + 1$
  (c)

(vi) (a) $x = -0.1$, $(-0.1, 0.99)$
  (b) $y = x^2 + 0.2x + 1$
  (c)

2 (i) (a) $(x + 2)^2 + 5$
  (b) $x = -2$; $(-2, 5)$
  (c)

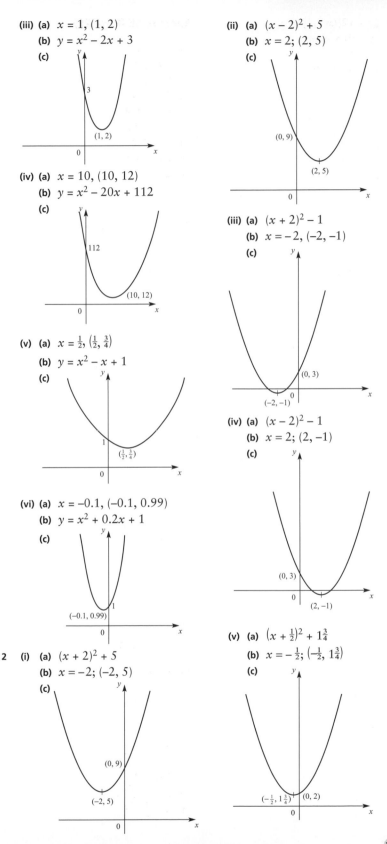

(ii) (a) $(x - 2)^2 + 5$
  (b) $x = 2$; $(2, 5)$
  (c)

(iii) (a) $(x + 2)^2 - 1$
  (b) $x = -2$, $(-2, -1)$
  (c)

(iv) (a) $(x - 2)^2 - 1$
  (b) $x = 2$; $(2, -1)$
  (c)

(v) (a) $\left(x + \frac{1}{2}\right)^2 + 1\frac{3}{4}$
  (b) $x = -\frac{1}{2}$; $\left(-\frac{1}{2}, 1\frac{3}{4}\right)$
  (c)

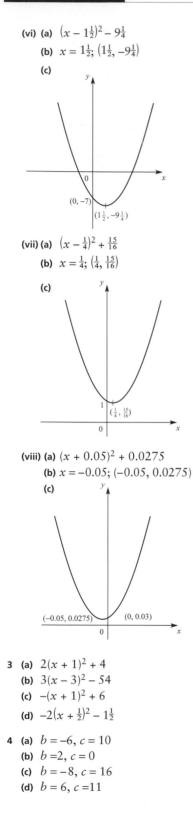

(vi) (a) $\left(x - 1\tfrac{1}{2}\right)^2 - 9\tfrac{1}{4}$

(b) $x = 1\tfrac{1}{2};\ \left(1\tfrac{1}{2}, -9\tfrac{1}{4}\right)$

(c)

$(0, -7)$

$\left(1\tfrac{1}{2}, -9\tfrac{1}{4}\right)$

(vii) (a) $\left(x - \tfrac{1}{4}\right)^2 + \tfrac{15}{16}$

(b) $x = \tfrac{1}{4};\ \left(\tfrac{1}{4}, \tfrac{15}{16}\right)$

(c)

$\left(\tfrac{1}{4}, \tfrac{15}{16}\right)$

(viii) (a) $(x + 0.05)^2 + 0.0275$

(b) $x = -0.05;\ (-0.05, 0.0275)$

(c)

$(-0.05, 0.0275)$ $(0, 0.03)$

3 (a) $2(x + 1)^2 + 4$

(b) $3(x - 3)^2 - 54$

(c) $-(x + 1)^2 + 6$

(d) $-2\left(x + \tfrac{1}{2}\right)^2 - 1\tfrac{1}{2}$

4 (a) $b = -6,\ c = 10$

(b) $b = 2,\ c = 0$

(c) $b = -8,\ c = 16$

(d) $b = 6,\ c = 11$

Exercise 1L (Page 32)

1 9

2 $\pm 2, \pm 3$

3 9

4 2, 3

5 81

6 $\pm 1, \pm 4$

7 1, 81

8 ± 5

9 $\tfrac{1}{8}, -1$

10 $\tfrac{1}{16}, 16$

Exercise 1M (Page 36)

1 $x = 1,\ y = 2$

2 $x = 0,\ y = 4$

3 $x = 2,\ y = 1$

4 $x = 1,\ y = 1$

5 $x = 3,\ y = 1$

6 $x = 4,\ y = 0$

7 $x = \tfrac{1}{2},\ y = 1$

8 $u = 5,\ v = -1$

9 $l = -1,\ m = -2$

10 $t_1 = 0,\ t_2 = 4$

11 (a) $5p + 8h = 10,\ 10p + 6h = 10$

(b) Paperbacks 40p, hardbacks £1

12 (a) $s + l = 17,\ 2s + 5l = 70$

(b) 5 short, 12 long

13 (a) $p = a + 5,\ 8a + 9p = 164$

(b) Apples 7p, pears 12p

14 (a) $t_1 + t_2 = 4;\ 110t_1 + 70t_2 = 380$

(b) 275 km motorway, 105 km country roads

15 $x = 1,\ y = -2$ or $x = 2.2,\ y = 0.4$

16 $x = 1,\ y = -3$ or $x = 6,\ y = -\tfrac{1}{2}$

17 $x = 3,\ y = 1$ or $x = 1,\ y = 3$

18 $x = 4,\ y = 2$ or $x = -20,\ y = 14$

19 $x = -3,\ y = -2$ or $x = 1\tfrac{1}{2},\ y = 2\tfrac{1}{2}$

20 $k = -1,\ m = -7$ or $k = 4,\ m = -2$

21 $t_1 = -10,\ t_2 = -5$ or $t_1 = 10,\ t_2 = 5$

22 $p = -3,\ q = -2$

23 $k = -6,\ m = -4$ or $k = 6,\ m = 4$

24 $p_1 = 1,\ p_2 = 1$

Exercise 1N (Page 38)

1 $x > -3$

2 $x \leq 8.5$

3 $x < 5$

4 $x \geq 2.5$

5 $x \leq 4$

6 $x < 4.5$

7 $x < 3$

8 $x \leq -\tfrac{1}{2}$

9 $x \leq 12$

10 $x > 0$

11 $1 < x < 3$

12 $x \leq -1,\ x \geq 2$

13 $x < -4,\ x > -3$

14 $-2 \leq x \leq 2$

15 $x < -1,\ x > 3$

16 $x < -1,\ x > 1\tfrac{1}{2}$

17 $-\tfrac{1}{3} \leq x \leq 4$

18 $x < -\tfrac{4}{3},\ x > \tfrac{1}{2}$

19 $-2 \leq x \leq \tfrac{3}{4}$

20 $x < -\tfrac{4}{3},\ x > \tfrac{4}{3}$

Exercise 1O (Page 42)

1 (a) 3 (b) 12 (c) 2 (d) 7

2 $2x^3 - 4$

3 $x^4 + 4x^3 + 6x^2 + 4x + 1$

4 $x^3 + 2x^2 + 5x + 7$

5 $-x^2 + 15x + 18$

6 $2x^4 + 8$

7 $x^4 + 4x^3 + 6x^2 + 4x + 1$

8 $x^4 - 5x^2 + 4$

9 $x^4 - 10x^2 + 9$

10 $x^{11} - 1$

11 $2x - 2$

12 $10x^2$

13 4

14 $2x^2 - 2x$

15 $-8x^3 - 8x$

16 $3x(5x + 4)$

17 $4x(2x^2 + 3xy + z)$

18 $4x(4x^2 + 3)$

19 $27x^3 + 54x^2 + 36x + 8$

20 $81 - 216x + 216x^2 - 96x^3 + 16x^4$

21 $x(x + 3)(x + 5)$

22 $x(x - 7)(x + 2)$

23 $x(2x + 1)(3x - 4)$

24 $(x + 1)(2x^2 + x - 6)$
   $(x + 1)(x + 2)(2x - 3)$
   $-1, -2, \frac{3}{2}$

25 $(x - 1)(x^2 - 2x - 4)$
   $1, 1 \pm \sqrt{5}$

Exercise 1P (Page 46)

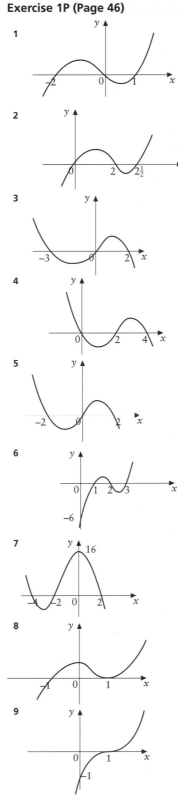

Exercise 1Q (Page 47)

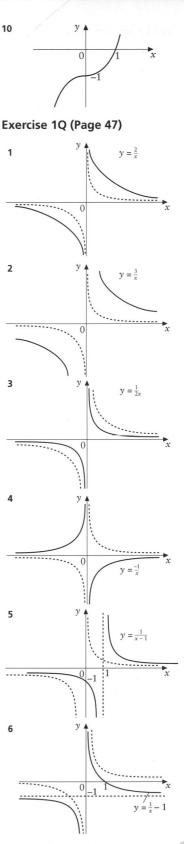

Exercise 1R (Page 50)

1 (a) $x = \pm 1$
   (b) $x = \pm 2$

2 (a) -0.6, 1.6
   (c) -1, 3
   (d) $-1 < x < 3$

3 (c) -3, 1

4 $1 < x < 3$

5 -3.3, 0.3

6 -3.1, 0.5, 2.7

7 -4.2, 1.2

8 (a) $2(x + 1)^2 - 9$
   (b)
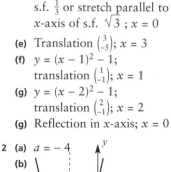
   (c) $k > -9$

9 (a) -0.8, 1.5, 4.3

10 -3.6, 0.6, 1

Exercise 1S (Page 59)

1 (a) Translation $\binom{0}{-2}$; $x = 0$
   (b) Translation $\binom{-4}{0}$; $x = -4$
   (c) Stretch, parallel to y-axis of
        s.f. 4 or stretch parallel to
        x-axis of s.f. $\frac{1}{2}$; $x = 0$
   (d) Stretch, parallel to y-axis of
        s.f. $\frac{1}{3}$ or stretch parallel to
        x-axis of s.f. $\sqrt{3}$; $x = 0$
   (e) Translation $\binom{3}{-5}$; $x = 3$
   (f) $y = (x - 1)^2 - 1$;
        translation $\binom{1}{-1}$; $x = 1$
   (g) $y = (x - 2)^2 - 1$;
        translation $\binom{2}{-1}$; $x = 2$
   (g) Reflection in x-axis; $x = 0$

2 (a) $a = -4$
   (b)
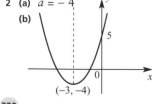

3 $p = 3$, $q = 2$

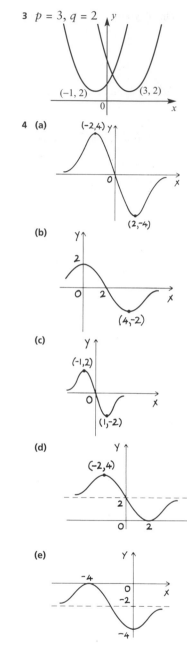

4 (a)

   (b)

   (c)

   (d)

   (e)

   (f)

5 (a)

   (b)

   (c)

6

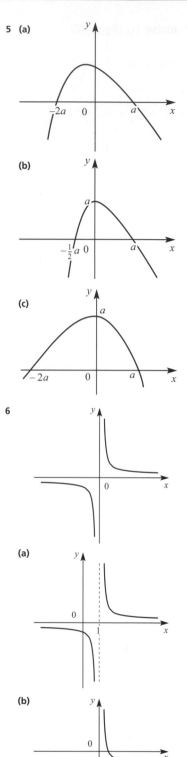

   (a)

   (b)

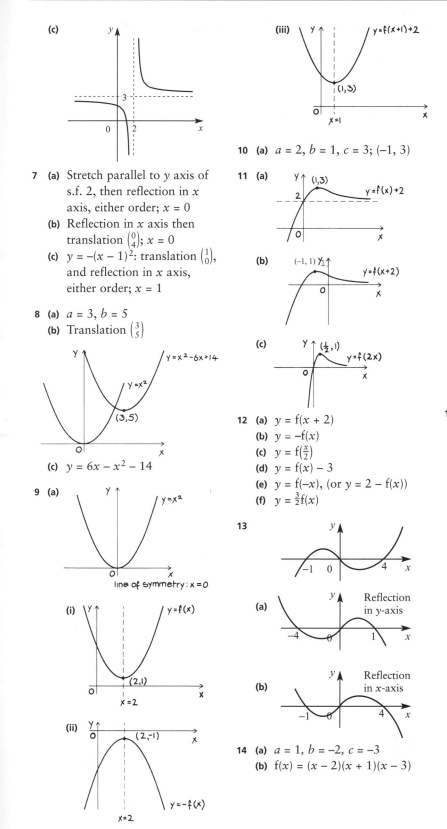

(c)

7 (a) Stretch parallel to y axis of s.f. 2, then reflection in x axis, either order; $x = 0$

(b) Reflection in x axis then translation $\binom{0}{4}$; $x = 0$

(c) $y = -(x - 1)^2$: translation $\binom{1}{0}$, and reflection in x axis, either order; $x = 1$

8 (a) $a = 3, b = 5$

(b) Translation $\binom{3}{5}$

(c) $y = 6x - x^2 - 14$

9 (a)

(i)

(ii)

(iii) $y = f(x+1)+2$ (1,3) $x=1$

10 (a) $a = 2, b = 1, c = 3; (-1, 3)$

11 (a) (1,3) $y = f(x)+2$

(b) $(-1, 1)$ $y = f(x+2)$

(c) $(\frac{1}{2}, 1)$ $y = f(2x)$

12 (a) $y = f(x + 2)$

(b) $y = -f(x)$

(c) $y = f\left(\frac{x}{2}\right)$

(d) $y = f(x) - 3$

(e) $y = f(-x)$, (or $y = 2 - f(x)$)

(f) $y = \frac{3}{2}f(x)$

13

(a) Reflection in y-axis

(b) Reflection in x-axis

14 (a) $a = 1, b = -2, c = -3$

(b) $f(x) = (x - 2)(x + 1)(x - 3)$

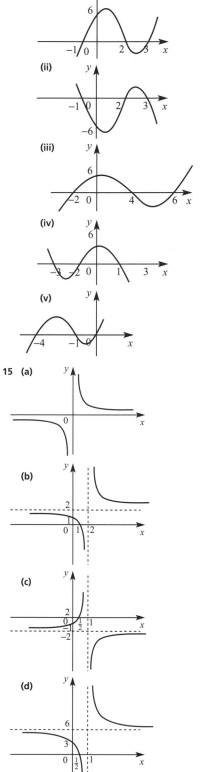

(c) (i)

(ii)

(iii)

(iv)

(v)

15 (a)

(b)

(c)

(d)

Exercise 1T (Page 62)

1 $\frac{3q^3}{5p^2}$

2 (a) 160　(b) $\frac{1}{2}$　(c) $\frac{1}{25}$

3 $5 - 2\sqrt{2}$

4 $29 - 12\sqrt{5}$

5 (a) $x^4 + (2 + b)x^3 + (a + 2b + 3)x^2$
　　　$+ (ab + 6)x + 3a$
　(b) $a = 4, b = 2, c = 4$

6 $k = -5$

7 (a) $f(2) = 0$
　(b) $(x - 2)(x + 2)(x + 5)$
　(c)

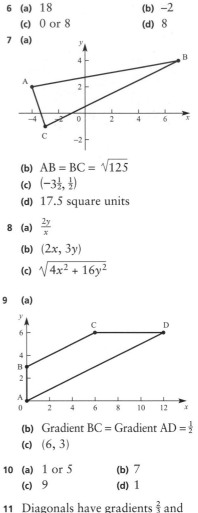

　(d) $-5 \leqslant x \leqslant 2, \ x \geqslant 2$

8 $f(x) = (x - 2)(2x^2 + 3x - 4)$
　Positive roots are $2, \dfrac{-3 + \sqrt{41}}{4}$

9 (a) $(x - 7)(x + 3)$
　(b) $-3, 7$

10 (a) $(5x - 1)(5x - 3)$
　(b) $\frac{1}{25}, \frac{9}{25}$

11 $x = 3, y = 2$ and $x = \frac{-3}{14}, y = \frac{37}{14}$

12 $x \leqslant -3, x \geqslant \frac{1}{2}$

13 $\frac{1}{16}, 64$

14 (a) $-5, 4$
　(b) $x < -5, x > 4$

15 (a) $f(2) = 0$
　(b) $2, \dfrac{-3 \pm \sqrt{5}}{2}$

16 (a) $k = 3$　　　(b) $x = 1.2$

17 (b) $k \leq 0, k \geq \frac{8}{25}$　(c) $0, \frac{8}{25}$

18 (a) $x = -\frac{1}{2}, y = \frac{5}{2}$
　(b) 8

19 (a) $-k \pm \sqrt{k^2 + 7}$
　(b) $\sqrt{k^2 + 7} \geq \sqrt{7} \Rightarrow$ real
　　　$\sqrt{k^2 + 7} > k \Rightarrow$ distinct
　(c) $\pm 3 - \sqrt{2}$

20 (a) $x < 2$　(b) $\frac{1}{2} < x < 5$
　(c) $\frac{1}{2} < x < 2$

CHAPTER 2

Exercise 2A (Page 72)

1 (i) (a) -2
　　　(b) $(1, -1)$
　　　(c) $\sqrt{20}$
　　　(d) $\frac{1}{2}$
　(ii) (a) -3
　　　(b) $(3\frac{1}{2}, \frac{1}{2})$
　　　(c) $\sqrt{10}$
　　　(d) $\frac{1}{3}$
　(iii) (a) 0
　　　(b) $(0, 3)$
　　　(c) 12
　　　(d) Infinite
　(iv) (a) $\frac{10}{3}$
　　　(b) $(3\frac{1}{2}, -3)$
　　　(c) $\sqrt{109}$
　　　(d) $-\frac{3}{10}$
　(v) (a) $\frac{3}{2}$
　　　(b) $(3, 1\frac{1}{2})$
　　　(c) $\sqrt{13}$
　　　(d) $-\frac{2}{3}$
　(vi) (a) Infinite
　　　(b) $(1, 1)$
　　　(c) 6
　　　(d) 0

2 5

3 1

4 (a) $AB:\frac{1}{2}, BC:\frac{3}{2}, CD:\frac{1}{2}, DA:\frac{3}{2}$
　(b) Parallelogram
　(c)

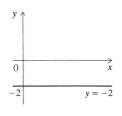

5 (a) 6
　(b) $AB = \sqrt{20}, BC = \sqrt{5}$
　(c) 5 square units

6 (a) 18　　　　(b) -2
　(c) 0 or 8　　(d) 8

7 (a)

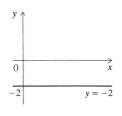

　(b) $AB = BC = \sqrt{125}$
　(c) $(-3\frac{1}{2}, \frac{1}{2})$
　(d) 17.5 square units

8 (a) $\frac{2y}{x}$
　(b) $(2x, 3y)$
　(c) $\sqrt{4x^2 + 16y^2}$

9 (a)

　(b) Gradient BC = Gradient AD $= \frac{1}{2}$
　(c) $(6, 3)$

10 (a) 1 or 5　(b) 7
　(c) 9　　　　(d) 1

11 Diagonals have gradients $\frac{2}{3}$ and
　$-\frac{3}{2}$ so are perpendicular.
　Mid-points of both diagonals are
　$(4, 4)$ so they bisect each other.

　52 square units

Exercise 2B (Page 81)

1 (a)

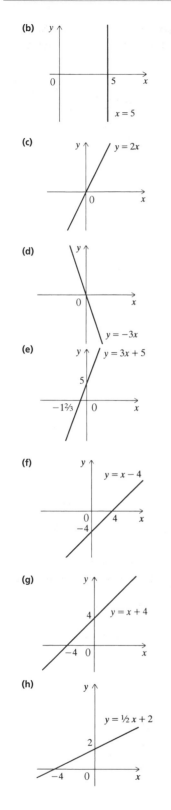

(b) $x = 5$

(c) $y = 2x$

(d) $y = -3x$

(e) $y = 3x + 5$, $-1\frac{2}{3}$, 5

(f) $y = x - 4$, 4, -4

(g) $y = x + 4$, 4, -4

(h) $y = \frac{1}{2}x + 2$, 2, -4

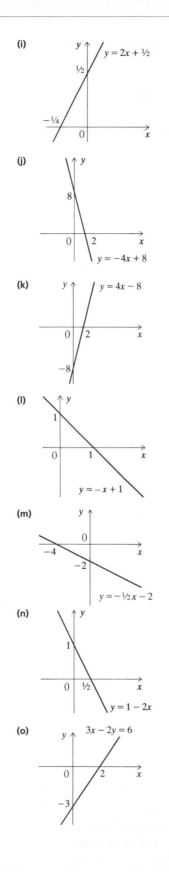

(i) $y = 2x + \frac{1}{2}$, $\frac{1}{2}$, $-\frac{1}{4}$

(j) $y = -4x + 8$, 8, 2

(k) $y = 4x - 8$, 2, -8

(l) $y = -x + 1$, 1, 1

(m) $y = -\frac{1}{2}x - 2$, -4, -2

(n) $y = 1 - 2x$, 1, $\frac{1}{2}$

(o) $3x - 2y = 6$, 2, -3

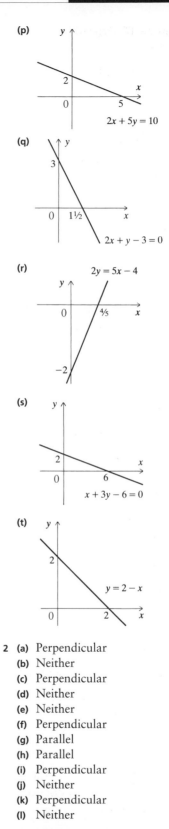

(p) $2x + 5y = 10$, 2, 5

(q) $2x + y - 3 = 0$, 3, $1\frac{1}{2}$

(r) $2y = 5x - 4$, $\frac{4}{5}$, -2

(s) $x + 3y - 6 = 0$, 2, 6

(t) $y = 2 - x$, 2, 2

2 (a) Perpendicular
(b) Neither
(c) Perpendicular
(d) Neither
(e) Neither
(f) Perpendicular
(g) Parallel
(h) Parallel
(i) Perpendicular
(j) Neither
(k) Perpendicular
(l) Neither

3 (a) $x = 7$
  (b) $y = 5$
  (c) $y = 2x$
  (d) $x + y = 2$
  (e) $x + 4y + 12 = 0$

4 (a) $y = x$
  (b) $x = -4$
  (c) $y = -4$
  (d) $x + 2y = 0$
  (e) $x + 3y - 12 = 0$

5 (a) $y = 2x + 3$
  (b) $y = 3x$
  (c) $2x + y + 3 = 0$
  (d) $y = 3x - 14$
  (e) $2x + 3y = 10$
  (f) $y = 2x - 3$
  (g) $2y = 3x - 10$
  (h) $5y = 2x + 20$
  (i) $2x + y + 3 = 0$
  (j) $y = 2x + 11$

6 (a) $x + 3y = 0$
  (b) $x + 2y = 0$
  (c) $x - 2y - 1 = 0$
  (d) $2x + y - 2 = 0$
  (e) $3x - 2y - 17 = 0$
  (f) $x + 4y - 24 = 0$
  (g) $3x + 2y = 18$
  (h) $y = 4x + 10$
  (i) $3y = 2x - 12$
  (j) $x + 4y = 14$

7 (a) $3x - 4y = 0$
  (b) $y = x - 3$
  (c) $x = 2$
  (d) $3x + y - 14 = 0$
  (e) $x + 7y - 26 = 0$
  (f) $y = -2$
  (g) $3x + y = 6$
  (h) $x + 3y + 1 = 0$
  (i) $x + 5y = 8$
  (j) $24y = 36x - 1$

8 (a)
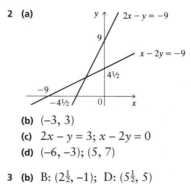
  (b) AC: $x + 3y - 12 = 0$,
    BC: $2x + y - 14 = 0$
  (c) AB $= \sqrt{20}$, BC $= \sqrt{20}$,
    area = 10 square units

(d) $\sqrt{10}$

9 (a)
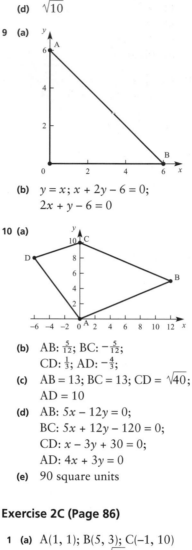
  (b) $y = x$; $x + 2y - 6 = 0$;
    $2x + y - 6 = 0$

10 (a)

  (b) AB: $\frac{5}{12}$; BC: $-\frac{5}{12}$;
    CD: $\frac{1}{3}$; AD: $-\frac{4}{3}$;
  (c) AB = 13; BC = 13; CD $= \sqrt{40}$;
    AD = 10
  (d) AB: $5x - 12y = 0$;
    BC: $5x + 12y - 120 = 0$;
    CD: $x - 3y + 30 = 0$;
    AD: $4x + 3y = 0$
  (e) 90 square units

Exercise 2C (Page 86)

1 (a) A(1, 1); B(5, 3); C(−1, 10)
  (b) BC = AC $= \sqrt{85}$

2 (a)

  (b) (−3, 3)
  (c) $2x - y = 3$; $x - 2y = 0$
  (d) (−6, −3); (5, 7)

3 (b) B: $(2\frac{1}{2}, -1)$; D: $(5\frac{1}{2}, 5)$

4 (a)
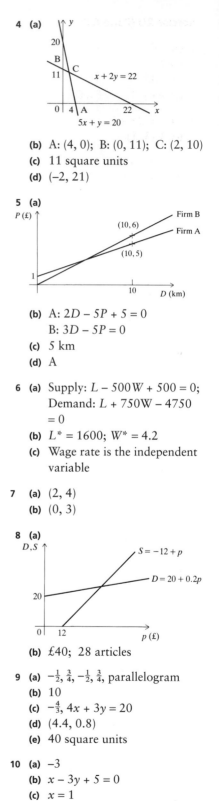
  (b) A: (4, 0); B: (0, 11); C: (2, 10)
  (c) 11 square units
  (d) (−2, 21)

5 (a)

  (b) A: $2D - 5P + 5 = 0$
    B: $3D - 5P = 0$
  (c) 5 km
  (d) A

6 (a) Supply: $L - 500W + 500 = 0$;
    Demand: $L + 750W - 4750 = 0$
  (b) $L^* = 1600$; $W^* = 4.2$
  (c) Wage rate is the independent variable

7 (a) (2, 4)
  (b) (0, 3)

8 (a)

  (b) £40; 28 articles

9 (a) $-\frac{1}{2}, \frac{3}{4}, -\frac{1}{2}, \frac{3}{4}$, parallelogram
  (b) 10
  (c) $-\frac{4}{3}$, $4x + 3y = 20$
  (d) (4.4, 0.8)
  (e) 40 square units

10 (a) −3
  (b) $x - 3y + 5 = 0$
  (c) $x = 1$
  (d) (1, 2)
  (f) 3.75 square units

Exercise 2D (Page 89)

1 (a) 10
   (b) $(1, 5)$
   (c) $-3/4$
   (d) $4x - 3y + 11 = 0$

2 (a) $y = 2x - 5$
   (b) $C(3, 1)$

3 (a) $y = x - 3$
   (b) $2y = -x + 3$ $(3, 0)$

4 (a) 13
   (b) $\left(8, \frac{11}{2}\right)$
   (c) $12y = 5x + 26$
   (d) $\left(\frac{21}{2}, -\frac{1}{2}\right) \left(\frac{11}{2}, \frac{23}{2}\right)$

5 (a) $3x + 5y - 14 = 0$
   (b) $y + 2x - 7 = 0$

6 (a) $y = -4x + 21$
   (b) $y = 3x$
   (c) $S(3, 9)$

7 (b) $x + 4y - 56 = 0$
   (d) $2\sqrt{17}$

8 (a) $\sqrt{3}$
   (b) $y = \sqrt{3}x + 2\sqrt{3}$
   (d) 6
   (e) $60°$

9 (a) $7x + 5y - 18 = 0$
   (b) $\frac{162}{35}$

10 (a) $2y = -x + 6$
   (b) $4y = x + 9$
   (c) 1.12
   (d) 18.75

11 (a)

   (b) $\left(\frac{4}{3}, -\frac{1}{3}\right)$
   (c) $12x - 3y - 17 = 0$

12 (a) $\frac{1}{2}$
   (b) 6

   (c) $4\sqrt{5}$
   (d) 10
   (e) $2x + y - 16 = 0$
   (f) $(4, 8)$

13 (a) $\left(\frac{5}{2}, 1\right)$
   (b) $6x - 11y - 4 = 0$
   (c) $11x + 6y - 112 = 0$
   (d) $\left(7\frac{7}{15}, 4\frac{44}{45}\right)$

14 (a) $x + 2y - 16 = 0$
   (b) $4x + y = 0$
   (c) $\left(\frac{6}{7}, \frac{53}{7}\right)$

CHAPTER 3

Exercise 3A (Page 100)

1 (a) $19, 22, 25, 28$. Arithmetic with 1st term $a = 7$ and common difference $d = 3$.
   (b) $4, 3, 2, 1$. Arithmetic: $a = 8, d = -1$
   (c) $810, 2430, 7290, 21\,870$. Geometric with 1st term $a = 10$ and common ratio $r = 3$.
   (d) $4, 2, 1, \frac{1}{2}$. Geometric: $a = 64, r = \frac{1}{2}$
   (e) $2, 2, 2, 5$. Periodic with period $p = 4$.
   (f) $4, 2, 1, 2$. Periodic: $p = 8$
   (g) $-32, 64, -128, 256$ Geometric, oscillating, non-periodic: $a = 1, r = -2$
   (h) $4, 6, 8, 2$. Periodic: $p = 4$
   (i) $3.3, 3.1, 2.9, 2.7$ Arithmetic: $a = 4.1, d = -0.2$
   (j) $1.4641, 1.61051, 1.771\,561, 1.948\,717\,1$. Geometric: $a = 1, r = 1.1$

2 (a) $3, 5, 7, 9$
   (b) $6, 12, 24, 48$
   (c) $4, 8, 14, 24$
   (d) $1, \frac{1}{2}, \frac{1}{3}, \frac{1}{4}$
   (e) $1, 3, 6, 10$
   (f) $1, 4, 27, 256$
   (g) $0, 3, 8, 15$
   (h) $\frac{1}{2}, \frac{1}{6}, \frac{1}{12}, \frac{1}{20}$
   (i) $4, 6, 4, 6$
   (j) $512, 256, 128, 64$

3 (a) $12, 15, 18, 21$
   (b) $-5, 5, -5, 5$
   (c) $72, 36, 18, 9$
   (d) $4, 6, 4, 6$
   (e) $2, 3, 8, 63$
   (f) $1, 2, 6, 42$
   (g) $2, 2, 2, 2$
   (h) $2, 7, 17, 37$
   (i) $3, \frac{1}{4}, \frac{4}{5}, \frac{5}{9}$
   (j) $10^8, 10^4, 10^2, 10$

4 (a) $0, 1, 1, 2, 3$. The sequence of differences, after the first term, is the same as the sequence itself.
   (b) $13, 21, 34$

5 (a) $2, 6, 2, 6, 2, 6$
   (b) Oscillating periodic sequence, $p = 2$.
   (c) (i) $2, 8, -4, 20, -28, 68$ Diverging oscillating sequence.
       (ii) $3\frac{1}{2}, 4\frac{1}{4}, 3\frac{7}{8}, 4\frac{1}{16}, 3\frac{31}{32}, 4\frac{1}{64}$ Oscillating sequence, converging towards 4.

6 $1, 1, 2, 3$ Fibonacci

7 (a) $0, -2, 2, 2$
   (b) -1
   (c) (i) diverges
       (ii) converges
       (iii) periodic

8 (a) 55
   (b) 87
   (c) 40
   (d) 24

9 (a) $\sum_{r=1}^{6} r^3$ (b) $\sum_{r=1}^{15} r(r+1)$
   (c) $\sum_{r=1}^{9} \frac{1}{2^{r-1}}$ (d) $\sum_{r=1}^{20} (-1)^{r+1} r^2$

10 (a) $1\frac{2}{3}, 1\frac{2}{13}, 1\frac{2}{63}$
   (b) 1

Exercise 3B (Page 103)

1 $\frac{1}{2}, \frac{1}{8}, \frac{1}{32}, \frac{1}{128}, 0$

2 $11, 3, 1$. Cannot have the root of a negative number.

3 1.5, 1.75, 1.875, 1.9375

4 **(a)** $u_1 = 0 \implies$ 0 fixed point

$\quad\quad u_1 = 1 \implies$ 1 fixed point

$\quad\quad u_1 = 2 \implies$ diverges

(b) $u_2 = u_1^2,\ u_3 = u_1^4$

(c) $u_n = u_1^{2n-1}$

5 3969

6 **(a)** 4, 6, 8, 10, 12

(b) 8, 12, 16, 20, 24

(c) $u_n = 4(n + 1),\ u_{100} = 404$

7 **(a)** $\frac{1}{2}, \frac{1}{4}, \frac{1}{8}, \frac{1}{16}, \frac{1}{32}$

(b) 2

8 **(a)** $x_0 = 1$ $-5, 19, 355, 126\,019$

$\quad\quad x_0 = 2$ $-2, -2, -2, -2$

$\quad\quad x_0 = 3$ 3, 3, 3, 3

(b) $\overline{x^2 - x - 6} = 0, -2, 3$

(c) Solutions to quadratic
= fixed points.

9 **(a)** 0.3875

(b) $5x = 2$

(c) 0.4; 0.4

10 **(a)** $\frac{1}{3}u_1, \frac{1}{9}u_1, \frac{1}{27}u_1, \frac{1}{81}u_1$

(b) $\frac{1}{3^{n-1}}u_1$

(c) $\frac{40}{81}$

Exercise 3C (Page 110)

1 **(a)** Yes: $d = 2, a_7 = 39$

(b) No

(c) No

(d) Yes: $d = 4, a_7 = 27$

(e) Yes: $d = -2, a_7 = -4$

2 **(a)** 10

(b) 37

3 **(a)** 4

(b) 34

4 **(a)** 5

(b) 850

5 **(a)** 16, 18, 20

(b) 324

6 **(a)** 15

(b) 1170

7 **(a)** First term 4, common
difference 6

(b) 12

8 **(a)** 3

(b) 165

9 **(a)** 23p

(b) £1.44

10 **(a)** 5000

(b) 5100

(c) 10 100

(d) The 1st sum, 5000, and the
2nd sum, 5100, add up to
the third sum, 10 100. This is
because the sum of the odd
numbers plus the sum of the
even numbers between 50
and 150 is the same as the
sum of all the numbers
between 50 and 150.

11 **(a)** 22 000

(b) The 17th term is the first
negative term.

12 **(a)** $u_n = 4 + 3n$, 23

(b) $S_n = \frac{1}{2}n(11 + 3n)$, 63

13 **(a)** 10, 7   35 650

(b) 5, 9 $u_n = 1 + 4n$

14 **(a)** £16 500   **(b)** 8

Exercise 3D (Page 112)

1 **(a)** 13, 17, 21   **(b)** 2130

2 3, 12, 27, 48, 75   9, 15, 21, 27
$6n + 3$ 7800

3 **(a)** $a = 36, d = -2$ **(b)** 37

4 £30.80   £837

5 **(a)** 55   5050

(b) 4, 9, 16

(c) $(n + 1)^2$

6 **(a)** 560, 626, 698 population
growth is greater than the
number caught.

(b) fourth year = 35
fifth year = no fish left.

(c) 50

7 **(a)** $a = -12, d = 4$

(b) 240

(c) 27

8 **(a)** $a = 1.5\ \ S = 150$

(b) 30

9 **(a)** $a = 25, d = -3$

(b) -3810

10 **(a)** 26 733   **(b)** 53 467

11 **(a)** $d = 5\,\text{cm}$

(b) $a = 45\,\text{cm}$

12 3367
Sum of numbers from 1 to 100
inclusive, excluding multiples of 3

13 **(a)** 71 071   **(b)** 71 355

14 **(a)** £2450 **(b)** £59 000 **(c)** £30

15 **(a) (i)** 455   **(ii)** 13 230

(b) 11

CHAPTER 4

Exercise 4A (Page 121)

2

$f(x)$	$f'(x)$
x^2	$2x$
x^3	$3x^2$
x^4	$4x^3$
x^5	$5x^4$
x^6	$6x^5$
$\vdots$	
x^n	nx^{n-1}

Exercise 4B (Page 127)

1 **(a)** $5x^4$ **(b)** $8x$

(c) $6x^2$ **(d)** $11x^{10}$

(e) $40x^9$ **(f)** $15x^4$

(g) 0   **(h)** 7

(i) $6x^2 + 15x^4$ **(j)** $7x^6 - 4x^3$

(k) $2x$ **(l)** $3x^2 + 6x + 3$

(m) $3x^2$ **(n)** $x + 1$

(o) $6x + 6$

2 (a) $-2x^{-3}$

(b) $\frac{2}{3}x^{-\frac{1}{3}}$

(c) $-12x^{-5}$

(d) $-\frac{2}{x^2}$

(e) $x^{-\frac{1}{4}}$

(f) $\frac{1}{3}x^{-\frac{2}{3}}$

(g) $\frac{1}{2}x^{-\frac{1}{2}} - \frac{1}{2}x^{-\frac{3}{2}}$

(h) $\frac{2}{3}x - \frac{6}{x^3}$

(i) $-\frac{3}{x^2} - \frac{1}{3x^{\frac{4}{3}}}$

(j) $9x^{\frac{1}{2}}$

3 (a) $12x + 2$

(b) $12x + 11$

(c) $3x^2 + 10x + 2$

(d) $2x - 6$

(e) $x^{-\frac{1}{2}} + 3x^{\frac{1}{2}}$

(f) $x^{-\frac{1}{2}} + 1$

(g) $4x^{\frac{1}{3}} + 2x^{-\frac{1}{3}}$

(h) $-8x^{-3}$

4 (a) $6x - \frac{5}{x^2}$

(b) $\frac{1}{2}x^{-\frac{1}{2}} - \frac{1}{2}x^{-\frac{3}{2}}$

(c) $-\frac{1}{2x^2} + \frac{3}{x^3} - \frac{6}{x^4}$

(d) $\frac{10}{9}x^{\frac{2}{3}} + \frac{8}{9}x^{-\frac{1}{3}} + \frac{1}{3}x^{-\frac{4}{3}}$

(e) $-\frac{1}{2}x^{-\frac{3}{2}} - x^{-2}$

(f) $2 - 9x^{-4}$

(g) $-\frac{1}{2}x^{-\frac{3}{2}} + \frac{1}{2}x^{-\frac{1}{2}}$

(h) -1

5 (a) $8\pi r$ (b) $4\pi r^2$

(c) 2π

6 (a) $\frac{1}{2} + \frac{3}{x^2}$

7 (a) 4

(b) 5

(c) 7

(d) 17

(e) $\frac{1}{12}$

(f) 11

(g) $-\frac{1}{8}$

(h) $\frac{1}{6}$

(i) $\frac{7}{2}$

(j) $\frac{3}{16}$

8 (a) $(2, 4)$

(b) $(-1, 0)$

(c) $(\frac{1}{2}, -\frac{3}{4})$

(d) $(\frac{1}{16}, \frac{9}{4})$

(e) $(\frac{1}{2}, 6), (-\frac{1}{2}, -6)$

(f) $(1, 5), (-1, -5)$

(g) $(1, -13), (2, -8)$

(h) $(-1, -9), (3, 7)$

(i) $(1, 2)$

(j) $(-1, -8), (3, 8)$

Exercise 4C (Page 129)

1 (a) $3x^2, 6x$

(b) $5x^4, 20x^3$

(c) $8x, 8$

(d) $-2x^{-3}, 6x^{-4}$

(e) $\frac{3}{2}x^{\frac{1}{2}}, \frac{3}{4}x^{-\frac{1}{2}}$

(f) $4x^3 + 6x^{-4}, 12x^2 - 24x^{-5}$

(g) $3x^{-\frac{2}{3}}, -2x^{-\frac{5}{3}}$

(h) $32x + 24, 32$

(i) $1 + x^{-2}, -2x^{-3}$

(j) $6x^2 + 10x - 5, 12x + 10$

2 (a) $20x^4 - 12x^2, 4928$

(b) $-\frac{3}{2}x^{-\frac{5}{2}}, -\frac{3}{16}$

(c) $\frac{48}{x^4} - \frac{1}{4}, -\frac{1}{16}$

(d) $30x^4, 7680$

(e) $0, 0$

(f) $-\frac{1}{4}x^{-\frac{3}{2}} + \frac{3}{4}x^{-\frac{1}{2}}, \frac{11}{32}$

(g) $8x^{-3} + 3x^{-\frac{5}{2}}, \frac{7}{32}$

(h) $\frac{3}{4}x^{-\frac{1}{2}} + \frac{15}{4}x^{-\frac{7}{2}}, \frac{207}{512}$

3 (a) $0, 4$ (b) $2, 4$

(c) $\frac{dy}{dx} > 0 \Rightarrow x < 2$

$\frac{dy}{dx} < 0 \Rightarrow x > 2$

(d)

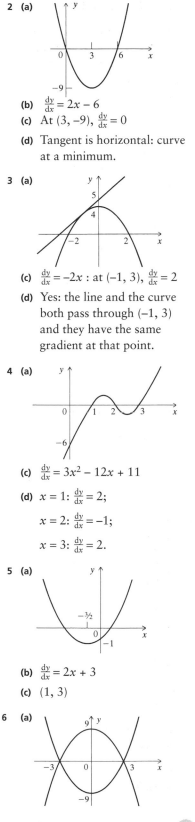

4 (a) $4.57, 19.59$

(b) 2.34

5 $-\frac{7}{2}$

Exercise 4D (Page 132)

1 (a)

(b) $(-2, 0), (2, 0)$

(c) $\frac{dy}{dx} = 2x$

(d) At $(-2, 0), \frac{dy}{dx} = -4$;

at $(2, 0), \frac{dy}{dx} = 4$

2 (a)

(b) $\frac{dy}{dx} = 2x - 6$

(c) At $(3, -9), \frac{dy}{dx} = 0$

(d) Tangent is horizontal: curve at a minimum.

3 (a)

(c) $\frac{dy}{dx} = -2x$: at $(-1, 3), \frac{dy}{dx} = 2$

(d) Yes: the line and the curve both pass through $(-1, 3)$ and they have the same gradient at that point.

4 (a)

(c) $\frac{dy}{dx} = 3x^2 - 12x + 11$

(d) $x = 1: \frac{dy}{dx} = 2$;

$x = 2: \frac{dy}{dx} = -1$;

$x = 3: \frac{dy}{dx} = 2$.

5 (a)

(b) $\frac{dy}{dx} = 2x + 3$

(c) $(1, 3)$

6 (a)

(b) $\frac{dy}{dx} = 2x$

(c) At $(2, -5)$, $\frac{dy}{dx} = 4$;
at $(-2, -5)$, $\frac{dy}{dx} = -4$

(d) At $(2, 5)$, $\frac{dy}{dx} = -4$;
at $(-2, 5)$, $\frac{dy}{dx} = 4$

7 (a)

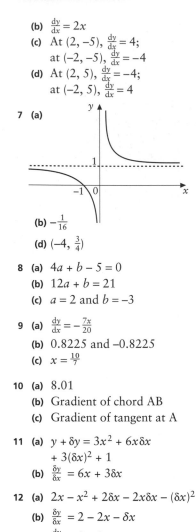

(b) $-\frac{1}{16}$

(d) $(-4, \frac{3}{4})$

8 (a) $4a + b - 5 = 0$

(b) $12a + b = 21$

(c) $a = 2$ and $b = -3$

9 (a) $\frac{dy}{dx} = -\frac{7x}{20}$

(b) 0.8225 and -0.8225

(c) $x = \frac{10}{7}$

10 (a) 8.01

(b) Gradient of chord AB

(c) Gradient of tangent at A

11 (a) $y + \delta y = 3x^2 + 6x\delta x$
$+ 3(\delta x)^2 + 1$

(b) $\frac{\delta y}{\delta x} = 6x + 3\delta x$

12 (a) $2x - x^2 + 2\delta x - 2x\delta x - (\delta x)^2$

(b) $\frac{\delta y}{\delta x} = 2 - 2x - \delta x$

(c) $\frac{dy}{dx} = 2 - 2x$

Exercise 4E (Page 136)

1 (a) $\frac{dy}{dx} = 6 - 2x$

(b) 4

(c) $y = 4x + 1$

2 (a)

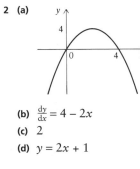

(b) $\frac{dy}{dx} = 4 - 2x$

(c) 2

(d) $y = 2x + 1$

3 (a)

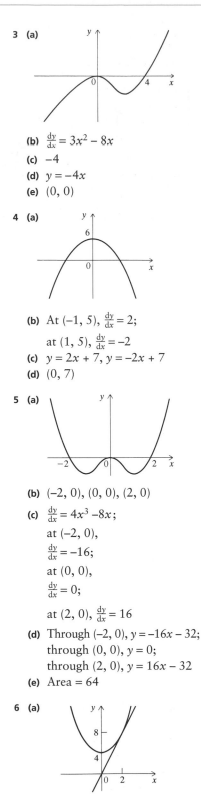

(b) $\frac{dy}{dx} = 3x^2 - 8x$

(c) -4

(d) $y = -4x$

(e) $(0, 0)$

4 (a)

(b) At $(-1, 5)$, $\frac{dy}{dx} = 2$;
at $(1, 5)$, $\frac{dy}{dx} = -2$

(c) $y = 2x + 7$, $y = -2x + 7$

(d) $(0, 7)$

5 (a)

(b) $(-2, 0)$, $(0, 0)$, $(2, 0)$

(c) $\frac{dy}{dx} = 4x^3 - 8x$;
at $(-2, 0)$,
$\frac{dy}{dx} = -16$;
at $(0, 0)$,
$\frac{dy}{dx} = 0$;
at $(2, 0)$, $\frac{dy}{dx} = 16$

(d) Through $(-2, 0)$, $y = -16x - 32$;
through $(0, 0)$, $y = 0$;
through $(2, 0)$, $y = 16x - 32$

(e) Area $= 64$

6 (a)

(c) $y = 4x$ is the tangent to the
curve at $(2, 8)$.

7 (a)

(b) At $(-2, -1)$, $\frac{dy}{dx} = -4$;
at $(2, -1)$, $\frac{dy}{dx} = 4$

(c) At $(-2, -1)$: $4y = x - 2$;
at $(2, -1)$: $4y = -x - 2$

(d) $(0, -\frac{1}{2})$

8 (a) $f(x) = x(x + 1)(x + 2)$;
$x = 0, -1$ or -2

(b) $\frac{dy}{dx} = 3x^2 + 6x + 2$

(c) At $(-2, 0)$: $2y = -x - 2$;
at $(-1, 0)$: $y = x + 1$;
at $(0, 0)$: $2y = -x$

(d) The normals do not intersect
in one point: two of them are
parallel and the other one
crosses them.

9 (a) $y = 6x + 28$

(b) $(3, 45)$

(c) $6y = -x + 273$

10

(a) -4

(b) $4y = x + 15$

11 (a)

A$(1, 0)$; B$(2, 0)$ or vice versa

(b) At $(1, 0)$, $\frac{dy}{dx} = -1$;
at $(2, 0)$, $\frac{dy}{dx} = 1$

(c) At $(1, 0)$, tangent is $y = -x + 1$,
normal is $y = x - 1$;
at $(2, 0)$, tangent is $y = x - 2$,
normal is $y = -x + 2$.

(d) A square.

12 (a) $(1, -7)$ and $(4, -4)$

(b) $\frac{dy}{dx} = 4x - 9$. At $(1, -7)$,

tangent is $y = -5x - 2$; at

$(4, -4)$, tangent is $y = 7x - 32$.

(c) $(2.5, -14.5)$

(d) No.

13 (a) $\frac{dy}{dx} = 3x^2 - 1$

(b) $y = 6$

(c) $\frac{dy}{dx} = 2$

(d) $y = 2x + 8$

(e) $Q(1, 6)$

(f) $x + 2y - 13 = 0$

Exercise 4F (Page 139)

1 (a) $\frac{1}{3}x^{-\frac{2}{3}} - \frac{1}{3}x^{-\frac{4}{3}}$

(b) $3 - \frac{2}{x^2}$

2 (a) $3x^2 + 2x - 1$

(b) $\frac{3}{4}x^{\frac{1}{2}} - x^{-\frac{1}{2}} - x^{-\frac{3}{2}}$

3 (a) $f'(x) = -8x^{-3} + \frac{1}{4}x^{-\frac{1}{2}}$

(b) 4

(c) $\frac{5}{64}$

4 (a) $A = 16 \quad B = 24 \quad C = 9$

(b) $-16x^{-2} - 12x^{-\frac{3}{2}}$

(c) $\frac{17}{16}$

5 (a)

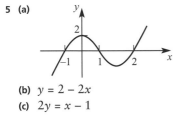

(b) $y = 2 - 2x$

(c) $2y = x - 1$

6 (a) $x = 8, y = 12$

8 (a) $\frac{dy}{dx} = \frac{3}{2}x^{\frac{1}{2}} - \frac{48}{x^2}$

$\frac{d^2y}{dx^2} = \frac{3}{4}x^{-\frac{1}{2}} + \frac{96}{x^3}$

(b) $x = 4, y = 20$

(c) $16y = 93x - 416$

9 (a) $4 - (x - 1)^2$

(b) vertex $(1, 4)$

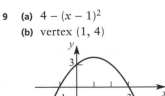

(c) 2

(d) $P(0, 3)$

(e) $x + 2y - 6 = 0$

10 (a) $\frac{dy}{dx} = 3x^2 - 10x + 5$

(b) $x = \frac{1}{3}$

(c) $y = 2x - 7$

(d) $y = \frac{7}{2}\sqrt{5}$

CHAPTER 5

Exercise 5A (Page 145)

1 (a) $\frac{1}{3}x^3 + c$

(b) $\frac{1}{4}x^4 + c$

(c) $x^5 + c$

(d) $2x^2 + c$

(e) $7x + c$

(f) $-\frac{4}{x} + c$

(g) $-\frac{5}{2x^2} + c$

(h) $\frac{8}{3}x^{\frac{3}{2}} + c$

(i) $\frac{15}{4}x^{\frac{4}{3}} + c$

(j) $6x^{\frac{1}{2}} + c$

2 (a) $\frac{2}{3}x^{\frac{3}{2}} + 2x^{\frac{1}{2}} + c$

(b) $\frac{1}{2}x^4 - \frac{2}{x^2} + c$

(c) $x^3 + x^2 + 4x + c$

(d) $\frac{3}{2}x^4 + \frac{4}{3}x^3 + 4x^2 - 10x + c$

(e) $\frac{1}{3}x^3 + \frac{1}{2}x^2 + c$

(f) $\frac{2}{5}x^{\frac{5}{2}} - \frac{2}{3}x^{\frac{3}{2}} + c$

(g) $3x - \frac{5}{x} + c$

(h) $\frac{2}{3}x^{\frac{3}{2}} - 2x^{\frac{1}{2}} + c$

(i) $\frac{1}{3}x^3 + 2x^2 + 4x + c$

(j) $\frac{2}{5}x^{\frac{5}{2}} - \frac{2}{3}x^{\frac{3}{2}} - 4x^{\frac{1}{2}} + c$

Exercise 5B (Page 146)

1 (a) $x^3 + c$

(b) $x^5 + x^7 + c$

(c) $2x^3 + 5x + c$

(d) $\frac{x^4}{4} + \frac{x^3}{3} + \frac{x^2}{2} + x + c$

(e) $x^{11} + x^{10} + c$

(f) $x^3 + x^2 + x + c$

(g) $\frac{x^3}{3} + 5x + c$

(h) $5x + c$

(i) $2x^3 + 2x^2 + c$

(j) $\frac{x^5}{5} + x^3 + x^2 + x + c$

2 (a) $-3x^{-1} + c$

(b) $\frac{3}{4}x^{\frac{4}{3}} + c$

(c) $-\frac{1}{4}x^{-4} + c$

(d) $4x^{\frac{3}{2}} + c$

(e) $\frac{3}{4}x^4 + \frac{3}{2}x^2 + c$

(f) $\frac{16}{3}x^3 + 4x^2 + x + c$

(g) $\frac{1}{5}x^{\frac{5}{3}} - \frac{1}{4}x^{\frac{2}{3}} + c$

(h) $\frac{4}{5}x^{\frac{5}{2}} - \frac{16}{3}x^{\frac{3}{2}} + c$

(i) $6x + \frac{2}{3}x^{\frac{3}{2}} - \frac{1}{2}x^2 + c$

(j) $-\frac{5}{x} + x^3 + c$

3 (a) $-\frac{10}{3}x^{-3} + c$

(b) $x^2 - x^{-3} + c$

(c) $2x + \frac{1}{4}x^4 - \frac{5}{2}x^{-2} + c$

(d) $2x^3 + 7x^{-1} + c$

(e) $4x^{\frac{5}{4}} + c$

(f) $-\frac{1}{3x^3} + c$

(g) $\frac{2}{3}x^{\frac{3}{2}} + c$

(h) $\frac{2}{5}x^5 + \frac{4}{x} + c$

Exercise 5C (Page 148)

1 (a) $y = 2x^3 + 5x + c$

(b) $y = 2x^3 + 5x + 2$

2 (a) $y = 2x^2 + 3$ **(b)** 5

3 (a) $y = 2x^3 - 6$

4 (b) $t = 4$. Only 4 is applicable here.

5 (a) $y = 5x + c$

(b) $y = 5x + 3$

(c)

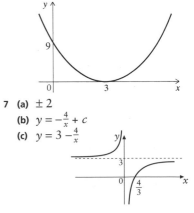

6 (a) $y = x^2 - 6x + 9$

(b) The curve passes through $(1, 4)$

7 (a) ± 2

(b) $y = -\frac{4}{x} + c$

(c) $y = 3 - \frac{4}{x}$

8 (a) $y = \frac{2}{3}x^{\frac{3}{2}} - 2x^{\frac{1}{2}} + c$

  (b) $y = \frac{2}{3}x^{\frac{3}{2}} - 2x^{\frac{1}{2}} + 3$

9 (a) $y = \frac{1}{3}x^3 + 2x - \frac{1}{x} + \frac{5}{3}$

  (b) $x = -1, x = 1$

10 (a) $y = \frac{2}{5}x^{\frac{5}{2}} + \frac{8}{3}x^{\frac{3}{2}} + 8x^{\frac{1}{2}} + c$

   (b) $y = \frac{2}{5}x^{\frac{5}{2}} + \frac{8}{3}x^{\frac{3}{2}} + 8x^{\frac{1}{2}} - \frac{1}{15}$

Exercise 5D (Page 150)

1 (a) $\frac{1}{3}x^3 + \frac{3}{2}x^2 - 4x + c$

  (b) $\frac{3}{4}x^{\frac{4}{3}} + \frac{3}{2}x^{\frac{2}{3}} + c$

2 (a) $\frac{1}{4}x^4 + 3x^{\frac{4}{3}} + x^{-2} + c$

  (b) $\frac{1}{3}x^6 + \frac{6}{5}x^5 + c$

3 (a) $2x - \frac{5}{2}x^2 + x^3 - \frac{3}{2}x^4 + c$

  (b) $\frac{1 - x^2}{x^3} + c$

4 $\frac{16}{3}x^{\frac{3}{2}} - 8x^{\frac{1}{2}} + c$

5 $\frac{4}{5}x^5 + \frac{1}{12}x^{-3} + c$

6 (a) $y = x^2 + 4x + c$

  (b) $y = x^2 + 4x - 5$

  (c) -9

7 (a) $y = \frac{1}{3}x^3 - 2x - \frac{1}{x} + c$

  (b) $y = \frac{1}{3}x^3 - 2x - \frac{1}{x} - \frac{2}{3}$

  (c) $-2\frac{1}{2}$

8 (a) $f(x) = x^3 - 4x^2 + 3x$

  (b) $0, 1, 3$

  (c)

9 $f(x) = \frac{1}{3}x^3 - 2x - \frac{1}{x} - \frac{8}{3}$

10 (a) $y = 5x - \frac{1}{x} + c$

   (b) $12\frac{1}{2}$

ANSWERS (CORE 2)

CHAPTER 6

Exercise 6A (Page 157)

1 $x + 5$
2 $x + 4$
3 $x - 7$
4 $2x + 3$
5 $3x - 7$
6 $x^2 - 2x - 3$
7 $x^2 - x - 4$
8 $x^2 + 3x + 5$
9 $x^2 + 3x$
10 $2x^2 - 5x + 5$
11 $2x^2 + 5$
12 $3x^2 - 2x + 2$
13 $4x^2 - 2x - 7$
14 $5x^2 - 3x - 6$
15 $x^2 + 3x + 4$
16 $x^3 + x^2 + x + 1$
17 $x^3 - 3x^2 + 2x - 5$
18 $2x^3 - 5x^2 + 7x + 3$
19 $3x^3 + 2x^2 - 4x - 3$
20 $x^3 + 2x^2 + 3x + 4$

Exercise 6B (Page 163)

1 (a) $(x - 2)(x - 3)(x + 3)$
   (b) $(x + 2)(x - 3)(x - 4)$
   (c) $(x + 1)(x - 3)(x + 5)$
   (d) $(x + 4)(x - 4)(x - 6)$
   (e) $-(x + 5)(x - 3)(x + 4)$
   (f) $(2x - 1)(x - 1)(x + 3)$
   (g) $(3x + 1)(x - 2)(x + 4)$
   (h) $(2x - 1)(3x + 1)(x - 5)$
   (i) $-(4x + 3)(x + 3)(x - 3)$
   (j) $(2x - 1)(3x - 1)(4x - 1)$

2 (a) $0, 0, -8, -18, -24, -20, 0$
   (b) $(x + 3)(x + 2)(x - 3)$
   (c) $-3, -2$ or 3

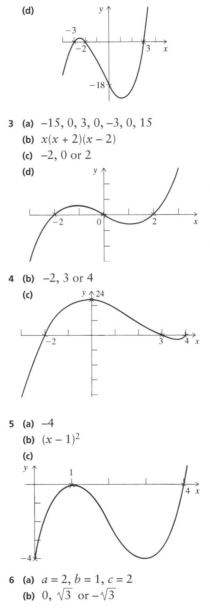

(d)

3 (a) $-15, 0, 3, 0, -3, 0, 15$
   (b) $x(x + 2)(x - 2)$
   (c) $-2, 0$ or 2
   (d)

4 (b) $-2, 3$ or 4
   (c)

5 (a) -4
   (b) $(x - 1)^2$
   (c)

6 (a) $a = 2, b = 1, c = 2$
   (b) $0, \sqrt{3}$ or $-\sqrt{3}$

7 (a) $(x^2 - 4)(x^2 - 1)$
   (b) $(x + 2)(x - 2)(x^2 + 1)$
   (c) 2 real roots: -2 and 2

8 (a) $\pm 6, \pm 3, \pm 2, \pm 1$,
   (b) $-1, 2$ or 3

9 (a) $(x - 1)(x - 2)(x + 2)$
   (b) $(x + 1)(x^2 - x + 1)$
   (c) $(x - 2)(x^2 + 2x + 5)$
   (d) $(x + 2)(x^2 - x + 3)$

10 (a) $\left(\frac{8}{x^2}\right)m$
    (d) $4, 1.464$

Exercise 6C (Page 168)

1 (a) 7
   (b) 26
   (c) $-2\frac{1}{2}$
   (d) -9
   (e) 37
   (f) $\frac{1}{4}$
   (g) 3
   (h) $3\frac{5}{16}$
   (i) 50
   (j) 0

2 (a) $k = -6$
   (c) $(2x + 1)(x - 3)(x + 2)$
   (d)

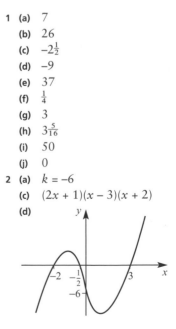

3 $a = 2, b = -4$

331

4 (a) $a = 8, b = 1, c = -42$
 (b) $(x + 3)(x + 7)(x - 2)$
 (c) $-7, -3, 2$

5 $2x^2 + 3x + 9, 13$

6 -16

7 93

8 7

9 $a = 2, b = -3$

10 (a) $R = P(a)$
 (b) -7
 (c) 6
 (d) -3

Exercise 6D (Page 169)

1 $x^2 + 3x - 1$

2 $2x^2 - 3x + 1$

3 $3x^3 + 2x^2 - 4x + 5$

4 $x^2 + 2x - 4$

5 (a) $x^2 + 6x + 8$
 (b) $x^2 + 5x + 4$

6 $k = -5$

7 (a) $f(2) = 0$
 (b) $(x - 2)(x + 2)(x + 5)$
 (c)

y

-5   -2   2
         -20
x

 (d) $-5 \leqslant x \leqslant 2, x \geqslant 2$

8 $f(x) = (x - 2)(2x^2 + 3x - 4)$
 Positive roots are $2, \frac{-3 + \sqrt{41}}{4}$.

9 (a) $(x - 7)(x + 3)$
 (b) $-3, 7$

10 (a) $(5x - 1)(5x - 3)$
 (b) $\frac{1}{25}, \frac{9}{25}$

11 (a) $x^2 + 4x + 14, 44x - 42$
 (b) (i) 2 **(ii)** 90

12 (a) -5
 (b) $-\frac{25}{3}$

13 (b) $b = 4, c = -6$

14 $a = -4, b = 1, c = 10$

15 (a) $a = 2, b = -13$
 (b) $(2x + 1)(x + 2)(x - 3)$
 (c) $2x^2 + 3x - 7, -20$

CHAPTER 7

Exercise 7A (Page 175)

1 (a) $(x - 1)^2 + y^2 = 16$
 (b) $(x - 2)^2 + (y + 1)^2 = 9$
 (c) $(x - 1)^2 + (y + 3)^2 = 25$
 (d) $(x + 2)^2 + (y + 5)^2 = 1$

2 (a) (i) $(0, 0)$ **(ii)** 3
 (b) (i) $(0, 2)$ **(ii)** 5
 (c) (i) $(3, -1)$ **(ii)** 4
 (d) (i) $(-2, -2)$ **(ii)** 2
 (e) (i) $(-4, 0)$ **(ii)** $\sqrt{8}$
 (f) (i) $(-4, 8)$ **(ii)** 7
 (g) (i) $(\frac{1}{2}, -1)$ **(ii)** 1
 (h) (i) $(\frac{9}{2}, \frac{5}{2})$ **(ii)** 5
 (i) (i) $(3, 4)$ **(ii)** $2\sqrt{3}$
 (j) (i) $(-2, -\frac{12}{5})$ **(ii)** $\frac{17}{5}$

3 $(x - 1)^2 + (y - 7)^2 = 169$

4

y

(4, 5)
5

0   4   x

5 $r = 2; (-1, 2); 2$

6 (a) $(2, 11); \sqrt{10}$
 (b) $(x - 2)^2 + (y - 11)^2 = 10$

7

y

(4, 4)
4

0   4   x

$(x - 4)^2 + (y - 4)^2 = 16$

8 $(x - 5)^2 + (y - 4)^2 = 25$ or
  $(x - 5)^2 + (y + 4)^2 = 25$

9 $4x - 3y + 13 = 0$

10

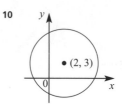

 (b) $2\sqrt{7}$

Exercise 7B (Page 182)

1 (a) $y = 2x - 1$
 (b) $y + x + 3 = 0$
 (c) $3y = 4x - 17$
 (d) $9y = 4x + 16$
 (e) $3y = x + 8$

2 (a) $2x + y = 3$
 (b) $y = x + 7$
 (c) $3x + 4y = 19$
 (d) $2x + y = 14$
 (e) $3x + y = 16$

3 (a) $y = x - 1$
 (b) $(x - 1)^2 + y^2 = 8$
  $(x - 5)^2 + (y - 4)^2 = 8$

4 Centre $(2, 4)$
 radius 7

5 $2\sqrt{21}$

6 $(x - 9)^2 + (y - 4)^2 = 169$

8 (a) $x + 2y = 18$ $y + 3x = 34$
 (d) $x^2 + y^2 - 20x - 8y + 66 = 0$

9 Centre $(4, 2)$
 $x^2 + y^2 - 8x - 4y - 5 = 0$

10 $k = -1 \pm \sqrt{3}$

Exercise 7C (Page 183)

1 $x^2 + y^2 - 6x + 8y + 13 = 0$

2 (a) $(-4, 2), 5$
 (b) $x^2 + y^2 + 2x - 8y + 4 = 0$

3 (a) $(5, -3)$
 (b) 7

4 (a) $(1, 3)$, $\sqrt{26}$
  (b) $(0, -2)$, $(6, 4)$

5 (a) $x^2 + y^2 - 6x - 8y - 43 = 0$
  (b) $(2 \pm 2\sqrt{2})$, $(5 \pm 2\sqrt{2})$
  (c) 8

6 (a) $\sqrt{29}$
  (c) $2x + 5y = 32$

7 (a) $(4, 8)$, 17
  (b) $-\frac{x-4}{y-8}$
  (d) $x = 21$

8 (a) $(3, -4)$, 10
  (b) $(9, 4)$

9 (a) $(-1, 2)$
  (b) 6
  (d) $x^2 + y^2 - 8x - 12y + 16 = 0$

10 (a) $y = x - 5$
  (b) $4x + 5y + 7 = 0$
  (c) $(2, -3)$
  (d) $x^2 + y^2 - 4x + 6y - 28 = 0$

12 (a) $XY = \sqrt{26}$
    $YZ = \sqrt{26}$
    $XZ = 2\sqrt{13}$
  (b) $(x - 4)^2 + (y - 2)^2 = 13$

13 (b) $x^2 + y^2 - 16x - 8y + 16 = 0$

14 (a) $(6, 10)$, 5

15 (a) $(2, -5)$ and $(15, 8)$
  (d) $x^2 + y^2 - 17x - 3y - 10 = 0$

CHAPTER 8

Exercise 8A (Page 189)

1 (a) Yes: $r = 2$, $u_7 = 320$
  (b) No
  (c) Yes: $r = -1$, $u_7 = 1$
  (d) Yes: $r = 1$, $u_7 = 5$
  (e) No
  (f) Yes: $r = \frac{1}{2}$, $u_7 = \frac{3}{32}$
  (g) No

2 (a) 2, $u_{10} = 1024$
  (b) $\frac{1}{2}$, $u_{10} = \frac{1}{512}$
  (c) -2, $u_{10} = -2560$
  (d) $-\frac{1}{3}$, $u_{10} = -\frac{1}{6561}$
  (e) 11, $u_{10} = 2.59 \times 10^{10}$
  (f) 0.4, $u_{10} = 0.00131$
  (g) 1.5, $u_{10} = 2306.6$

3 (a) 9
  (b) 10th term

4 (a) 12
  (b) 11
  (c) 10
  (d) 7
  (e) 8

5 (a) $a = 6$, $r = 5$, $u_8 = 468\ 750$

6 (a) -2560

7 (a) 4
  (b) 81 920

8 (a) $u_n = 8 \times (\frac{1}{2})^{n-1}$
    14

9 (a) 4
  (b) £670.04

10 (a) $\frac{1}{16}$ cm
  (b) $1\frac{15}{16}$ cm
  (c) No

Exercise 8B (Page 192)

1 (a) 2046
  (b) 3.996
  (c) 174 762.5
  (d) 8.8439
  (e) $-29\ 524$
  (f) 0.799 999
  (g) 85.25

2 (a) 3072
  (b) 6142.5

3 (a) $a = \frac{2}{3}$, $r = 3$
  (b) 13 122
  (c) $19\ 682\frac{2}{3}$

4 (a) $a = 4$
  (b) 7.999 99

5 (a) -1705

6 (a) 384
  (b) 765

7 (a) 9
  (b) 4088

8 (a) 6
  (b) 267 (to 3 s.f.)

9 $-261\ 888$

10 12

Exercise 8C (Page 195)

1 (a), (b), (d), (e), (f)

2 (a) 4
  (b) $\frac{9}{4}$
  (c) 25
  (d) 4
  (e) $\frac{2}{9}$
  (f) $\frac{4}{3}$
  (g) $\frac{1}{1-r}$

3 1, 1.5, 1.75, 1.875, 1.9375, 2

4 (a) $\frac{1}{2}$
  (b) 8

5 (a) $\frac{1}{10}$
  (b) $\frac{7}{9}$

6 (a) 0.9
  (b) 45th
  (c) 1000
  (d) 44

7 (a) 0.2
  (b) 1

8 (a) 12
  (b) 11

9 (a) $\frac{2}{3}$
  (b) 4
  (c) 11

10 (a) $r = \frac{1}{2}$, $a = \frac{5}{2}$
  (b) 4.960 937 5

Exercise 8D (Page 200)

1. (a) $x^4 + 12x^3 + 54x^2 + 108x + 81$
   (b) $x^5 - 10x^4 + 40x^3 - 80x^2 + 80x - 32$
   (c) $1 + 8x + 24x^2 + 32x^3 + 16x^4$
   (d) $8 - 36x + 54x^2 - 27x^3$
   (e) $64 + 192x + 240x^2 + 160x^3 + 60x^4 + 12x^5 + x^6$
   (f) $16x^4 - 96x^3 + 216x^2 - 216x + 81$

2. (a) $1 + 7x + 21x^2 + 35x^3$
   (b) $x^8 - 4x^7 + 7x^6 - 7x^5$
   (c) $64 + 96x + 60x^2 + 20x^3$

3. (a) 6
   (b) 3
   (c) 70

4. (a) $1 + 5x + 10x^2 + 10x^3 + \ldots$
       $1 - 5x + 10x^2 - 10x^3 + \ldots$
   (b) $1 - 5x^2$
   (c) Same because $[(1-x)(1+x)]^5 = (1-x^2)^5$

5. $1 + 3x + 6x^2 + 7x^3 + 6x^4 + 3x^5 + x^6$

Exercise 8E (Page 206)

1. (a) $x^4 + 4x^3 + 6x^2 + 4x + 1$
   (b) $1 + 7x + 21x^2 + 35x^3 + 35x^4 + 21x^5 + 7x^6 + x^7$
   (c) $x^5 + 10x^4 + 40x^3 + 80x^2 + 80x + 32$
   (d) $64x^6 + 192x^5 + 240x^4 + 160x^3 + 60x^2 + 12x + 1$
   (e) $32x^5 - 240x^4 + 720x^3 - 1080x^2 + 810x - 243$
   (f) $8x^3 + 36x^2y + 54xy^2 + 27y^3$

2. (a) 6
   (b) 15
   (c) 20
   (d) 15
   (e) 1
   (f) 220
   (g) 220
   (h) 1365
   (i) 1

3. (a) 56
   (b) 210
   (c) 673 596
   (d) $-823\,680$
   (e) 70

4. (a) $1 + 4x + 6x^2 + 4x^3 + x^4$
   (b) 1.008
   (c) 0.0024%

5. (a) $6x + 2x^3$

6. (a) $32 - 80x + 80x^2 - 40x^3 + 10x^4 - x^5$
   (b) 31.208
   (c) 0.000 13%

7. (a) $x^6 + 6x^4 + 15x^2 + 20 + \frac{15}{x^2} + \frac{6}{x^4} + \frac{1}{x^6}$
   (b) $16x^4 - 16x^2 + 6 - \frac{1}{x^2} + \frac{1}{16x^4}$
   (c) $1 + \frac{10}{x} + \frac{40}{x^2} + \frac{80}{x^3} + \frac{80}{x^4} + \frac{32}{x^5}$

8. (a) £2570
   (b) True value = £2593.74
   (c) 0.92%

9. (b) $x = 0, -1$ and -2

10. (a) $1 + 24x + 264x^2 + 1760x^3$
    (b) 1.268 16
    (c) 0.0064%

Exercise 8F (Page 207)

1. (a) 2
   (b) 3
   (c) 3069

2. (a) 68th swing is the first less than 1°.
   (b) 241° (to nearest degree)

3. (a) Height after nth impact $= 10 \times (\frac{2}{3})^n$
   (b) 59.0 m (to 3 s.f.)

4. (a) Common ratio $r = 1.08$
   (b) £29 985
   (c) £1496 (to the nearest £)

5. (a) 2.5, 1.25, 0.625, 0.3125
   (b) 4.999 923 706

6. (a) $\frac{1}{3}$
   (b) 121.5
   (c) 13

7. (b) 6
   (c) 18

8. (a) 1 hour 21 minutes
   (b) 3 hours 57 minutes

9. (b) 8.95×10^{-4}
   (c) 200
   (d) $\frac{1000}{3}$

10. (b) $r = 0.6, a = 5$
    (c) 12.5

11. $81x^4 - 36x^2 + 6 - \frac{4}{9x^2} + \frac{1}{81x^4}$

12. 4096, 24 576, 67 584, 112 640

13. (a) $1 - 44x + 880x^2 - 10\,560x^3$
    (b) 0.637 44

14. (a) $81 + 216x + 216x^2 + 96x^2 + 16x^4$
    (b) $81 - 216x + 216x^2 - 96x^2 + 16x^4$
    (c) 1154

15. (a) 8
    (b) $4\frac{3}{8}$

Exercise 9A (Page 215)

1. (a) 7.27 cm
   (b) 21.9 cm
   (c) 6.38 cm
   (d) 17.2 cm

2. (a) 46.1°
   (b) 52.6°
   (c) 57.7° or 122.3°
   (d) 53.4° or 126.6°

3. (a) 182 km
   (b) 6.60 km

4. $x = 14.5$ cm, QR = 18.7 cm

5. (a) 56.4° or 123.6°
   (b) 0.534 m or 1.198 cm

Exercise 9B (Page 217)

1 (a) 7.48 cm
   (b) 9.95 cm
   (c) 11.4 cm
   (d) 3 cm

2 (a) 26.4°
   (b) 38.0°
   (c) 133.4°
   (d) 125.1°

3 (a) 156°
   (b) 59.9°

4 29.0°, 46.6°, 104.5°

5 (a) 7 cm

Exercise 9C (Page 219)

1 (a) 16.9 cm²
   (b) 16.7 cm²
   (c) 13.5 cm²
   (d) 2.30 cm²

2 (a) 17.4 cm²
   (b) 35.0 cm²
   (c) 16.1 cm²
   (d) 15.3 cm²

3 (a) $5, 13, 9\sqrt{2}$
   (b) 22.4°, 75.7°, 81.9°
   (c) 31.5 square units

4 (a) $x = 6$
   (b) 8, 13.5 cm
   (c) 13°, 17°

5 1.92 cm²

Exercise 9D (Page 222)

1 (a) $\frac{\pi}{4}$
   (b) $\frac{\pi}{2}$
   (c) $\frac{2\pi}{3}$
   (d) $\frac{5\pi}{12}$
   (e) $\frac{5\pi}{3}$
   (f) 0.4 rad
   (g) $\frac{5\pi}{2}$
   (h) 3.65 rad
   (i) $\frac{5\pi}{6}$
   (j) $\frac{\pi}{25}$

2 (a) 18°
   (b) 108°
   (c) 114.6°
   (d) 80°
   (e) 540°
   (f) 300°
   (g) 22.9°
   (h) 135°
   (i) 420°
   (j) 77.1°

Exercise 9E (Page 224)

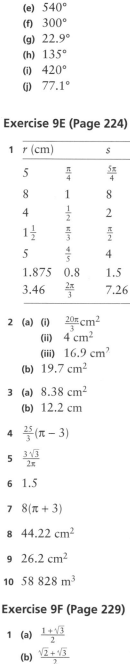

1 r (cm)	s		A(cm²)
5	$\frac{\pi}{4}$	$\frac{5\pi}{4}$	$\frac{25\pi}{8}$
8	1	8	32
4	$\frac{1}{2}$	2	4
$1\frac{1}{2}$	$\frac{\pi}{3}$	$\frac{\pi}{2}$	$\frac{3\pi}{8}$
5	$\frac{4}{5}$	4	10
1.875	0.8	1.5	1.41
3.46	$\frac{2\pi}{3}$	7.26	4π

2 (a) (i) $\frac{20\pi}{3}$cm²
         (ii) 4 cm²
         (iii) 16.9 cm²
   (b) 19.7 cm²

3 (a) 8.38 cm²
   (b) 12.2 cm

4 $\frac{25}{3}(\pi - 3)$

5 $\frac{3\sqrt{3}}{2\pi}$

6 1.5

7 $8(\pi + 3)$

8 44.22 cm²

9 26.2 cm²

10 58 828 m³

Exercise 9F (Page 229)

1 (a) $\frac{1+\sqrt{3}}{2}$
   (b) $\frac{\sqrt{2}+\sqrt{3}}{2}$
   (c) $1 + \sqrt{3}$

2 (a) $2 - \sqrt{3}$
   (b) $2 - \sqrt{3}$

3 1

Exercise 9G (Page 236)

1 (a)

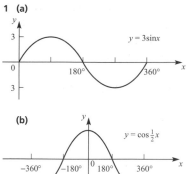

   (b)
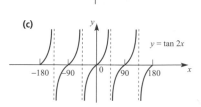

   (c)
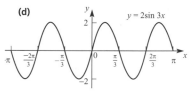

   (d)
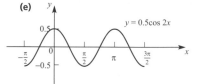

   (e)
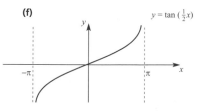

   (f)
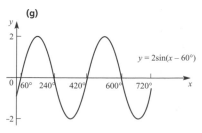

   (g)

(h)

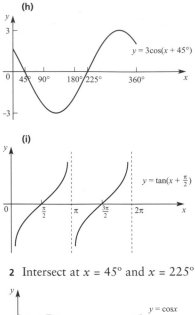

(i)

2 Intersect at $x = 45°$ and $x = 225°$

3 (b) (i) False
    (ii) True
    (iii) False
    (iv) True

4 Possible equations are
   (a) $y = 2\cos 4x$
   (b) $y = \tan\left(x - \frac{\pi}{2}\right)$
   (c) $y = 3\sin\left(\frac{1}{2}x\right)$
   (d) $y = 3\cos(2x - 30°)$

5 (a)

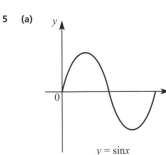

(b)

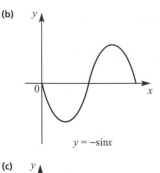

$y = -\sin x$

(c)

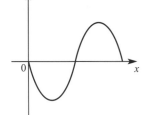

$y = \sin(-x)$

(d)

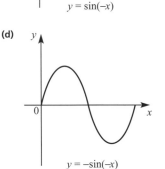

$y = -\sin(-x)$

Comment: $-\sin(-x) = \sin x$,
$-\sin x = \sin(-x)$.

Exercise 9H (Page 240)

1 (a) $30°, 150°$
   (b) $60°, 300°$
   (c) $26.6°, 206.6°$
   (d) $225°, 315°$
   (e) $150°, 210°$
   (f) $120°, 300°$

2 (a) $10°, 50°, 130°, 170°$
   (b) $73.7°, 646.3°$
   (c) $-151.8°, -61.8°, 28.2°,$
        $118.2°$
   (d) $-81.3°, -8.7°, 98.7°, 171.3°$
   (e) $0, \pm\frac{\pi}{2}, \pm\pi, \pm\frac{3\pi}{2}, \pm 2\pi$
   (f) $\frac{\pi}{2}, \frac{7\pi}{2}$
   (g) $\frac{\pi}{2}, \frac{5\pi}{6}$
   (h) $-18.6°, -101.4°$
   (i) $-4.1°, 85.9°$

(j) $53.1°, 126.9°, 233.1°, 306.9°$
(k) $\pm\frac{\pi}{6}, \pm\frac{5\pi}{6}, \pm\frac{7\pi}{6}, \pm\frac{11\pi}{6}$
(l) $-90°, -30°, 90°, 150°, 270°$

3 (a), (b)

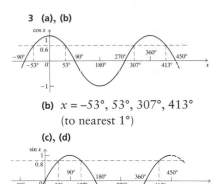

(b) $x = -53°, 53°, 307°, 413°$
      (to nearest 1°)

(c), (d)

(d) $x = 53°, 127°, 413°$
      (to nearest 1°)
(e) For $0 \le x \le 90°$, $\sin x = 0.8$
      and $\cos x = 0.6$ have the same
      root. For $90° \le x \le 360°$,
      $\sin x$ and $\cos x$ are never both
      positive.

4 (a) $-60°$
   (b) $-155.9°$
   (c) $54.0°$

5 (a) α between $0°$ and $90°$, $360°$
      and $450°$, $720°$ and $810°$, etc.
   (b) No: since $\tan\alpha = \frac{\sin\alpha}{\cos\alpha}$, all
      must be positive or one
      positive and two negative.
   (c) No: $\sin\alpha = \cos\alpha \Rightarrow \alpha = 45°$,
      $225°$ etc. but $\tan\alpha = \pm 1$
      for these values of α, and
      $\sin\alpha = \cos\alpha = \frac{1}{\sqrt{2}}$

Exercise 9I (Page 243)

2 (a) $60°, 180°, 300°$
   (b) $-90°, 19.5°, 160.5°$
   (c) $\pm\frac{\pi}{6}, \pm\frac{5\pi}{6}, \pm\frac{7\pi}{6}, \pm\frac{11\pi}{6}$
   (d) $60°, 109.5°, 250.5°, 300°$
   (e) $\pm\frac{\pi}{3}, \pm\frac{5\pi}{3}$
   (f) $-90°, 270°$
   (g) $-90°, 41.8°, 138.2°$
   (h) $\frac{\pi}{4}, \frac{\pi}{2}, \frac{5\pi}{4}, \frac{3\pi}{2}$
   (i) $\frac{\pi}{6}, \frac{5\pi}{6}, \frac{7\pi}{6}, \frac{11\pi}{6}$
   (j) $19.5°, 160.5°$

3 (b) $\sin Q = \frac{8}{17}$, $\cos Q = \frac{15}{17}$, $\tan Q = \frac{8}{15}$

4 $-\frac{2\sqrt{6}}{5}$

5 $\cos A = \frac{5}{7}$, $\sin A = \frac{2\sqrt{6}}{7}$

Exercise 9J (Page 244)

1 (a) 8 cm
 (b) 35.3°

2 (a) 325.5 cm
 (b) 146.2°

3 5

4 1

6 (b) Area = 6π

7 (a) $2\sqrt{3}$cm
 (b) $\frac{3\pi}{2}$ cm²

8 (a) 12θ m²
 (c) 19 m
 (d) 40 cm

9 (a) 8.75cm²
 (b) 0.70cm²

10 (a) $(0,1)$
 (b) $\frac{17\pi}{24}$, $\frac{23\pi}{24}$, $\frac{41\pi}{24}$, $\frac{47\pi}{24}$

11 (a) 60°, 150°, 240°, 330°
 (b) (i) C $(340, 0)$
 (ii) $p = 1.5$, $q = 60$

12 70.5°, 180°, 289.5°

13 (a) 120°, 240°
 (b) 133.3°, 226.7°

14 36.5°, 83.5°, 156.5°

15

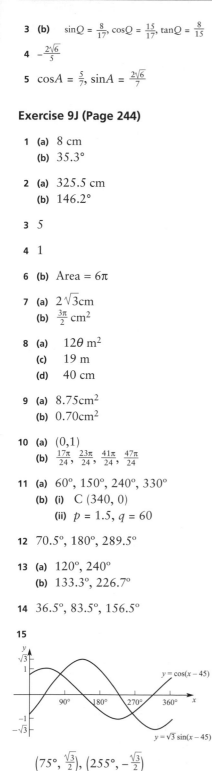

$\left(75°, \frac{\sqrt{3}}{2}\right), \left(255°, -\frac{\sqrt{3}}{2}\right)$

Chapter 10

Exercise 10A (Page 251)

1 (a)
(b)
(c)
(d)

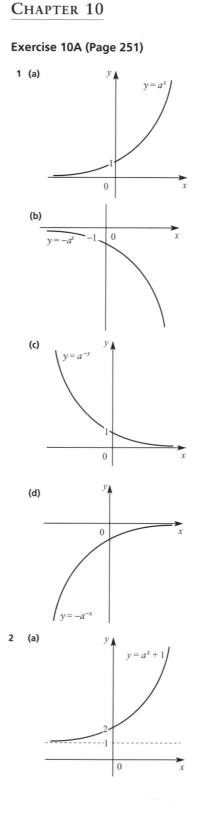

2 (a)

3 (a)

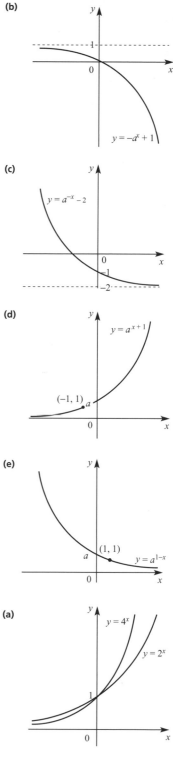

(b) $4^x = (2^2)^x = 2^{2x}$
stretch scale factor $\frac{1}{2}$ parallel
to x-axis, y-axis invariant

4 (a)

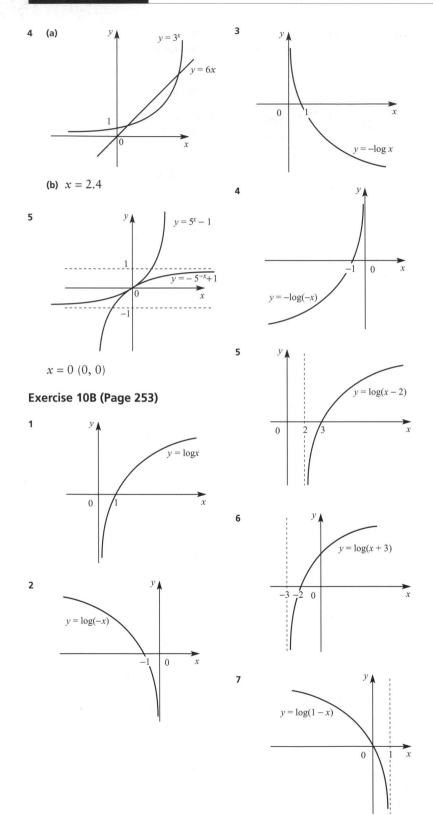

$y = 3^x$

$y = 6x$

(b) $x = 2.4$

5

$y = 5^x - 1$

$y = -5^{-x} + 1$

$x = 0 \ (0, 0)$

Exercise 10B (Page 253)

1

$y = \log x$

2

$y = \log(-x)$

3

$y = -\log x$

4

$y = -\log(-x)$

5

$y = \log(x - 2)$

6

$y = \log(x + 3)$

7

$y = \log(1 - x)$

8

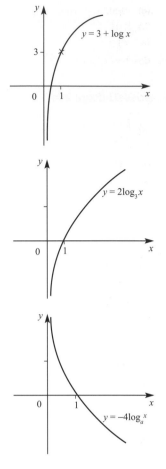

$y = 3 + \log x$

9

$y = 2\log_3 x$

10

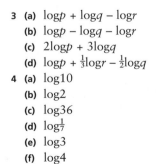

$y = -4\log_a x$

Exercise 10C (Page 257)

1 (a) $\log_3 81 = 4$
   (b) $\log_2 256 = 8$
   (c) $\log_a l = b$
   (d) $\log_y x = z$

2 (a) $10\,000 = 10^4$
   (b) $216 = 6^3$
   (c) $b = c^q$
   (d) $y = n^x$

3 (a) $\log p + \log q - \log r$
   (b) $\log p - \log q - \log r$
   (c) $2\log p + 3\log q$
   (d) $\log p + \frac{1}{3}\log r - \frac{1}{2}\log q$

4 (a) $\log 10$
   (b) $\log 2$
   (c) $\log 36$
   (d) $\log \frac{1}{7}$
   (e) $\log 3$
   (f) $\log 4$

(g) $\log 4$
(h) $\log \frac{1}{3}$
(i) $\log \frac{1}{2}$
(j) $\log 12$

5 (a) 3
(b) -4
(c) $\frac{1}{2}$
(d) 0
(e) 4
(f) -4
(g) $\frac{3}{2}$
(h) $\frac{1}{4}$
(i) $\frac{1}{2}$
(j) -3

6 (a) 19.93
(b) -9.97
(c) 9.01
(d) 48.32
(e) 1375
(f) 4.64
(g) 6.11
(h) 10.48
(i) 8.58
(j) -5

7 (a) -2, 1.58
(b) 2

8 $\log_3 \dfrac{x+1}{2x}$

$\frac{1}{17}$

9 1

10 (a) $P = P_0 \times 10^{-\frac{t}{500}}$
(b) $P_0 = 1000$

Exercise 10D (Page 259)

1 (a)

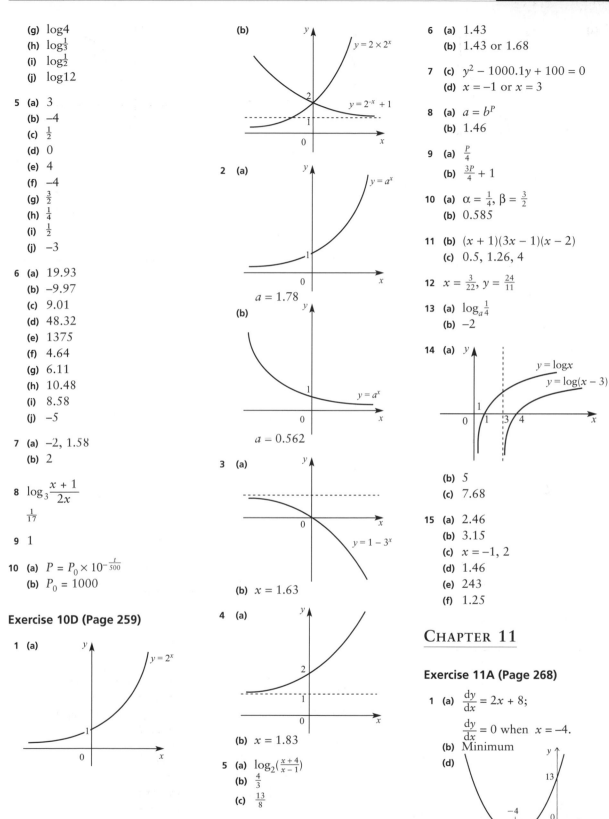

(b)

2 (a)
$a = 1.78$
(b)
$a = 0.562$

3 (a)
(b) $x = 1.63$

4 (a)
(b) $x = 1.83$

5 (a) $\log_2 \left(\frac{x+4}{x-1} \right)$
(b) $\frac{4}{3}$
(c) $\frac{13}{8}$

6 (a) 1.43
(b) 1.43 or 1.68

7 (c) $y^2 - 1000.1y + 100 = 0$
(d) $x = -1$ or $x = 3$

8 (a) $a = b^P$
(b) 1.46

9 (a) $\frac{P}{4}$
(b) $\frac{3P}{4} + 1$

10 (a) $\alpha = \frac{1}{4}$, $\beta = \frac{3}{2}$
(b) 0.585

11 (b) $(x+1)(3x-1)(x-2)$
(c) 0.5, 1.26, 4

12 $x = \frac{3}{22}$, $y = \frac{24}{11}$

13 (a) $\log_a \frac{1}{4}$
(b) -2

14 (a)
(b) 5
(c) 7.68

15 (a) 2.46
(b) 3.15
(c) $x = -1$, 2
(d) 1.46
(e) 243
(f) 1.25

CHAPTER 11

Exercise 11A (Page 268)

1 (a) $\dfrac{dy}{dx} = 2x + 8$;

$\dfrac{dy}{dx} = 0$ when $x = -4$.
(b) Minimum
(d)

2 (a) $\dfrac{dy}{dx} = 2x + 5$;

$\dfrac{dy}{dx} = 0$ when $x = -2\frac{1}{2}$

(b) Minimum

(c) $y = -4\frac{1}{4}$

(d)
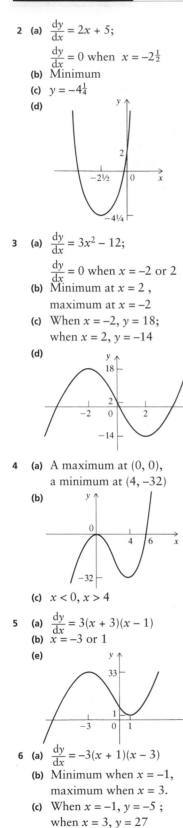

3 (a) $\dfrac{dy}{dx} = 3x^2 - 12$;

$\dfrac{dy}{dx} = 0$ when $x = -2$ or 2

(b) Minimum at $x = 2$,
maximum at $x = -2$

(c) When $x = -2$, $y = 18$;
when $x = 2$, $y = -14$

(d)

4 (a) A maximum at $(0, 0)$,
a minimum at $(4, -32)$

(b)

(c) $x < 0$, $x > 4$

5 (a) $\dfrac{dy}{dx} = 3(x + 3)(x - 1)$

(b) $x = -3$ or 1

(e)

6 (a) $\dfrac{dy}{dx} = -3(x + 1)(x - 3)$

(b) Minimum when $x = -1$,
maximum when $x = 3$.

(c) When $x = -1$, $y = -5$;
when $x = 3$, $y = 27$

(d)
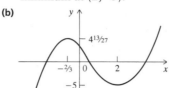

7 (a) Maximum at $\left(-\frac{2}{3}, 4\frac{13}{27}\right)$,
minimum at $(2, -5)$.

(b)
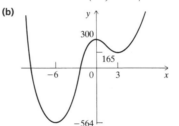

8 (a) Maximum at $(0, 300)$,
minimum at $(3, 165)$,
minimum at $(-6, -564)$

(b)

9 (a) $\dfrac{dy}{dx} = 3(x^2 + 1)$

(b) There are no turning points.

(d)

x	-3	-2	-1	0	1	2	3
y	-36	-14	-4	0	4	14	36

(e)

10 (a) $\dfrac{dy}{dx} = 4x^3 - 16x$

(b) Maximum at $(0, 16)$,
minima at $(2, 0)$ and $(-2, 0)$.

(c)
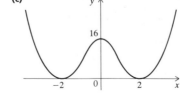

(d) $x^4 - 8x^2 + 16 = (x^2 - 4)^2$
$= (x - 2)^2 (x + 2)^2$

(e) The factorised form shows
that the graph touches the
x-axis twice (repeated roots
at $x = -2$ and $+2$). The
unfactorised version shows
behaviour for large positive
and negative x-values, and
shows where the graph
crosses the y-axis. By
symmetry this must be a
turning point.

11 (a) $\dfrac{dy}{dx} = 6x^2 + 6x - 72$

(b) $y = 18$

(c) $\dfrac{dy}{dx} = 48$

(d) $(-4, 338)$ and $(3, -5)$

12 (a) $y = 0 \Rightarrow x^2 = -4 \Rightarrow$ no
solution

(b) As $x \to 0$, $y \to \infty$

(c) Minimum at $(2, 4)$
Maximum at $(-2, -4)$

(d)
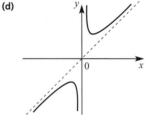

13 Maximum at $\left(-8, 1\frac{1}{3}\right)$
Minimum at $\left(8, -1\frac{1}{3}\right)$

14 (a) $x = 4$, $y = 3$

15 $(3, 9)$ Minimum.

Exercise 11B (Page 274)

1 (a) $\dfrac{dy}{dx} = 12x^2 (x - 1)$

(b) $x = 0$ or 1

(c) Point of inflection when
$x = 0$; minimum when $x = 1$

(d) When $x = 0$, $y = 0$;
when $x = 1$, $y = -1$

(e) A: $(0, 0)$; B: $(1, -1)$; C: $\left(\frac{4}{3}, 0\right)$

2 (a) $\dfrac{dy}{dx} = (3x + 1)(x - 1)$

(b) $x = -\frac{1}{3}$ or 1

(c) Maximum at $x = -\frac{1}{3}$, minimum at $x = 1$

(d) When $x = -\frac{1}{3}$, $y = 4\frac{5}{27}$; when $x = 1$, $y = 3$

(e)
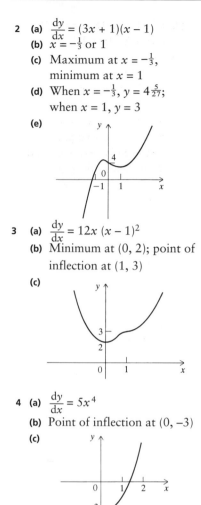

3 (a) $\dfrac{dy}{dx} = 12x\,(x - 1)^2$

(b) Minimum at $(0, 2)$; point of inflection at $(1, 3)$

(c)
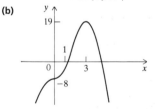

4 (a) $\dfrac{dy}{dx} = 5x^4$

(b) Point of inflection at $(0, -3)$

(c)
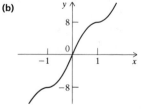

5 (a) Point of inflection at $(0, -8)$; maximum at $(3, 19)$

(b)

6 (a) Points of inflection at $(-1, -8)$ and $(1, 8)$

(b)

7 (a) $\dfrac{dy}{dx} = (3x - 7)(x - 1)$

(b) Maximum at $(1, 0)$; minimum at $(2\frac{1}{3}, -1\frac{5}{27})$

(c)

8 (a) $\dfrac{dy}{dx} = 4x(x - 1)(x - 2)$

(b) Minimum at $(0, 0)$; maximum at $(1, 1)$; minimum at $(2, 0)$

(c)

9 (a) $p + q = -1$

(b) $3p + 2q = 0$

(c) $p = 2$ and $q = -3$

10 (a) $\dfrac{dy}{dx} = 12x^2(x + 1)$

(b) $(0, 0)$ and $(-1, -1)$

(c) Minimum at $(-1, -1)$; point of inflection at $(0, 0)$

(d)

Exercise 11C (Page 278)

1 (a) $y = 60 - x$

(b) $A = 60x - x^2$

(c) $\dfrac{dA}{dx} = 2(30 - x)$. Dimensions 30 m by 30 m, area 900 m²

2 (a) $V = 4x^3 - 48x^2 + 144x$

(b) $\dfrac{dV}{dx} = 12(x - 2)(x - 6)$

3 (a) $y = 8 - x$

(b) $S = 2x^2 - 16x + 64$

(c) 32

4 (a) $2x + y = 80$

(b) $A = 80x - 2x^2$

(c) $x = 20$, $y = 40$

5 (a) $x(1 - 2x)$

(b) $V = x^2 - 2x^3$

(c) $\dfrac{dV}{dx} = 2x(1 - 3x)$

(d) All dimensions $\frac{1}{3}$ m (a cube); volume $\frac{1}{27}$ m³

6 (a) $(4 - 2x)$ cm

(b) $(16 - 16x + 4x^2)$ cm²

(d) $x = 1.143$

(e) $A = 6.857$

7 (a) $z = \dfrac{1000}{x}$

(b) $(x + 24)$ cm; $\left(\dfrac{1000}{x} + 12\right)$ cm

(d) $\dfrac{dA}{dx} = 12 - \dfrac{24\,000}{x^2}$; min. area $= 2361$ cm²

8 (b) $x = 3$ cm

9 (a) $h = \dfrac{100}{\pi r^2}$

(c) 94.7 cm² (3 s.f.)

10 (b) 10.5 cm, 3.5 cm, 7 cm

Exercise 11D (Page 281)

1 (a) $12x^2 - 12 + \dfrac{3}{x^2}$

(c) $f'(x) = 3\left(2x - \frac{1}{x}\right)^2 > 0$

2 (b) $\dfrac{10}{3}$ cm

(d) 268 cm³ (3 s.f.)

(e) 22.2%

3 (a) $S = \dfrac{2000}{r} + 2\pi r^2$ cm²

(b) 5.42 cm (3 s.f.)

(d) 554 cm² (3 s.f.)

4 (b) $x < -2$, $x > 2$

5 (a) $x = 4$, $y = 20$

6 (a) $(0, 15)(-5, 0)(\frac{3}{2}, 0)$

(b)
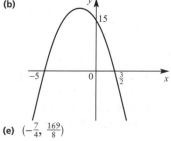

(e) $\left(-\frac{7}{4}, \frac{169}{8}\right)$

341

7 (a) $7\frac{31}{32}$
(b) $f(x) = \frac{1}{3}x^3 - 2x - \frac{1}{x} - \frac{8}{3}$
(c) $f'(x) = (x - \frac{1}{x})^2 > 0$

8 (a) $4x^3 - 16x$
(b) $(-2, -13), (0, 3), (2, -13)$
(c) $(\pm 2, -13)$ Minimum turning point
$(0, 3)$ Maximum turning point
(d)

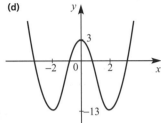

9 (b) $0 < x < 20$
(c) 7.362 (4 s.f.)
(d) 6564 (4 s.f.)

10 (c) 1200 cm^2

CHAPTER 12

Exercise 12A (Page 286)

1 (a) 3
(b) 9
(c) 27
(d) $10\frac{1}{2}$
(e) 12
(f) 15
(g) 114
(h) $\frac{1}{6}$
(i) $2\frac{9}{20}$
(j) 0
(k) $-105\frac{3}{4}$
(l) 5

2 (a) $\frac{1}{16}$
(b) 9.3
(c) 126.6
(d) $3\frac{1}{3}$
(e) $36\frac{2}{3}$
(f) 9
(g) $-9\frac{3}{4}$
(h) 725.5
(i) $\frac{1}{8}$

(j) $2\frac{2}{3}$
(k) 42.34
(l) -88

3 (a) $2\frac{1}{4}$
(b) $\frac{3}{4}$
(c) 41.4
(d) $-2\frac{2}{3}$
(e) 3.68
(f) $10\frac{2}{3}$

Exercise 12B (Page 290)

1 (a) A: $(2, 4)$; B: $(3, 6)$
(b) 5
(d) In this case the area is not a trapezium since the top is curved.

2 (a)

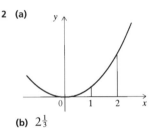

(b) $2\frac{1}{3}$

3 (a)

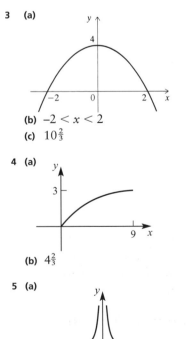

(b) $-2 < x < 2$
(c) $10\frac{2}{3}$

4 (a)

(b) $4\frac{2}{3}$

5 (a)

(b) $\frac{2}{3}$
(c) $\frac{2}{3}$
(d) Discontinuity at $x = 0$

6 (a)

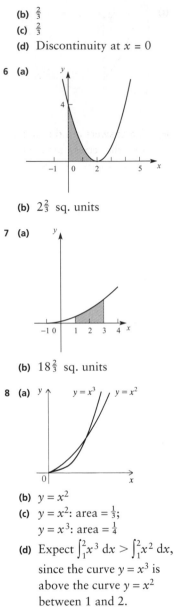

(b) $2\frac{2}{3}$ sq. units

7 (a)

(b) $18\frac{2}{3}$ sq. units

8 (a) $y = x^3$, $y = x^2$
(b) $y = x^2$
(c) $y = x^2$: area $= \frac{1}{3}$; $y = x^3$: area $= \frac{1}{4}$
(d) Expect $\int_1^2 x^3 \, dx > \int_1^2 x^2 \, dx$, since the curve $y = x^3$ is above the curve $y = x^2$ between 1 and 2. Confirmation: $\int_1^2 x^3 \, dx = 3\frac{3}{4}$ and $\int_1^2 x^2 \, dx = 2\frac{1}{3}$

9 (a)

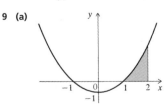

(b) $1\frac{1}{3}$

(c)

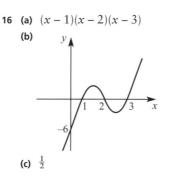

(d) $1\frac{1}{3}$

(e) The answers are the same, since the second area is a translation of the first.

10 (a)

(b) 24

11 (a)

(b) $7\frac{1}{3}$
(c) $7\frac{1}{3}$, by symmetry
(d) $7\frac{1}{3}$

12 (a)

(b) $\int_0^4 (x^2 - 2x + 1)\,dx$ larger, as area between 3 and 4 is larger than area between -1 and 0.
(c) $\int_{-1}^3 (x^2 - 2x + 1)\,dx = 5\frac{1}{3}$;
$\int_0^4 (x^2 - 2x + 1)\,dx = 9\frac{1}{3}$

13 (a) and **(b)**

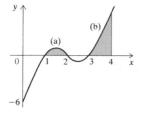

(c) (i) $\frac{1}{4}$ **(ii)** $2\frac{1}{4}$
(d) 0.140625. The maximum lies before $x = 1.5$.

14 (a) $\dfrac{dy}{dx} = -2x + 2$
(b) max $y = 4$
(c) 6m^2

15 (a) 9 m
(b) Yes: distance in 3rd second was $6\frac{1}{3}$ m; distance in first 2 seconds was $2\frac{2}{3}$ m.

16 (a) $(x - 1)(x - 2)(x - 3)$
(b)

(c) $\frac{1}{2}$

Exercise 12C (Page 295)

1 $20\frac{1}{4}$

2 9

3 $2\frac{1}{6}$

4 1

5 $\frac{4}{15}$

6 $2\frac{1}{16}$

7 $\frac{1}{6}$

8 $8\frac{1}{6}$

9 $11\frac{1}{12}$

10 $8\frac{1}{6}$

11 (a) $\dfrac{dy}{dx} = 20x^3 - 5x^4$; $(0, 0)$ and $(4, 256)$
(b) $520\frac{5}{6}$ square units
(c) 0

Exercise 12D (Page 298)

1 (a) A: $(-3, 9)$; B: $(3, 9)$

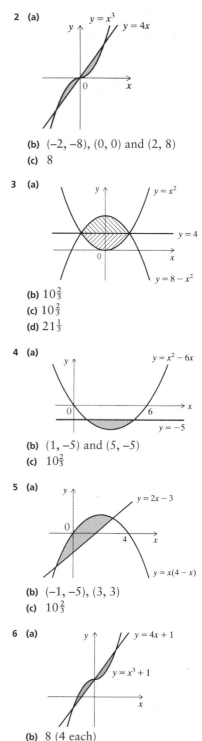

2 (a)

(b) $(-2, -8)$, $(0, 0)$ and $(2, 8)$
(c) 8

3 (a)

(b) $10\frac{2}{3}$
(c) $10\frac{2}{3}$
(d) $21\frac{1}{3}$

4 (a)

(b) $(1, -5)$ and $(5, -5)$
(c) $10\frac{2}{3}$

5 (a)

(b) $(-1, -5)$, $(3, 3)$
(c) $10\frac{2}{3}$

6 (a)

(b) 8 (4 each)

7 (a)

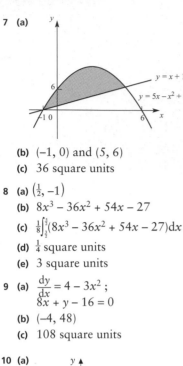

(b) $(-1, 0)$ and $(5, 6)$

(c) 36 square units

8 (a) $\left(\frac{1}{2}, -1\right)$

(b) $8x^3 - 36x^2 + 54x - 27$

(c) $\frac{1}{8}\int_{\frac{3}{2}}^{\frac{5}{2}}(8x^3 - 36x^2 + 54x - 27)dx$

(d) $\frac{1}{4}$ square units

(e) 3 square units

9 (a) $\dfrac{dy}{dx} = 4 - 3x^2$;
$8x + y - 16 = 0$

(b) $(-4, 48)$

(c) 108 square units

10 (a)

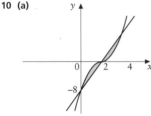

(b) 8

Exercise 12E (Page 305)

1 2.6286, 2.5643, 2.5475;
overestimate

2 5.1667, 5.1167, 5.1032;
overestimate

3 1.2622, 0.9436, 0.8395;
overestimate

4 1.3965

5 (a)

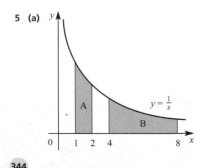

(b) for A and B

(i) 0.7083

(ii) 0.6970

(iii) 0.6941

(c) A and B are of equal area

6 (a) 458 m

(b) A curve is approximated by a straight line. The speeds are only given to 1 d.p.

7 (a) 3.1311

(b) 3.1389, 3.1409

(c) 3.14

8 (a) 7.32775

(b) overestimate

9 (b) 12.6597, too small

(c) $2\frac{1}{3}$

(d) 0.055%

10 (a) 1.05101, 1.53862, 0.53565

(b) $1 + 5x^2 + 10x^4 + 10x^6 + 5x^8 + x^{10}$, 0.52715

(c) (i) Too high – trapezia above graph.

(ii) Too low – remaining terms would increase the value.

Exercise 12F (Page 308)

1 (a) $20\frac{2}{3}$

(b) $\frac{1}{18}$

2 (a) $\frac{2}{3}$

(b) $(1\frac{1}{2}, 6\frac{1}{4})$

(c)

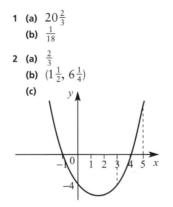

(d) Part of the area is below x-axis and part is above.

3 (a)

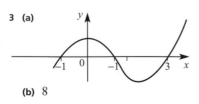

(b) 8

4 (a) $A = 9$, $B = 12$, $C = 4$

(b) $6x^{\frac{3}{2}} + 12x + 8x^{\frac{1}{2}} + c$

5 (a) 493

(b) $A = 4$, $B = 9$, $C = -12$

(c) 6.375

6 (a) $A(3, 0)$

(b) $B(1, 4)$

(c) 6.75

7 (a) $A(1, 2)$ $B(4, 5)$

(b) $4\frac{1}{2}$

8 (a) $A(-2, 11)$ $B(3, 6)$

(b) $20\frac{5}{6}$

9 1.588

10 3.135, more strips

11 (a) 14.72

(b)

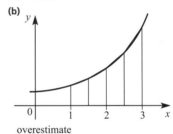

overestimate

(c) $14\frac{2}{3}$

12 0.845

13 133.9

14 $p = 1.357$, $q = 1.382$
2.5885

15 (a) $a = 0.268$, $b = 0.577$, $c = 1$

(b) 0.348

Index

C1 C2